Connected Vehicle Systems

Yunpeng Wang, Daxin Tian, Zhengguo Sheng, and Jian Wang

Connected Vehicle Systems: Communications, Data, and Control

CRC Press
Taylor & Francis Group
Boca Raton London New York

CRC Press is an imprint of the
Taylor & Francis Group, an **informa** business

CRC Press
Taylor & Francis Group
6000 Broken Sound Parkway NW, Suite 300
Boca Raton, FL 33487-2742

International Standard Book Number-13: 978-1-138-03587-4 (Hardback)

Visit the Taylor & Francis Web site at
http://www.taylorandfrancis.com

and the CRC Press Web site at
http://www.crcpress.com

MIX
Paper from responsible sources
FSC www.fsc.org FSC® C013056

Printed and bound in Great Britain by
TJ International Ltd, Padstow, Cornwall

Contents

List of Figures vii

List of Tables xiii

Preface xv

Contributors xvii

PART I: INTRAVEHICLE COMMUNICATIONS 1

1 A Survey of Automotive Networking Applications and Protocols 3
Victor Ocheri, Zhengguo Sheng, and Falah Ali

2 CAN-Based Media Access Control Protocol 21
Zhengguo Sheng, Amir Kenarsari-Anhari, Nima Taherinejad, and Yunpeng Wang

3 Worst-Delay Analysis of Intravehicle Communication Buses Using HomePlug GP 47
Zhengguo Sheng, Roberto P. Antonioli, Morgan Roff, and Yunpeng Wang

PART II: INTERVEHICLE COMMUNICATIONS 57

4 Physical Channel Modeling and Sharing 59
Jian Wang and Yunlei Zhang

5 Short-Range Communication Technology in the Internet of Vehicles 69
Daxin Tian, Jianshan Zhou, Yingrong Lu, Yunpeng Wang, and Zhengguo Sheng

6 Cross-Layer Multihop Broadcasting 95
Jian Wang and Fangqi Liu

PART III: VEHICULAR DATA APPLICATION AND MOBILITY CONTROL 113

7 **Road Recognition from Trajectories Collected by Freight Vehicles Based on Least Squares Support Vector Machine Fitting** 115
Daxin Tian, Yue Yang, Xiaolei Ma, Yunpeng Wang, and Zhengguo Sheng

8 **DSRC for Vehicular Positioning** 131
Yunpeng Wang, Xuting Duan, Daxin Tian, Xuejun Zhang, and Yingrong Lu

9 **Vehicular Positioning Enhancement** 159
Yunpeng Wang, Xuting Duan, Daxin Tian, Xuejun Zhang, and Min Chen

10 **A Lightweight, Adaptive Security Mechanism** 187
Jian Wang and Zemin Sun

11 **A Social-Network-Enabled Green Transportation System Framework Driven by Connected Vehicle Innovations** 203
Wei Shu, Guohui Zhang, Min-You Wu, and Jia-Liang Lu

12 **A Mobility Model for Connected Vehicles Induced by the Fish School** 213
Daxin Tian, Keyi Zhu, Jianshan Zhou, and Yunpeng Wang

References 247

Index 269

List of Figures

1.1 Class A application zone. 5
1.2 Class B multiplex application. 5
1.3 Class C applications. 6
1.4 Automotive network architecture. 10
1.5 A roadmap of automotive communication development. 12
1.6 Block diagram of the architecture of the automotive network test bed platform. 13
1.7 An implementation of a minimum viable test bed for automotive communication using Raspberry Pi and Arduino boards. 14
1.8 Distribution of data bit rate on the Ethernet connection over the period of testing. 16
1.9 Partitioning of total data throughput on the Ethernet connection into overhead and data traffic. 16
1.10 Proportion of the traffic on the Ethernet link used by overhead bits as well as data bits transmission. 17
1.11 Graph showing the average latency in signal transmission over the Ethernet link of the test bed. 17
1.12 Graph showing the traffic over the Wi-Fi connection between the onboard unit and RSU of the test bed. 18
1.13 Graph showing proportion of traffic used by overhead in the Wi-Fi connection of the test bed. 18
1.14 Pie chart showing the proportion of traffic that comes from transmission of overhead bits. 18
1.15 Graph showing the average RTT over the Wi-Fi connection. 19
2.1 Magnitude of $S_{21}(f)$ for different connections from HEV VCU to potential communication nodes [222], where VCU is Vehicle Control Unit, ESS is Energy Storage System, TMDMOC is Traction Motor Digital Module Controller, and DC/DC stands for DC-to-DC converter unit of the vehicle. 24
2.2 View of a single transmission cycle. 26
2.3 Illustration of the collision resolution algorithm performed on each channel. 27
2.4 Channel selection probabilities when $M = 10$ channels are available for multiple choices of β. 30
2.5 Block diagram of a robust sensing module. 32

2.6 ECDF of the received signal before and after preprocessing. 33
2.7 Illustration of the protocol operation with imperfect sensing. 34
2.8 Success probability versus number of contending nodes for different values of γ, $N = 50$, $k = 6$, $M = 2$. 36
2.9 Average success probability versus number of channels for different number of time slots (k), $N = 50$, $\gamma = 0.6$. 37
2.10 ρ_1 versus number of channels for different number of time slots (k), $N = 50$, $\gamma = 0.6$. 37
2.11 Average success probability versus total number of nodes connected to the harness (N), $\gamma = 0.6$. 38
2.12 Probability mass function of the number of transmission cycles required to transmit the first packet (a) all packets (b) when there are 25 contenders, $N = 50$, $\gamma = 0.6$, $k = 4$, $M = 3$. 39
2.13 Probability mass function of the number of transmission cycles needed to transmit the first packet (a) all packets (b) when there are five contenders, $N = 50$, $\gamma = 0.6$, $k = 4$, $M = 3$. 40
2.14 Success probability of the proposed protocol, and CDR versus number of contending nodes, $M = 3$, $N = 50$, $\gamma = 0.6$. 41
2.15 Throughput ρ_1 of the proposed protocol and CDR versus number of contending nodes, $M = 3$, $N = 50$, $\gamma = 0.6$. 42
2.16 Average success probability versus threshold λ of the energy detector, $N = 50$, $\gamma = 0.6$, $k = 6$, $M = 2$, $u = 10$. 42
2.17 Success probability versus number of contending nodes for multiple values of r, where the good channel is indexed 1, $k = 6$, $M = 2$, $N = 50$, $\gamma = 0.6$. 43
2.18 Success probability versus number of contending nodes for multiple values of r, where the bad channel is indexed 1, $k = 6$, $M = 2$, $N = 50$, $\gamma = 0.6$. . 43
2.19 Success probability versus number of contending nodes for multiple values of r, where channels are equally good, $k = 6$, $M = 3$, $N = 50$, $\gamma = 0.6$. . . 44
2.20 Success probability versus number of contending nodes for multiple values of r, where channels are equally bad, $k = 6$, $M = 5$, $N = 50$, $\gamma = 0.6$. . . . 44
3.1 Frame structure of the proposed MAC protocol 50
3.2 Topology of extended HomePlug Green PHY network 51
3.3 Average delay for the regular HomePlug GP messages 52
3.4 Average delay for the proposed HomePlug GP messages 52
3.5 Percentage of collisions in the regular HomePlug 53
3.6 Percentage of collisions in the modified HomePlug 54
4.1 Dynamic spectrum access taxonomy. 60
4.2 Overlay system model of three secondary users. 63
4.3 State probabilities of channel occupancy in overlay. 67
4.4 Secondary system throughput comparison between two and three secondary users. 67
4.5 Throughput comparison between two and three secondary users in overlay scheme. 68
5.1 WAVE stack. 71
5.2 Channel access mechanisms. 72
5.3 V2V communication system architecture with DSRC devices. 74
5.4 The V2V communication system model. 75
5.5 Varying activity α in simulations. 80

5.6 Evaluation of the impact of random noise variability on the dynamics system. 81
5.7 The handover decision framework based on the extended attractor selection model. 82
5.8 The variation of the adopted sigmoid function with different parameter settings. 84
5.9 The real testing scenario. 85
5.10 Performance comparison: Transmission delay versus relative distance. . . . 86
5.11 Performance comparison: Packet loss rate versus relative distance. 86
5.12 A simulation scenario. 87
5.13 The variation of the decision vector states of the different applications: (a) the voice application, (b) one video application, and (c) another video application. 90
5.14 The simulation results under different vehicular terminals: (a) the results of the activity α and (b) the results of the fairness. 91
5.15 The simulation results under different initial velocities: (a) the results of the activity α and (b) the results of the fairness. 92
6.1 Relation between the probability of sending packet p_0 and packet transmission delay T_{delay}. 100
6.2 Relation between F_{value} and n. 102
6.3 Flow diagram of the decision-making process. 104
6.4 Relation between the number of multicast n and S under different vehicle densities. 105
6.5 Relation between the number of multicast n and E_{decide} under different vehicle densities. 105
6.6 Relation between the number of mutilcast n and S under different multicast ranges r. 106
6.7 Relation between the number of multicast n and E_{decide} under different multicast ranges r. 107
6.8 Relation between the number of multicast n and S under different packet's lengths L_{P}. 108
6.9 Relation between the number of multicast n and E_{decide} under different packet's lengths L_{P}. 108
6.10 Relation between the number of multicast n and S under different transmission data rates R_{d}. 109
6.11 Relation between the number of multicast n and E_{decide} under different transmission data rates R_{d}. 109
6.12 Relation between the number of multicast n and S under different vehicle speeds v. 110
6.13 Relation between the number of multicast n and E_{decide} under different vehicle speeds v. 110
7.1 Ideal partitioning. 118
7.2 Definition of distance. 119
7.3 Potential problem. 120
7.4 Map generated by data and OSM map. 122
7.5 Merged map. 123
7.6 Borders and centerline of the roads. 125
7.7 Nearest point on the road for each node. 126

7.8 Fitting result of LS-SVM and B-spline. 127
7.9 Leftmost marked part. 128
7.10 Middle marked part. 128
7.11 Rightmost marked part. 129
7.12 Deviation of lines fitted by the two methods and the original lines. 129
7.13 Final fitting centerlines of the test area 130
8.1 The hybrid TOA/DOA method. 135
8.2 Maximum throughputs and TUL of 802.11p. 137
8.3 Minimum delays and DLL of 802.11p. 138
8.4 Negotiated deceleration rates used by communication vehicles. 139
8.5 Highway capacity with varied PLRs at different speeds (delay = DLL, payload size = 1000 bytes). 140
8.6 Details of highway capacity with varied PLRs at different speeds (delay = DLL, payload size = 1000 bytes). 140
8.7 Highway capacity with varied bitrates at different speeds (payload size = 1000 bytes, PLR = 0). 141
8.8 Details of highway capacity with varied bitrates at different speeds (payload size = 1000 bytes, PLR = 0). 141
8.9 Highway capacity with varied PLRs at different speeds (delay = 50 ms, payload size = 1000 bytes). 142
8.10 Details of highway capacity with varied PLRs at different speeds (delay = 50 ms, payload size = 1000 bytes). 142
8.11 The mobility of multiple vehicles in different traffic scenarios. 143
8.12 DSRC test bed and its performance evaluation system. 144
8.13 The evaluation results at different speeds and distances. 145
8.14 Problem definition and the initial idea for the solution. 146
8.15 Impinging signals on uniform linear antenna array and the hybrid TOA/DOA positioning method. 147
8.16 Angular scope discretization. 149
8.17 Plots of the number of received signals ($d = /2$; $K = 2$): (a) $M \in [5{:}5{:}25]$, $N \in [10{:}10{:}100]$; (b) $M \in [5{:}5{:}25]$, SNR $\in [0{:}5{:}20]$ dB; (c) $N \in [10{:}10{:}100]$, SNR $\in [0{:}5{:}20]$ dB. 152
8.18 MT-BCSbased DOAs. 153
8.19 Actual and estimated DOAs. 154
8.20 Multiple snapshots DOA estimation ($M = 10$; $N = 10$; $d = /2$; $K = 2$) Plots of $\overline{RMSE}$ when SNR[0:5:20]. 155
8.21 RMSE of hybrid TOA/DOA-based location estimation. 156
8.22 CDF of the location error (CRB for TOA/DOA for one single RSU). 156
8.23 The absolute position error related to the two simulated cars obtained by comparison methods. 157
9.1 Vehicular positioning system with information fusion of the DFS and GPS measurements from both itself and the other neighboring vehicles. . . 164
9.2 Initial scenario. 166
9.3 CV-IMM-KF and GPS performance in positioning error. 168
9.4 Positioning enhanced by the DFS and the RSSI measurements. 168
9.5 The schematic of the proposed vehicular cooperative localization method. . 172
9.6 One trial demo of vehicular positioning estimation with both the DFS and the RSSI measurements. 182

9.7 The performance of the GPS and the proposed method, and the fundamental limits bounded by the R.SPEB and the R.mSPEB. 182
9.8 The enhancements for different CV penetrations—DFS only. 183
9.9 The neighbors number and achieved performance—DFS only. 183
9.10 The enhancements for different CV penetrations—RSSI only. 184
9.11 The neighbors' number and achieved performance—RSSI only. 184
9.12 The enhancements for the DFS and the RSSI measurements combination, and for the DFS measurements only. 185
10.1 The communication scenario. 189
10.2 Effect of the information security strength on the node utility and the optimal transmit power. 197
10.3 Effect of the relative density of eavesdropper nodes on the node utility and the optimal transmit power, where $\alpha = 0.8$. 198
10.4 Effect of the total node density on the node utility and the optimal transmit power, where $\alpha = 0.8$. 198
10.5 Effect of the transmit power of other nodes on the node utility and the optimal transmit power, where $\alpha = 0.8$. 199
10.6 Effect of the wireless channel bandwidth on the node utility and the optimal transmit power, where $\alpha = 0.8$. 200
10.7 Effect of the key length on the node utility and the optimal transmit power, where $n = 1.63 \times 10^{1053}$times$/\mu$s. 200
10.8 Effect of the attacker's capability on the node utility and the optimal transmit power, where (a) $\lambda = 128$ bits, (b) $\lambda = 192$ bits, and (c) $\lambda = 256$ bits. . . . 201
11.1 Cyberspace and physical space. 206
11.2 Hyper simulator platform system architecture and modulated components. V2V, vehicle-to-vehicle; V2I, vehicle-to-infrastructure. 211
12.1 The interplay range of vehicle i. 216
12.2 Bump function with different parameters. 218
12.3 The variation of the potential functions with different parameter settings. . . 220
12.4 The interaction range. 228
12.5 Sigmoid function with different parameters. 229
12.6 The traveling trajectory of vehicles. 233
12.7 The road situation for the simulation. 234
12.8 The mobility of multiple vehicles in different traffic scenarios. 236
12.9 The results of P_o and P_v under different initial velocity settings. 240
12.10 The results of P_o and P_v under different vehicle numbers. 241
12.11 The collision probability in different traffic scenarios. 242
12.12 The results of flow rate. 244
12.13 Flow rate. 245

List of Tables

1.1 Comparisons of Functional Classes 6
1.2 Comparison of Automotive Protocols 11
2.1 Parameters Used for Performance Analysis 28
2.2 Parameters Used for Performance Analysis 28
3.1 Original and Proposed Values of Contention Window (CW) as a Function of Backoff Procedure Event Counter (BPC) and Priorities 49
3.2 Summary of Differences 50
3.3 Minimum and Maximum Intermessage Times in the Regular and Proposed Real-Time HomePlug GP 51
4.1 Parameters 66
5.1 Network Conditions Settings 88
5.2 Applications Settings 88
5.3 Model Settings 89
5.4 Settings on the Weights 89
6.1 Parameters 104
8.1 Parameters of IEEE 802.11p 136
8.2 Procedure 1: The DOAs, $\hat{\theta}^i$, Estimation 149
8.3 Parameters Setting 149
8.4 $\hat{\eta}^{opt}$ in Different SNRs ($M = 10$, $N = 10$, $K = 4$) 150
8.5 DOA Estimation Based on MT-BCS and MSE Values 151
8.6 DOA Estimation Based on MT-BCS ($K = 4$) and MSE values 153
9.1 CV-IMM-KF and GPS Error Comparison 167
10.1 Parameters and Values 197
12.1 Parameters 235

Preface

In recent years, connected vehicles have attracted the interest of the research community; e.g., wireless communications, intelligent vehicle technology, intelligent transportation systems, mobile computing, sensors, multimedia, etc. The vehicular systems can help improve the efficiency and safety of a system. Some potential applications include emergency warning system for vehicles, automatic driving assistance systems, fleet management, smart road, and dynamic traffic light, just to name a few. These applications have been recognized by governmental organizations and vehicle manufacturers. The U.S. Federal Communications Commission has allocated 75 MHz of spectrum band around 5.9 GHz for Dedicated Short-Range Communications (DSRC), which is used for vehicle-to-vehicle (V2V) and vehicle-to-infrastructure (V2I) communications to support vehicular networking applications. Besides standardization efforts such as VSCC in North America, Car-2-Car consortium in Europe, and IEEE 802.11p working group, a number of research projects including NoW, CVIS, Fleetnet, SafeSpot, and SEVECOM have addressed vehicular telematics research around the world. Therefore, there is much interest in better understanding the properties of the related system and developing new application systems.

Connected Vehicle Systems: Communications, Data, and Control is the book to present and discuss the recent advances in theory and practice in connected vehicle systems; it will cover emerging research that aims at dealing with the challenges in designing the essential functional components of connected vehicles. The major topics will cover intra- and intervehicle communications, trace and position data analysis, security and privacy, and mobility control. It starts with (Chapter 1) a summary of automotive applications and provides an outline of the main standards used in the automotive industry, in particular, the networks and their protocols. In-vehicle communications are emerging to play an important role in the continued development of reliable and efficient x-by-wire applications in new vehicles. Chapter 2 presents the advancement of power line communications, which can provide a very low-cost and virtually free platform for in-vehicle communications. The latest Home Plug Green PHY (HPGP) has been promoted by major automotive manufacturers for communications with electric vehicles; Chapter 3 gives the results of its hard delay performance in supporting mission-critical in-vehicle applications. Vehicular ad hoc networks (VANETs) physical channel suffers from serious multipath fading and Doppler spreading; the physical channel behaviors are discussed in Chapter 4. Safety-related applications are geared primarily toward avoiding the risk of car accidents; they include cooperative collision warning, precrash sensing, lane-change warnings, and traffic violation warning—these applications all have real-time constraints. Chapter 5 discusses the short-range communications. Because of the special features of vehicular networks such as mobility, high speed, and self organization, V2V multihop broadcast has become a hot topic in recent years. A high-speed and efficient way to transmit data is discussed in Chapter 6. With the large-scale popularization of GPS equipment, the massive amount of vehicle trajectory data has become an important source for location service, digital map building, and moving trace monitoring. Chapter 7 focuses on road recognition; Chapters 8 and 9 present the DSRC positioning and enhancement methods. VANETs differ from the wired networks and behave in a highly dynamic context; e.g., frequently changing signal-to-noise ratio (SNR) and security risks, which undoubtedly affects the suffered security risk. Chapter 10 presents a lightweight and adaptive security mechanism. Chapter 11 demonstrates a prototype of social-network-enabled transportation system that enables communication between vehicles, monitoring, information gathering, assistant driving, and traffic flow control. Chapter 12 proposes a mobility model to describe the self-organized behavior of the vehicle swarm in VANETs.

Contributors

Falah Ali
University of Sussex
Brighton, United Kingdom

Roberto P. Antonioli
Federal University of Ceará
Fortaleza, Brazil

Min Chen
Huazhong University of Science and Technology
Wuhan, People's Republic of China

Xuting Duan
Beihang University
Beijing, People's Republic of China

Amir Kenarsari-Anhari
University of British Columbia
Vancouver, British Columbia, Canada

Jia-Liang Lu
Shanghai Jiaotong University
Shanghai, People's Republic of China

Yingrong Lu
Beihang University
Beijing, People's Republic of China

Xiaolei Ma
Beihang University
Beijing, People's Republic of China

Victor Ocheri
University of Sussex
Brighton, United Kingdom

Morgan Roff
Queen's University
Kingston, Ontario, Canada

Zhengguo Sheng
University of Sussex
Brighton, United Kingdom

Wei Shu
University of New Mexico
Albuquerque, New Mexico

Nima Taherinejad
Technical University of Vienna
Vienna, Austria

Daxin Tian
Beihang University
Beijing, People's Republic of China

Jian Wang
Jilin University
Changchun, People's Republic of China

Yunpeng Wang
Beihang University
Beijing, People's Republic of China

Min-You Wu
Shanghai Jiaotong University
Shanghai, People's Republic of China

Yue Yang
Beihang University
Beijing, People's Republic of China

Guohui Zhang
University of Hawaii
Honolulu, Hawaii

Xuejun Zhang
Beihang University
Beijing, People's Republic of China

Jianshan Zhou
Beihang University
Beijing, People's Republic of China

Keyi Zhu
Beihang University
Beijing, People's Republic of China

I

INTRAVEHICLE COMMUNICATIONS

Chapter 1

A Survey of Automotive Networking Applications and Protocols

Victor Ocheri
University of Sussex

Zhengguo Sheng
University of Sussex

Falah Ali
University of Sussex

CONTENTS

1.1 Automotive Applications 4
1.1.1 Class A 4
1.1.2 Class B 4
1.1.3 Class C 5
1.2 Automotive Networking Protocols 7
1.2.1 Local interconnect network 7
1.2.2 Controller area network 7
1.2.3 FlexRay 8
1.2.4 Media-oriented system transport 9
1.2.5 Ethernet 9
1.3 Technical Comparisons 10
1.4 Automotive Communication Network Test Bed Implementation 12
1.4.1 The onboard/V2X unit 14
1.4.2 The head/dashboard unit 14
1.4.3 The ZigBee gateway 15
1.4.4 The ZigBee units 15

1.4.5 The roadside unit 15
1.5 Basic Performance Outcomes from the Automotive Network Test Bed 15
1.5.1 Ethernet link test results 15
1.5.2 Wi-Fi link test results 17
1.6 Further Work Areas for the Test Bed 19

In this survey, we summarize automotive applications and provide an outline of the main standards used in the automotive industry, in particular, the networks and their protocols. Moreover, we analyze and provide some insights into how the standard solutions can cope with application characteristics.

1.1 Automotive Applications

The Society of Automotive Engineers (SAE) has classified the automotive applications into Classes A, B, and C with increasing order of criticality on real-time and dependability constraints [45,199].

- Class A, low speed ($<$10 kbps) for convenience features such as body and comfort.
- Class B, medium speed (between 10 and 125 kbps) for general information transfer, such as emission data and instrumentation.
- Class C, high speed ($>$125 kbps) for real-time control, such as traction control, brake by wire, etc.

1.1.1 Class A

Class A is the first SAE classification and maintains the lowest data rate, a rate that peaks at over 10 kbps and must support event-driven message transmission. Usage is for low-end, nonemission diagnostic, general-purpose communication. The implementation of Class A has significantly reduced the bulk of automotive wiring harnesses. Cost is generally about "x" adder per node. According to [143], a very rough estimate of \$0.50 to \$1 may be used for the value of "x" [143]. This cost includes any silicon involved (e.g., microprocessor module or transceiver, etc.), software, connector pin(s), service, etc.

Class A devices typically support convenience operations like actuators and "smart" sensors. Figure 1.1 provides a Class A application zone. Typical applications involving these networks include controlling lights, windshield wipers, doors (e.g., door locks, opening/closing windows), or seats (e.g., seat position motors, occupancy control). The latency requirement for Class A applications varies from 50 to 150 ms [199]. It is noted that a full list of Class A applications is by no means complete and will vary from application to application.

1.1.2 Class B

Class B supports data rates between 10 kbps and approximately 125 kbps and typically supports the vast majority of nondiagnostic, noncritical communications. The utilization of Class B can eliminate redundant sensors and other system elements by providing a means to transfer data (e.g., parametric data values) between nodes; Class B must support event-driven and some periodic message transmission plus sleep/wakeup. Its cost is around $2x$ per node.

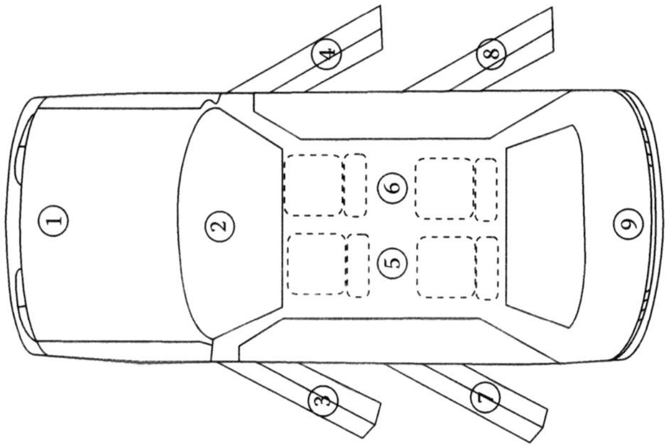

Figure 1.1 Class A application zone. (1, 9) controlling lights, (2) windshield wipers, (3,4, 7,8) doors (e.g., door locks, opening/closing windows), and (5, 6) seats. (From SAE, Class A Application Definition, J2507-1, 1997.)

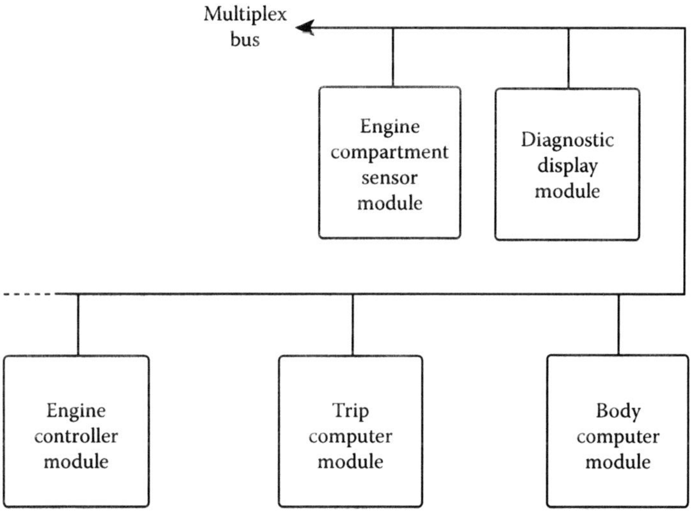

Figure 1.2 Class B multiplex application. (from SAE, Class A Application Definition, J2507-1, 1997.)

The shared information on a Class B network is not critical to the operation of all of the systems to which it is connected. The delay of a specific piece of information will not cause a critical failure in any of the systems. Therefore, the response window in the Class B network is not nearly as narrow as is that in Class C. In fact, the response time may be variable, depending on the application. Another characteristic of a Class B network is its interconnection of dissimilar systems. Figure 1.2 illustrates a Class B application.

1.1.3 Class C

Class C can support data rates between 125 kbps and 1 Mbps. Because of this level of performance, Class C facilitates distributed control via high data rate signals typically associated with critical and real-time control systems (perhaps in the range of a few seconds), such as control of engine and suspension. An unshielded twisted pair is the medium of choice rather than a shielded twisted pair or fiber optics. The cost is about 3*x* to 4*x* per node. However, the upper end to Class C utilization relies on expensive media, like fiber optics, that can push node costs much higher than estimated. Higher performance communication classifications in the range of 1 Mbps to 10 Mbps are expected in the future. Classifications like Class D, devoted to multimedia data (e.g., media-oriented system transport [MOST]), can be expected to push forward bandwidth and performance needs. Figure 1.3 illustrates typical Class C applications.

A hierarchical relationship exists between the classes of networks. By definition, Class C is a superset of Class B. Also, Class B is a superset of Class A. It should be noted that this is a functional relationship only. Therefore, it is important to distinguish between the function and the application of the multiplex network. Table 1.1 summarizes and compares the characteristics of the three application domains.

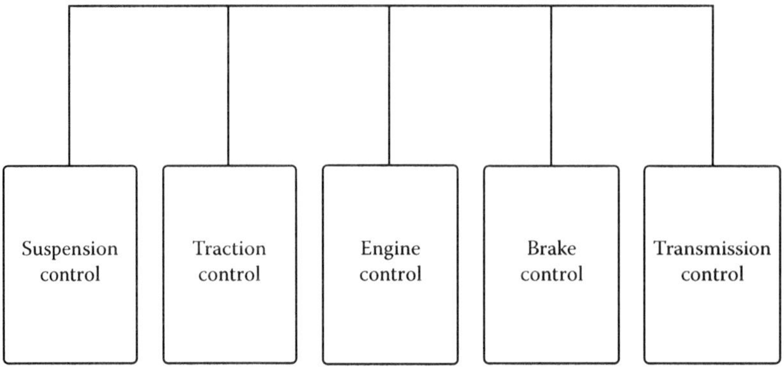

Figure 1.3 Class C applications.

Table 1.1 Comparisons of Functional Classes

	Class A	*Class B*	*Class C*
Purpose	*Sensor/actuator control*	*Information sharing*	*Real-time control*
Information	Real time	Varying window	Narrow window
Latency response time	Wide window (<150 ms)	Information sharing	Real-time control
Priority level	Low (for non-real time)	Low (for non-real time)	High (real time)
Information lost or corrupted	Nuisance	Nuisance or failure	Failure
Cost per unit	<$1	<$2	<$4
Application coverage	Body domain, i.e., comfort functions, dashboard, wipers, lights, doors, windows, seats, mirrors, climate control.	Multiplex bus can cover body domain and interactions among several function domains.	Powertrain (i.e., control of engine); chassis (i.e., control of suspension, steering, and braking). Telematics, multimedia, and human–machine interface (HMI)

It is not a straightforward matter to compare priority levels among functional classes, since each application domain may be controlled by an isolated subsystem using one of the bus protocols. The allocation of application priority is up to a system designer, but industry groups mutually agree on the significance of certain application domains. For example, the Powertrain Control bus is the high-priority information carrier, and the Body Control bus is the low-priority information carrier [213].

1.2 Automotive Networking Protocols

In this section, we introduce the main standards used in the automotive industry, in particular, the networks and their protocols. Due to stringent cost, real-time, and reliability constraints, specific communication protocols and networks have been developed to fulfill the needs of electronic control unit (ECU) multiplexing. The SAE has defined three distinct protocol classes, namely, Class A, Class B, and Class C. Class A protocol is defined for interconnecting actuators and sensors with a low bit rate (about 10 kbps). An example is a local interconnect network (LIN). Class B protocol supports a data rate as high as 125 kbps and is designed for supporting non real time control and inter ECU communication. A low speed controller area network (CAN) is an example of SAE Class B protocol. Class C protocol is designed for supporting real-time and critical applications. Networks like high-speed CAN belong to Class C, which support data rates as high as 1 Mbps. There are even more advanced protocols supporting high-speed communication, such as FlexRay and MOST.

1.2.1 Local interconnect network

LIN is an acronym for local interconnect network and is a low-cost serial bus network used for distributed body control electronic systems in vehicles. The standard is described in Ref. [136]. The LIN consortium includes VW/Audi, DaimlerChrysler, and Motorola. Its development was driven by academic institutions.

It is a single-master/multiple-slave architecture. One node, termed the master, possesses an accurate clock and drives the communication by polling the other nodes—the slaves—periodically, so there is no need for arbitration. A master can handle at most fifteen slaves (there are sixteen identifiers by class of data length). As it is time triggered, message latency is guaranteed. Silicon implementation is cheap, based on the common universal asynchronous receiver/transmitter (UART) and serial communications interface (SCI) hardware.

The LIN can be implemented using just a single wire, and it is commonly used as a sub-bus for CAN and FlexRay by building a hierarchical multiplex system. Speed is 20 kbps; hence, while it is considered to be most appropriate for SAE Class A applications, the speed is actually at the lower end of Class B. The data length can be 1/2/4/8 bytes.

Typical applications involving these networks include controlling doors (e.g., door locks, opening/closing windows) or controlling seats (e.g., seat-position motors, occupancy control).

1.2.2 Controller area network

CAN is without any doubt the most used in-vehicle network. It is a network protocol developed by Robert Bosch GmbH for vehicle systems, but is beginning to be used for linking distributed controllers, sensors, etc., in other fields. The integrated mechatronic subsystems based on low-cost networks are usually interconnected through a CAN backbone. The protocol has been adopted as a standard by the ISO, reference ISO11898 [89].

CAN is a priority-based bus that allows a bounded communication delay for each message priority. The media access control (MAC) protocol of CAN uses carrier sense multiple access/collision detection (CSMA/CD) with bit-by-bit nondestructive arbitration over the ID (identifier) field. The identifier is coded using 11 bits (CAN 2.0A) or 29 bits (CAN 2.0B), and it also serves as priority. CAN uses an Non-return-to-zero (NRZ)-bit-encoding scheme for making feasible the bit-by-bit arbitration with a logical AND operator. Higher priority messages always gain access to the medium during arbitration. Therefore, the transmission delay for higher priority messages can be guaranteed. However, the use of the bit-wise arbitration scheme intrinsically limits the bit rate of CAN as the bit time must be long enough to cover the propagation delay on the whole network. Up to 8 bytes of data can be carried by one CAN frame, and a cyclic redundancy check (CRC) of 16 bits is used for transmission error detection. It supports speeds of up to 1 Mbps, suitable for real-time control applications. CAN needs to be implemented using two wires, and its event-triggered nature is very efficient in terms of bandwidth usage.

Depending on the system requirements, CAN can support two classes of applications:

- Class B networks, between 10 and 125 kbps, used for body domain (low-speed CAN)
- Class C networks, between 125 kbps and 1 Mbps (high-speed CAN used for the powertrain and the chassis domains) for more real-time critical functions. If the distance is <40 m, the rate can be up to 1 Mbps.

CAN is not suited for safety-critical applications, such as some future x-by-wire systems; its main drawback is that a node has to diagnose itself. CAN is also not suitable for transmission of messages of large data sizes, although it does support fragmentation of data that is more than 8 bytes.

1.2.3 FlexRay

FlexRay is a protocol that combines time-triggered (primary) and event-triggered messaging. It is being developed by BMW and DaimlerChrysler with Philips and Motorola, and its purpose is to provide for x-by-wire applications with deterministic real-time and reliability communication. It is capable of a net data rate of 5 Mbps (10 Mbps gross). It is a protocol in bus architectures for safety–critical embedded systems and advanced control functions.

At the MAC level, FlexRay defines a communication cycle as the concatenation of a time-triggered (or static) window and an event-triggered (or dynamic) window. Time-triggered operation provides efficiency, determinism, and partitioning, but at the price of flexibility, whereas an event-triggered system responds to stimuli that are outside its control. To each communication window, whose size is set at design time, a different protocol applies. The time-triggered window uses a time division multiple access (TDMA) protocol. In the event-triggered part of the communication cycle, the protocol is flexible TDMA (FTDMA): the time is divided into so-called mini-slots. Each station possesses a given number of mini-slots (not necessarily consecutive), and it can start the transmission of a frame inside each of its own mini-slots. A mini-slot remains idle if the station has nothing to transmit. The communication cycles are executed periodically.

The highest priorities in FlexRay usually have slots allocated in the static (time-triggered) part, and lower priority sources in the dynamic (event-triggered) part, so a source has a slot in one or the other. This presumably reduces jitter for messages allocated in slots in the static portion (as compared to Byteight) as their timing is constant, unlike in Byteight, where vacant slots are shortened. FlexRay differs from SAFEbus and time-triggered architecture (TTA) in that the full schedule for the system is not installed in each node during construction. Instead, each node adapts to the full configuration of the system through exposure to messaging traffic.

The FlexRay network is very flexible with regard to topology and transmission support redundancy. It can be configured as a bus, a star, or multistars, and it is not mandatory that each station possesses replicated channels, even though this should be the case for x-by-wire applications. FlexRay should be classified as Class D networks (speed over 1 Mbps), which are x-by-wire applications that need predictability and fault tolerance.

1.2.4 Media-oriented system transport

MOST [44] is a multimedia fiber optic network developed in 1998 by MOST Cooperation (a consortium composed of carmakers, set makers, system architects, and key component suppliers). The basic application blocks supported by MOST are audio and video transfer, based on which end-user applications like radios, GPS navigation, video displays and amplifiers, and entertainment systems can be built.

The MOST protocol defines data channels and control channels. The control channels are used to set up what data channels the senders and receivers use. Once the connection is established, data can flow continuously, delivering streaming data (Audio/Video). MOST provides point-to-point audio and video data transfer with a data rate of 24.8 Mbps. Class D networks (speed over 1 Mbps) are devoted to multimedia data (e.g., MOST).

1.2.5 Ethernet

The Ethernet is a popular communication network technology used mainly in local area networks (LANs). Compared to the other protocols, the use of the Ethernet in cars is a relatively new development. Its specifications are detailed in the IEEE 802.3 standard. It was developed in the 1970s by Robert Metcalfe and David Boggs of the Xerox Corporation [60].

Several versions of the Ethernet exist beginning with the basic 10 Mbps version that uses a twisted-pair cable as a medium (10BASE-T). Other versions are the 100BASE-T Ethernet, which is capable of a transmission rate of 100 Mbps, and the 1000BASE-T Ethernet, also called the Gigabit Ethernet. The aforementioned versions are the most commonly used ones. For vehicular network purposes, the 10BASE-T and 100BASE-T are the most popular as Broadcom's BroadR-Reach 10/100 PHY Ethernet module (which is currently the de facto standard for Ethernet in automotive application) supports these Ethernet versions.

The most common transmission medium for the Ethernet is twisted-pair cables. Optical fibers are also attractive because of the immunity they provide from interference. One of the optical fibers gaining prominence for Ethernet connections is the plastic optical fiber (POB). In the past, coaxial cables were used. Of all these transmission media, twisted-pair cables are currently preferred due to their robust nature.

The Ethernet utilizes the CSMA/CD scheme in controlling access to a transmission medium. Essentially, nodes on an Ethernet network have to compete with each other to use the transmission medium. An Ethernet node that wants to transmit a message checks to see that there is no traffic in the shared transmission medium [60]. This is carrier sensing. If traffic is noticed in the medium, the node halts an attempt to transmit and waits for the medium to become idle before attempting again to transmit; otherwise it transmits on a free medium. Each Ethernet message frame transmitted on the network specifies a 48-bit source address, as well as a 48-bit destination address. These are addresses of the node transmitting and the node intended to be a recipient of the transmitted message, respectively. Once a transmitted message reaches the intended destination node, it is received via the Ethernet transceiver component of the Network Interface Card (NIC) in a node, after which the data is modified and made available to the node to process. A situation might arise where two nodes sense that the transmission medium is free and hence both transmit at the same time, resulting in collision. In this case, both nodes are notified of the collision. This is sensed as an increase in voltage by an NIC. The first node that senses the collision sends a signal to the other transmitting nodes indicating they should stop transmission. Each node would have to wait for a period time (a backoff period) before attempting to retransmit data. The backoff period is calculated by a Binary Exponential Backoff (BEB) algorithm [60]. The process is random and not the same for any two nodes.

1.3 Technical Comparisons

In this section, we mainly analyze and compare the performance of the automotive communication protocols discussed earlier and provide some future perspective on automotive communication development. Table 1.2 compares the main characteristics of automotive protocols.

In Figure 1.4, we give an example of automotive network architecture.

It is common in today's vehicles that the electronic architecture includes four different types of networks interconnected by gateways and up to 2500 signals (i.e., elementary information such as the speed of the vehicle) are exchanged by up to 70 ECUs.

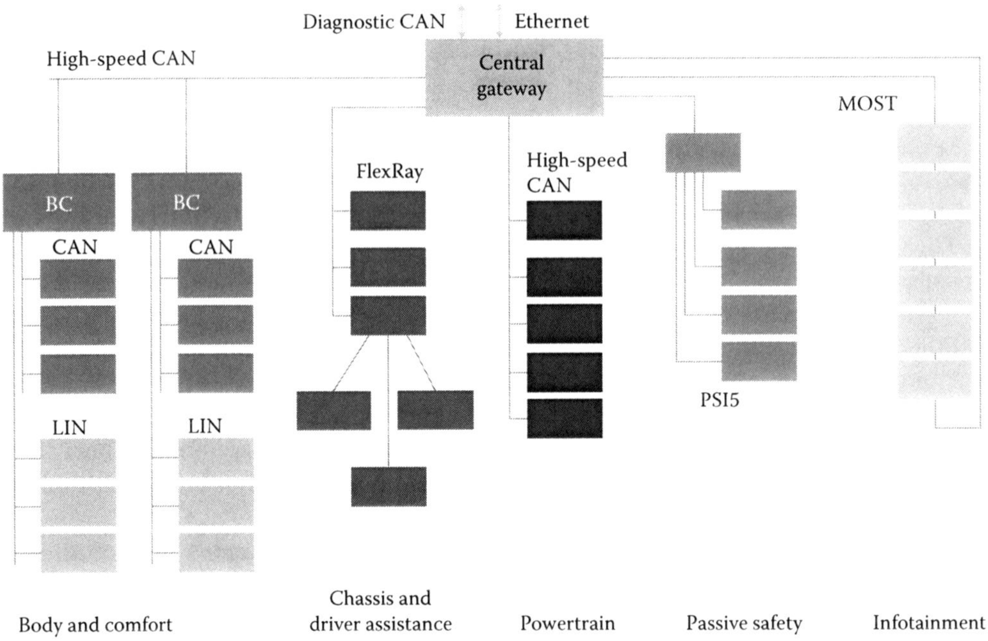

Figure 1.4 Automotive network architecture. (from H. Zinner, O. Kleineberg, C. Boiger, and A. Grzemba. Automotive requirements and definitions. Online presentation slides, 2012.)

Table 1.2 Comparison of Automotive Protocols

	Automotive Networking Protocols				
	LIN	*CAN*	*FlexRay*	*MOST*	*Ethernet*
Classification	Class A	Class B and C	Class D	Class D	Class D
Application	Body and comfort	Powertrain, driver assistance control (high speed); body and comfort (low speed)	Chassis, driver assistance, safety control	Infotainment: stream data and control	Infotainment, telematics, camera-based drivers assistance
Topology	Hierarchical bus	Hierarchical bus	Bus, star, multistar	Point-to-point	Star, point-to-point
Media	Single wire	Twisted-pair	Twisted pair or fiber	Optical	Twisted-pair
Bit encoding	NRZ	NRZ-5, MSb first	NRZ	BiPhase	Manchester Phase Encoding (MPE)
Schedule approach	Time triggered	Event triggered	Time and event triggered	Event triggered	Event triggered
Media access	Master/slave	Contention	TDMA with priority	Master/slave	Contention
Error detection	8-bit CS	CRC	24-bit CRC	CRC	CRC
Header length	2 Bits/Byte	11 or 29 Bits	40 Bits	Not specified	14–22 bytes
Data length	8 Bytes	0–8 Bytes	0–246 Bytes	Not specified	0–1500 bytes
In-message response	No	No	No	No	Not specified
Bit rate	20 kbps	10 kbps–1 Mbps	10 Mbps	25 Mbps	10 Mbps–100 Mbps
Maximum bus length	40 m	Not specified, typical 40 m	Not specified	Not specified	100 m
Maximum node	16	Not specified, typical 32	Not specified	24	Theorically 1024
Cost	Low	Medium	Medium	High	Not specified

Source: Adapted from C. A. Lupini. In-vehicle networking technology for 2010 and beyond. s.l. : SAE International, 2010.

At the end, we provide a road map of the automotive communication development in Figure 1.5. It is clear that the Ethernet technology will be a promising, universal, and low-cost solution for future automotive communications.

1.4 Automotive Communication Network Test Bed Implementation

In light of the trend toward the use of the Ethernet for automotive communication purposes, as indicated by the road map in Figure 1.5, and given increasing interest in the use of wireless communication technologies for automotive applications [62] such as in telematics [63], there is a need to integrate the Ethernet with legacy automotive communication technologies as well as to introduce wireless network technologies in the implementation of automotive networks. In an attempt to satisfy this need, a test bed was developed that utilizes CAN together with Ethernet, as well as the ZigBee wireless technology. The test bed also has Wi-Fi capability.

The aim of implementation of the test bed is to explore an automotive network design that integrates Ethernet, ZigBee, and Wi-Fi with the legacy CAN in an implementation of a prototype platform for automotive networks.

Working on the integration of these technologies has the potential of resulting in a communication network that is flexible enough to accommodate new requirements that might be associated with proposed new use cases or features for vehicles in the future.

The typical automotive network is hierarchical and consists of a backbone network and lower level networks [61]. The backbone network interconnects ECUs and the lower level networks via gateway units, [61] and is predominantly based on CAN. The lower level networks (or subnets) include networks, usually based on LIN protocol, that connect sensors, switches,

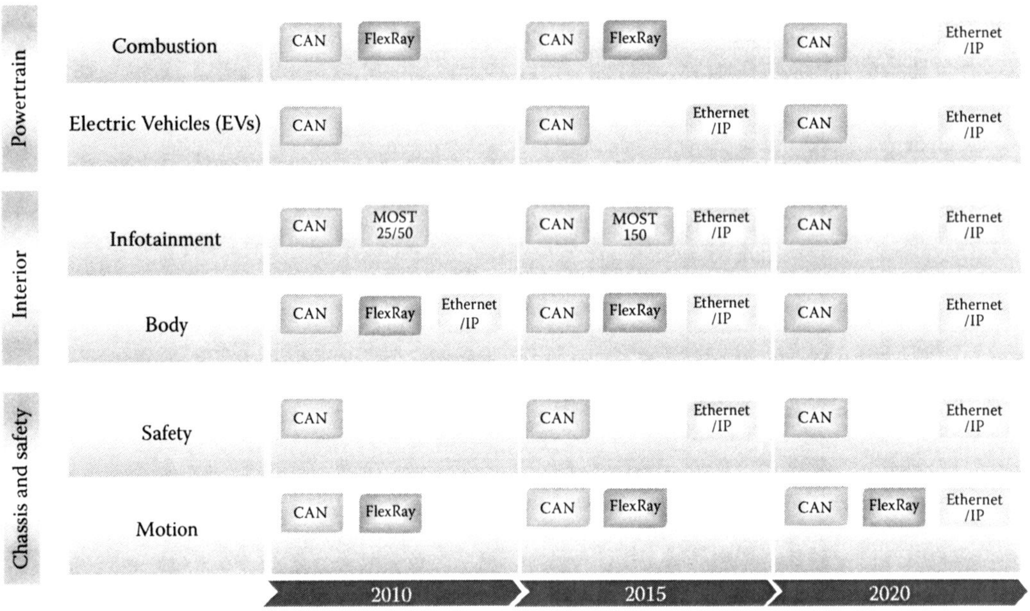

Figure 1.5 A road map of automotive communication development. (From H. Zinner, et al. Automotive requirements and definitions, http://www.ieee802.org/1/files/public/docs2012/new-avb-zinner-boiger-kleineberg-automotive-requirements-0112-v01.pdf, 2012.)

and vehicular actuators to the backbone network. Gateway units connect the subnets to the backbone network.

In the test bed design, key components considered include the following:

- A head/dashboard unit for monitoring purposes
- Gateway units for interconnecting different communication media
- Onboard unit(s) to handle external data
- Other arbitrary processing units that would be representative of other ECUs in a car and would be used for the purpose of generating artificial or real data. These units should be able to host sensor and actuator units for generating real data to be used in the test bed.

These components are linked together by appropriate communication links (wired or wireless) as shown in Figure 1.6.

In the design, shown in Figure 1.6, the nodes in the ZigBee network are representative of arbitrary processing units. The roadside unit (RSU), which is not a typical component in an in-vehicle network, is also taken into consideration for the purpose of demonstrating the functionality of the onboard unit.

On the basis of the design in Figure 1.6, a minimum viable automotive network test bed platform was implemented using the popular electronics prototyping boards Raspberry Pi and Arduino. The implementation is shown in Figure 1.7.

Add-on controller boards were attached to two of the four Raspberry Pi 3 boards to give them CAN communication capability. Also, the Arduino boards were adapted for ZigBee communication by using XBee shield (currently replaced by the Arduino wireless shield) together with XBee Coordinator and Router modules.

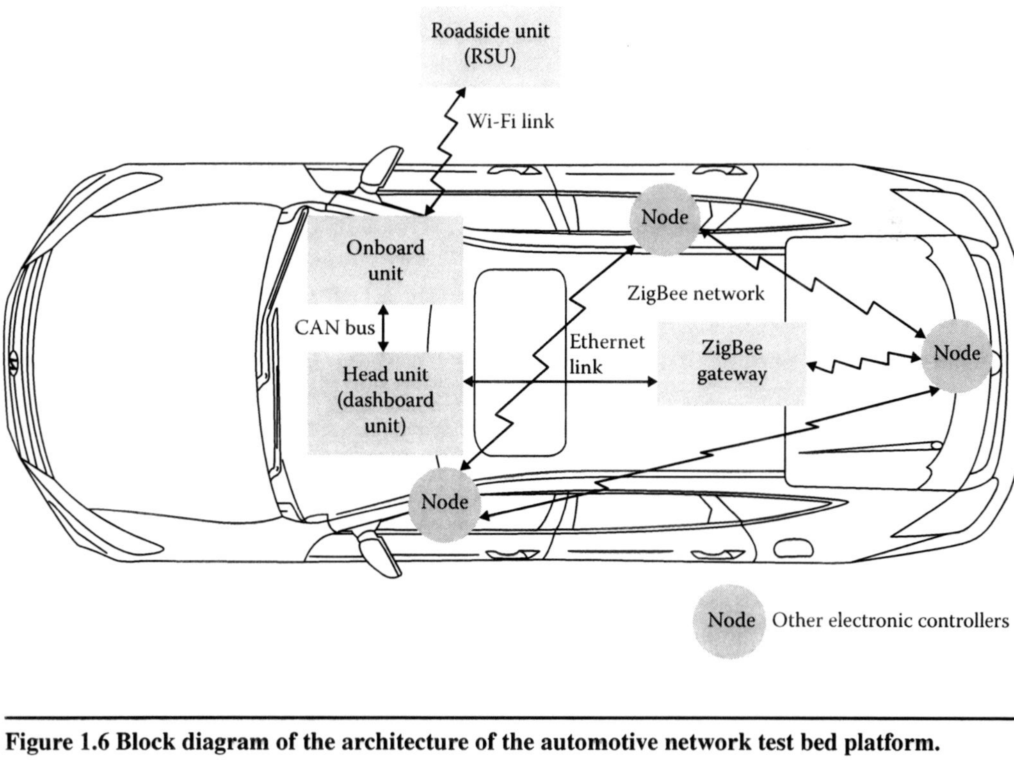

Figure 1.6 Block diagram of the architecture of the automotive network test bed platform.

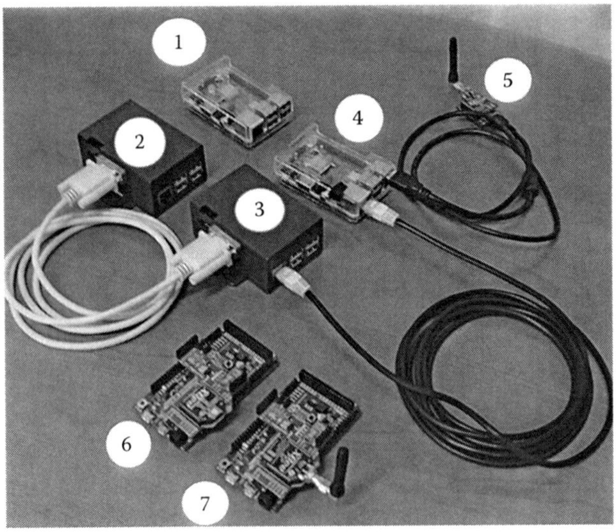

Figure 1.7 An implementation of a minimum viable test bed for automotive communication using Raspberry Pi and Arduino boards.

Details of each of the units of the test bed are provided in the next few sections.

1.4.1 The onboard/V2X unit

This unit is responsible for handling of communications with entities external to the test bed (which is representative of the in-car network). It is the Vehicle-to-Everything (V2X) communications unit.

In the test bed shown in Figure 1.7, the unit is marked with the number “2,” and it consists of a Raspberry Pi 3 with a CAN controller/transceiver add-on board as well as a CAN terminator that contains a 120 Ω resistor for terminating the CAN bus.

This unit is capable of communicating using the Transmission Control Protocol (TCP) and the CAN protocol. It was configured as a TCP client capable of sending out TCP requests. The data contained in the received response to each TCP request is loaded into a CAN frame, after which it is sent over the CAN bus (the white DB9 connector) to the head unit (marked with “3” in Figure 1.7) for the purpose of display and for further transmission to other entities in the test bed.

It is important to note that the onboard unit is connected to the CAN bus via a CAN terminator because it is situated at one end of the bus. In future attempts to expand the CAN bus of the test bed, nodes that are not connected to one of the two endpoints of the CAN bus should not be connected to the bus via a CAN terminator (or a terminating resistance).

1.4.2 The head/dashboard unit

The head unit is the monitoring and control unit of the test bed, just like the dashboard of a car. A visual display would normally be connected to this unit. In Figure 1.7, it is marked with the number “3.”

The head unit consists of the same hardware as the onboard unit, and it is also capable of TCP and CAN communication. It was configured as a TCP client so that it could send a TCP request over the Ethernet connection to the ZigBee gateway. Conversely, it can also collect data from a TCP response message received via the Ethernet. This data is then loaded into a CAN frame to be sent over the CAN bus.

It important to note that a CAN terminator is used for the head unit because it was connected to one end of the CAN bus, which in the case of the test bed consists of just two nodes (the onboard unit and the head unit).

1.4.3 The ZigBee gateway

This unit acts as an interface between the Ethernet connection and a ZigBee network. It is marked with the number "4" in Figure 1.7. The ZigBee gateway is responsible for adding ZigBee communication capability to the test bed. It consists of a Raspberry Pi 3 board with an XBEE ZigBee module connected to it via an XBEE-to-USB adapter.

The Zigbee gateway was configured as a TCP server that creates TCP responses using the data received from the ZigBee network as a result of TCP requests originating from the head unit. Also, it is able to extract data from a TCP request and send it serially over the Zigbee network.

1.4.4 The ZigBee units

The ZigBee units, which are marked "6" and "7" in Figure 1.7, are control units that are meant to be connected to various sensors and actuators. These units are capable of forming a ZigBee network as well as generating and receiving serial data. It consists of Arduino boards with XBEE ZigBee modules installed on them using an XBEE Arduino shield (expansion board).

1.4.5 The roadside unit

The RSU in the test bed is representative of an external entity that can communicate with an in-vehicle network. This unit is marked "1" in Figure 1.7.

In the test bed, it consists of a Raspberry Pi 3 that was configured as a Wi-Fi access point. It also has a TCP server running on it. The unit operates by responding to the recognized queries (in the form of a TCP request) that originate from the onboard unit of the test bed. An example of such a query is a request for traffic light state.

1.5 Basic Performance Outcomes from the Automotive Network Test Bed

The testing was focused on the Ethernet because it was the intermediary connection on the test bed. It is presumed that restrictions or limits imposed by the Ethernet connection dictate the performance of the core portion of the test bed (which is representative of an in-vehicle network of the test bed design). Also, a test was conducted on the Wi-Fi connection (representative of a V2X link) to evaluate V2X performance.

Data transmitted were randomly generated in both directions in the test bed. Also, the software Wireshark and Riverbed's Packet Analyzer was used to obtain the results shown subsequently.

1.5.1 Ethernet link test results

This section focuses on the throughput, latency, and overhead associated with a very basic transmission of arbitrary data.

The average data bit rate for transmission over the Ethernet link was found to be, between 1.8 and 2.0 kbps as shown in Figure 1.8. This is for the most basic transmission scenario that involves relay of data from two sources in opposite directions on the test bed. That is, from the onboard unit toward the ZigBee network and vice versa.

The given throughput value is a small portion of a 10 Mbps Ethernet connection, and it indicates the availability of room for more traffic associated with more nodes or data-intensive applications. This is one of the possible reasons that make the use of the Ethernet for automotive networks a really attractive proposition.

The overhead associated with this basic transmission over the Ethernet used close to 800 bps of the used bandwidth as shown in Figure 1.9.

Specifically, the overhead was about 40% of the total used bandwidth as is shown in Figure 1.10. The overhead is associated with the proportion of total throughput that is used by portions of an Ethernet frame excluding the payload data. This includes the preamble, CRC, and the interframe gap portions of the frame. This overhead for the Ethernet connection is getting close enough to about half of the total throughput, and hence care has to be taken to ensure that in any bid to make modification to the Ethernet for automotive applications, the overhead is not unnecessarily increased.

The round-trip time (RTT) was found to be between 300 and 350 s. This is shown in Figure 1.11. RTT specifies a measure of the latency in data transmission via the Ethernet link.

Figure 1.8 Distribution of data bit rate on the Ethernet connection over the period of testing.

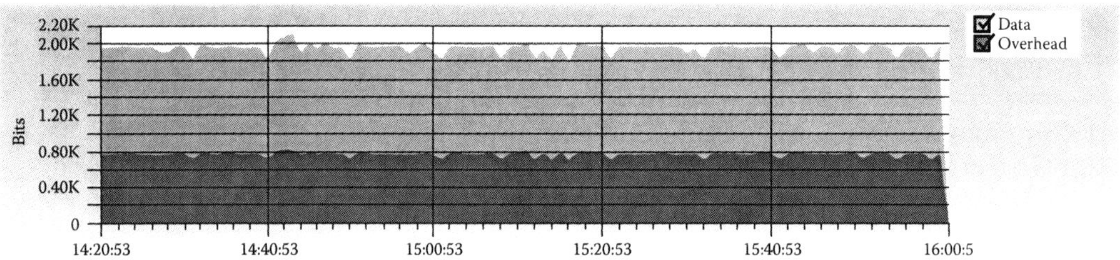

Figure 1.9 Partitioning of total data throughput on the Ethernet connection into overhead and data traffic.

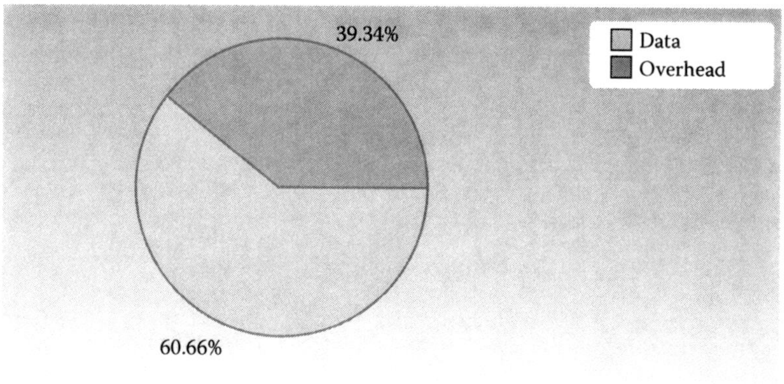

Figure 1.10 Proportion of the traffic on the Ethernet link used by overhead bits as well as data bits transmission.

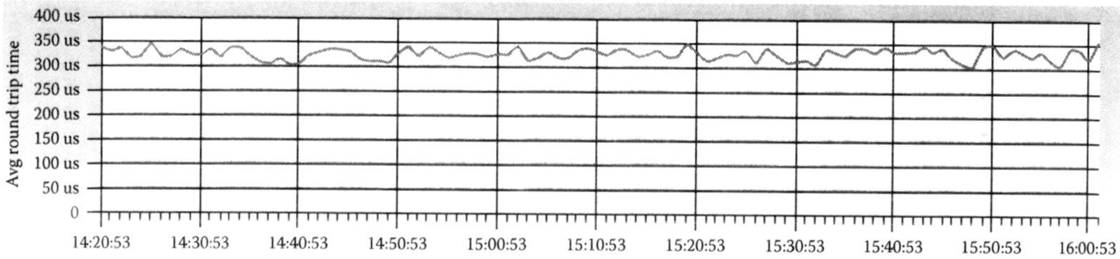

Figure 1.11 Graph showing the average latency in signal transmission over the Ethernet link of the test bed.

While the test bed may not have factored in conditions that contribute to latency in actual automotive network, this aforementioned RTT value at least provides a hint of the possibility of reducing latency in automotive network by the introduction of Ethernet connections.

1.5.2 Wi-Fi link test results

The results presented here assess the performance of the Wi-Fi link between the RSU and the onboard unit.

The Wi-Fi throughput was within the range of 8–14 kbps as shown in Figure 1.12. This is a small portion of a 54 Mbps Wi-Fi connection.

The overhead in the Wi-Fi transmission, which is due to portions of the transmitted message frames other than the data frame, was observed to amount to a traffic of about 3 kbps, as can be seen in Figure 1.13.

This implies that the overhead consists of a little over 18% of the total consumed bandwidth. The pie chart in Figure 1.14 shows the proportion of overhead in the Wi-Fi transmission throughput. This overhead value is noticeably less than that for the Ethernet connection, which

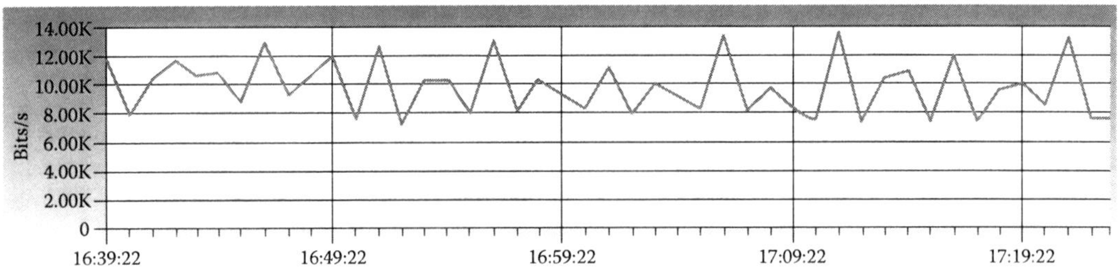

Figure 1.12 Graph showing the traffic over the Wi-Fi connection between the onboard unit and RSU of the test bed.

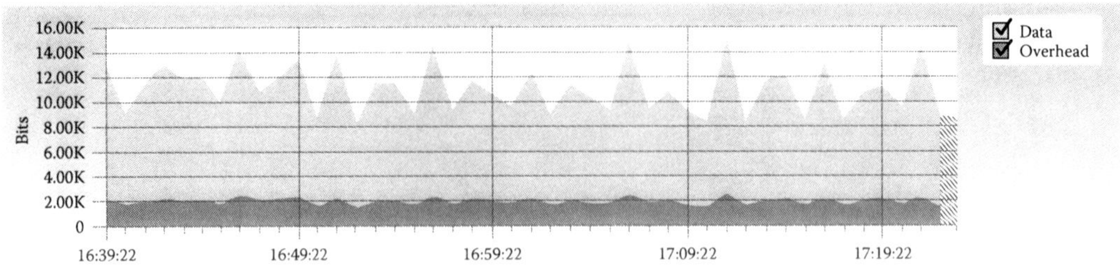

Figure 1.13 Graph showing proportion of traffic used by overhead in the Wi-Fi connection of the test bed.

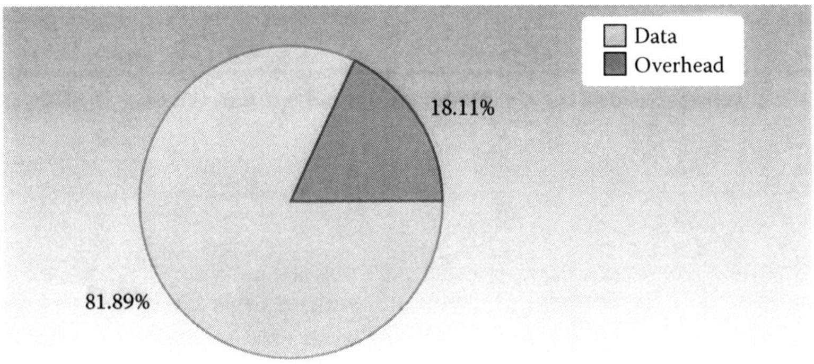

Figure 1.14 Pie chart showing the proportion of traffic that comes from transmission of overhead bits.

was at 40%, and it probably allows room for modification efforts that may attempt to adapt Wi-Fi for vehicular communication.

The associated latency in transmission over the Wi-Fi connection, as assessed using RTT, was 5–25 ms. This can be seen in Figure 1.15.

This latency is larger than the value observed for the Ethernet connection. This might have to do with the retransmissions as a result of transmission errors due to information loss.

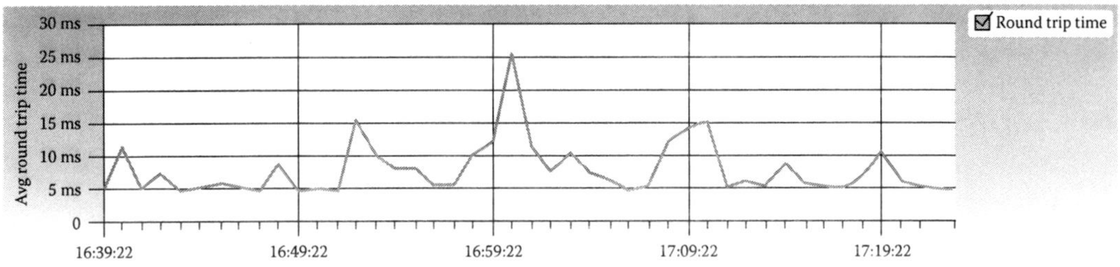

Figure 1.15 Graph showing the average RTT over the Wi-Fi connection.

1.6 Further Work Areas for the Test Bed

Certain areas in real automotive communication networks were not factored into the test bed discussion. But there is room to include them so as to take into consideration real-life automotive network conditions.

First, more nodes need to be included. For the CAN bus, an adapter could be used to include more CAN nodes (which could be representative of other arbitrary control units in cars). An Ethernet switch could be used to add more nodes to the Ethernet connection on the test bed.

Secondly, there is the need to make the data transmitted as close as possible to what is obtainable in automotive networks as opposed to the use of just random data.

Another area to consider is that, in the use of TCP in the automotive test bed implementation, there is the key challenge of ensuring that the server scripts are always ready before any TCP request is sent on the network by any TCP client script. Measures have to be taken to ensure that the system start-up sequence prioritizes the start of components running TCP servers above those housing TCP client operations.

In conclusion, for the design of automotive communication networks, the use of Ethernet and wireless communication technologies is clearly an interesting and viable option. The test bed discussed here hints at an approach. Also, it could be integrated with a host of other communication technologies, including legacy types (like CAN), as was demonstrated in the considered test bed implementation.

Chapter 2

CAN-Based Media Access Control Protocol

Zhengguo Sheng
University of Sussex

Amir Kenarsari-Anhari
University of British Columbia

Nima Taherinejad
Vienna University of Technology

Yunpeng Wang
Beihang University

CONTENTS

2.1 Introduction 22
2.2 Related Work 23
 2.2.1 Physical layer 23
 2.2.2 MAC layer 25
2.3 System Assumption and the Proposed Multichannel Contention Resolution Method 26
2.4 Performance Analysis under Perfect Sensing 27
 2.4.1 Throughput evaluation 31
2.5 Performance Analysis under the Presence of Sensing Errors 31
 2.5.1 Impulse noise filtering 31
 2.5.2 Impact of sensing errors 33
2.6 Numerical Results 36
2.7 Conclusions 45

2.1 Introduction

Over the past few years, we have witnessed an increasing interest in the use of power line communication (PLC) for home automation systems, automatic meter reading, real-time energy management systems, and many other applications. Recent research efforts have been focusing on the use of in-vehicle direct current (DC) power lines as a physical medium for data communications [13,19,51,201,222]. PLC is a promising method that can potentially reduce the complexity, cost, weight of the wiring harness, and fuel consumption of the vehicles.

The measurements of vehicular power line communication (VPLC) channels [19,51,201, 222], however, indicate that there are still a number of challenges for communications over power wires. These challenges include channel transfer functions varying in both time and frequency and experiencing deep notches [201,222], change of access impedance seen by communication devices varying in both time and access location [222,223], and the presence of nonstationary impulsive noise generated by various electrical devices connected to the VPLC networks [51]. These challenges make the impact of sensing errors on the performance of the in-vehicle communications a further important issue to be considered.

Recently, a media access control (MAC) protocol, *contention detection and resolution* (CDR), has been proposed for VPLC in Ref. [13]. To the best of our knowledge, this CDR protocol is the only existing well-established random access protocol designed for VPLC systems. The contention mechanism in CDR works as follows: Nodes use an n-bit random arbitration register (RAR) to randomize their access to the medium. Initially, a node waits until the power line is idle, followed by a random delay chosen uniformly from $[0, (n-1)\sigma_{\text{slot}}]$, where σ_{slot} is the duration of a single time slot, and n is the number of slots. After that, all nodes in contention switch between carrier sense (cs) and carrier transmission (ct) modes according to the content of their RARs and drop out of contention if they are listening to and hearing a carrier on the power line. The aim of the CDR is to have only one node remaining at end of contention to access the power line channel. Existing literature, e.g., Ref. [13], has analyzed the collision probability of the proposed MAC with the assumption of perfect sensing and compared the CDR protocol with carrier sense multiple access with collision avoidance (CSMA/CA) protocol to show the performance improvement of using CDR protocol for VPLC systems.

In this chapter, based on the challenges of VPLC mentioned earlier, we introduce an MAC protocol that provides access by resolving the contention using a combination of time and frequency multiplexing. In brief, the proposed protocol works as follows: First, each transmitter selects one of the available frequency channels based on a prespecified probability and then, on each channel, the contention is resolved over several slots in which nodes probabilistically send a carrier on the channel. At the end of the last slot, the receiver starts to scan the signal level from the first channel and locks and receives the packets from the first nonidle frequency channel. The use of multiple frequency channels is motivated by the fact that they can potentially provide robustness against interference and noise by periodically switching between frequency channels. During a slot, nodes sensing a busy channel will retire from contention, and nodes sending carriers on the same channel will move to the next slot. Different from our preliminary studies in Refs. [115,116], we aim to solve a joint optimization problem by deriving distributed channel selection and collision resolution such that they can reduce the chances of collision among transmitters. Moreover, we extend our work to investigate the effect of sensing errors on the MAC protocol performance and apply the optimal sensing detection method to improve the MAC protocol efficiency. In essence, our proposed MAC protocol provides fast collision resolution, demonstrates high efficiency under different traffic loads, and can be implemented directly in the hardware to enhance the performance of the VPLC system.

The rest of the chapter is organized as follows: In Section 2.2, we review the state-of-the-art developments of related work and emphasize the motivation and importance of our work. In Section 2.3, we describe our system assumption and present a brief description of the MAC protocol operation. Section 2.4 provides mathematical analysis of the proposed MAC protocol, under the assumption of perfect sensing, whereas in Section 2.5, we present the mathematical analysis of the protocol with the presence of sensing errors. We present numerical results in Section 2.6, and conclude in Section 2.7.

2.2 Related Work

With the emerging automated tasks in the vehicle domain, the development of in-vehicle communications is increasingly important and is subjected to new applications. Although both wired and wireless communications have been largely deployed for supporting diverse applications, most in-vehicle applications with time-critical nature, such as brake and engine controls, still prefer dedicated wired networks for reliable transmission. According to Ref. [11], the growth of electronic components in vehicles is in the order of n^2, where n is the number of electronic control units (ECUs). In other words, if each node is interconnected with all the others, the number of links grows by the square of n, which means that the wired strategy will be unable to cope with the increasing use of ECUs due to the problems of weight, cost, complexity, and reliability induced by the wires and the connectors. Motivated by the use of networks where the communications are multiplexed over a shared medium, VPLC has recently been considered by physical layer researchers as a low-cost and efficient way to deliver in-vehicle communications. This solution, considering its specific characteristics, consequently requires new defined protocols for managing communications and, in particular, for granting bus access. The rest of this section provides a deeper insight into the properties of the physical layer and challenges that impose on communication systems, as well as related works on communication protocols with respect to VPLC.

2.2.1 *Physical layer*

Understanding the characteristics of power wires in vehicles as a communication channel has been the drive for many measurement campaigns [18,52,75,135,222,223,233,272]. The findings show that vehicle power lines constitute a harsh and noisy transmission medium with both time- and frequency-selective channels, colored background noise, and periodic and aperiodic impulsive noises, see, e.g., Refs. [69,176]. These characteristics make obtaining a deterministic description of the channel and its noise an extremely complicated and cumbersome task. The tree-shaped topologies of the cable bundles, as well as the type, size, and length of cables in the bundle itself, are also quite different, which creates further diversity of channel characteristics among different vehicles. Therefore, a proper modeling of a channel would need quite a lot of detailed information on type and length of cables used to connect different nodes to their bundling, which can be a source of some characteristics especially cross-transmission interferences. In addition, the body of a vehicle, when used as the return path for many ground signals, affects the channel characteristics, which are different from one vehicle to another. Furthermore, the highly variable activation schedules of electrical functions such as windshield wipers or antilock braking systems (ABS), which produce sharp modifications in the circuits' load impedances over brief time intervals [135], can impose serious challenges on communication devices or even interrupt communications. These noise sources would also be largely different from one vehicle to another, since not all the vehicles use the same devices and in the same manner. In consequence, even if modeling the channel in a vehicle—however cumbersome—is achieved, the deterministic description and any derived conclusion thereof would be very specific to that certain model of vehicle and not useful for other vehicles. All these issues lead to a need for extra considerations in protocol design and necessitate a protocol that can address VPLC channel challenges in a rather more general manner.

To provide the reader with a better perspective on the nature and depth of the aforementioned challenges in the physical layer and why using a closed-form solution or deterministic model would not be a suitable path to take in order to design a communication system applicable as a general solution for all vehicles, we will demonstrate some of the measurement results obtained by our research team members working on the physical layer (Figure 2.1).

Characteristic measurements of the physical layer are often restricted by the access to vehicles for performing the measurement test. Existing measurements have been done for various vehicles, such as [16,222,223] for the internal combustion engine (ICE) vehicle and [18,19,222] for the electric or hybrid electric vehicle (HEV). In essence, we can learn that the channel attenuations are very link dependent and also vary from car to car. In a broader perspective, it can be observed that all the channels are fairly frequency selective, with random deep notches. Readers can refer to these papers for details of measurement setups, results, calculations, and discussion.

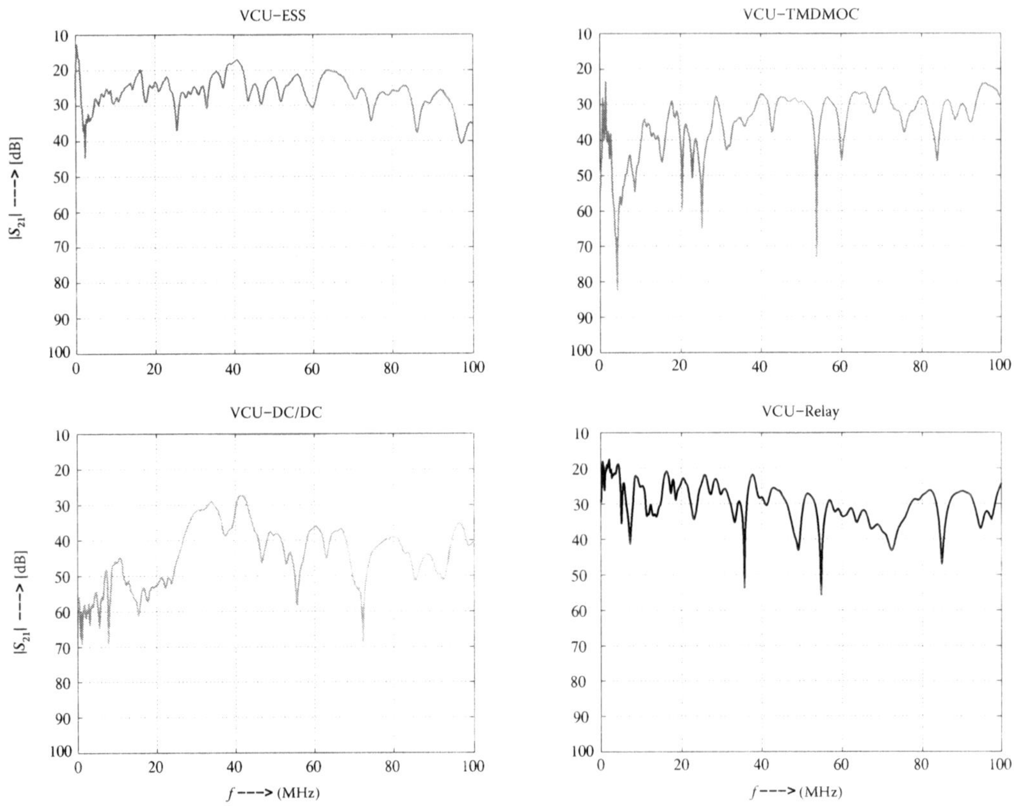

Figure 2.1 Magnitude of $S_{21}(f)$ for different connections from hybrid electric vehicle (HEV) VCU to potential communication nodes, where VCU is Vehicle Control Unit, ESS is Energy Storage System, TMDMOC is Traction Motor Digital Module Controller, and DC/DC stands for DC-to-DC converter unit of the vehicle. (From N. Taherinejad, et al. *Proceedings of IEEE International Symposium on Power Line Communications and Its Applications*, 440–445, Udine, Italy, April 2011.)

To further investigate the issue of noise on the signal being transmitted through the channel, a sample of measurements in the time domain showing the inflicted noise on the signal can be found in Ref. [51]. Specifically, we can observe different events happening that have inflicted noise upon the signal. Some of these noise events are rather periodical, which could be due to the operation of a certain device and hence could be rather predictable. However, others are neither periodical nor predictable. These noises are most likely due to the activation of different devices or loads inside the vehicle, as they have a different nature as well as a different effect on the signal. Therefore, a communication protocol that ascertains robustness and reliability of the communication despite all these challenges in physical layer plays an important role not only in establishing effectiveness but also in establishing the practicality of using power wires for communication purposes inside vehicles.

2.2.2 MAC layer

Beside the challenges in physical layer, the requirements of applications with time-critical nature can be fundamentally different from those of applications for which current MAC protocols are designed [89]. For example, energy is a valuable resource of sensor devices, and most of the existing MAC protocols are optimized to conserve energy, trade-off latency and throughput, etc. These protocols are typically not suitable when a vehicular application demands real-time requirements, because the energy sources are not as critically in shortage as in other scenarios. According to Ref. [128], some critical control messages in a car need to be delivered within a very small delay, such as 100 μs. The challenge in using existing solutions, such as IEEE 802.3 for Ethernet, as the universal in-vehicle network lies in meeting the real-time requirements of various time-critical car applications, since they all experience a longer delay[1] and do not provide the quality-of-service (QoS) guarantee on the minimum delay or bandwidth. The wireless interconnection of sensors and other devices within the vehicle, such as radio frequency in the IEEE 802.x based solutions [57,158], is also being investigated. Although there are advantages in using wireless transmission, such as lessening weight and physical network complexity, in-vehicle wireless devices still require connection to the electrical power source in the vehicle, which mitigates this advantage [231]. There are also concerns being raised about the security of wireless networks, such as the potential for eavesdropping on an in-vehicle network or of reverse engineering to jam false data are possible in a moving vehicle. This is particularly important, since the safety of the in-vehicle network is critical, and it is imperative to avoid security problems that lead to disastrous safety implications. In essence, if reliability or security gives higher priorities in in-vehicle communications, communication protocols need to be redesigned from the application's perspective.

There have been state-of-the-art works on in-vehicle MAC protocol design, since it is a fundamental issue in enabling channel access control, which makes it possible for several ECUs or network nodes to communicate in a multiple access network incorporating a shared medium, e.g., twisted-pair cable, and thus support upper layer protocols for application services. Local Interconnect Network (LIN) [189] is a low-cost serial bus network used for distributed body control electronic systems in vehicles. It is a single master/multiple slave architecture. One node, termed the master, possesses an accurate clock and drives the communication by polling the other nodes—the slaves—periodically. As it is time triggered, message latency is guaranteed. The LIN can be implemented using just a single wire. However, since the speed is only 20 kbps, it is considered to be the most appropriate for less time-critical applications, such as controlling doors (e.g., door locks, opening/closing windows) or seats (e.g., seat posi-

[1] A 1518-byte Ethernet message would take 122.08 μs to be forwarded in a 100 Mbps Ethernet switch.

tion motors, occupancy control). Controller Area Network (CAN) [32] is a priority-based bus that allows to provide a bounded communication delay for each message priority. The MAC protocol of CAN uses carrier sense multiple access/collision detection (CSMA/CD) with bit-by-bit nondestructive arbitration over the Identifier Field, which serves as priority. Therefore, the transmission delay for higher priority messages can be guaranteed. However, the use of a bit-wise arbitration scheme intrinsically limits the bit rate of CAN as the bit time must be long enough to cover the propagation delay on the whole network. It supports speeds of up to 1 Mbps, suitable for real-time control applications. CAN needs to be implemented using two wires, and the event-triggered nature is very efficient in terms of bandwidth usage. FlexRay [82] is a protocol that combines time-triggered (primary) and event-triggered messaging for point-to-point communications. It is being developed by BMW and DaimlerChrysler with Philips and Motorola, and its purpose is to provide x-by-wire applications with deterministic real-time and reliable communications. The FlexRay can support a net data rate of 5 Mbps (10 Mbps gross). It is a protocol in bus architectures for safety–critical embedded systems and advanced-control functions.

Our contribution in this paper is that we propose a contention resolution method for VPLC by leveraging both frequency and time domain-selections to resolve the bus-access collision. To the best of our knowledge, this is the first work that has considered time and frequency multiplexing in optimizing the VPLC performance. Hence, these results will potentially have a broad impact across a range of industry areas, including in-vehicle communications and control systems.

2.3 System Assumption and the Proposed Multichannel Contention Resolution Method

We consider a VPLC network in which N nodes are connected to the harness. Time is divided into fixed-size transmission cycles, where multiple-frequency channels can be used by the senders or receivers. Despite the use of multiple channels, we assume that each node includes one-signal feed and receive ports. We further assume that all nodes in the VPLC network are time synchronized. Figure 2.2 depicts the structure of a single transmission cycle. First, the contention between senders is resolved on the frequency domain, where each sender, at the beginning of the transmission cycle, picks up a channel randomly. Then, if more than one sender selects the same channel, the contention is resolved over number of slots by randomly performing one of the two following actions in each slot: a cs operation or a ct operation. At each time slot, the sender defers its transmission to the next transmission cycle if it senses that the channel is busy. But if the sender does not hear the carrier, it stays on the contention. At the

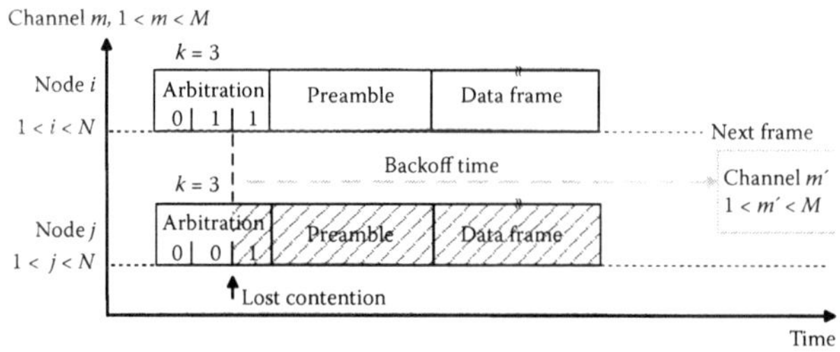

Figure 2.2 View of a single transmission cycle.

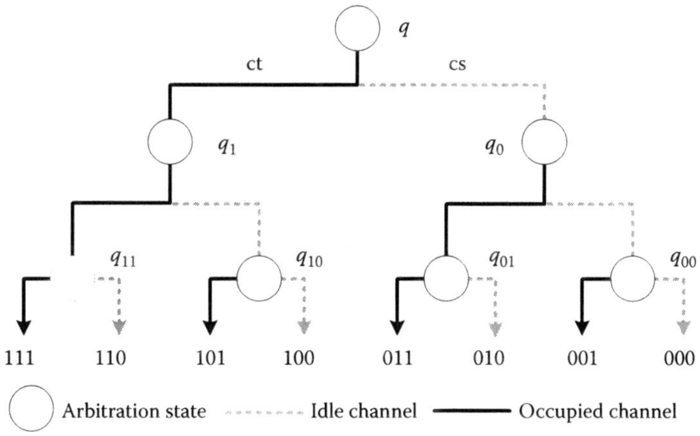

Figure 2.3 Illustration of the collision resolution algorithm performed on each channel.

end of the last slot, the remaining senders transmit a long preamble on their selected channel. After that, the receiver samples the signal level from the first channel, and locks and receives the packets from the first nonidle frequency channel. It is worth noting that the arbitration procedure relies on the fact that a sending node monitors the bus while transmitting. The signal must be able to propagate to the most remote node and return before the bit value is decided. This requires the bit time to be at least twice as long as the propagation delay.

An example of the collision resolution algorithm performed between contending nodes on each channel is given in Figure 2.3. Assume there are three time slots and that the left and right branches correspond to the ct and cs operations respectively. In the first slot, each node chooses ct with probability q, and only nodes who choose cs and sense that the channel is occupied will retire from contention. In the second slot, a node chooses ct with probability q_1 if it has emitted a carrier in the first slot, and with probability q_0 otherwise. This procedure repeats for the next slots. For the case of k contention slots, if we describe the whole process with a set of k binary digits, where bit 1 and 0 correspond to the ct and cs operations respectively, we can conclude that a node with the largest value (i.e., highest priority) wins the contention. In the following sections, we focus on the probabilistic analysis of the proposed MAC protocol and provide insights into designing protocol parameters to cope with contention and sensing errors. Tables 2.1 and 2.2 list the parameters used for performance analysis.

2.4 Performance Analysis under Perfect Sensing

Consider a system scenario where at a given time, n nodes try to transmit packets over the DC power line. We assume the value of n is not known to the nodes, but its probability mass function is known to all nodes in the network and can be expressed as

$$p_N(n) = \frac{1}{\zeta(\gamma)n^{\gamma}}, \tag{2.1}$$

where $n \in \{2, \ldots, N\}$, N is the number of nodes connected to the DC-bus, and $\zeta(\gamma) = \sum_{z=2}^{N} \frac{1}{z^{\gamma}}$, where γ is the shape parameter of the distribution. This distribution is widely applied to model self-similar packet arrivals [20]. We would like to remark, however, that the analysis in this paper are valid for any other distribution of interest.

Table 2.1 Parameters Used for Performance Analysis

Parameter	*Description*
M	Number of channels
N	Number of nodes connected to the DC power line
n	Number of nodes trying to transmit packets over the DC power line
k	Number of time slots to solve contention
T	Number of transmission circles
$p_N(n)$	Probability mass function of n
$p_m, 1 \leq m \leq M$	Probability that the mth channel is selected by a sender
$\mathbf{p} = (p_1, p_2, \ldots, p_M)$	Channel selection distribution
$q^{(m)}$	Probability vector to resolve the contention on the mth channel
$\mathbf{q} = [q^{(1)}, q^{(2)}, \ldots, q^{(M)}]^T$	Probability vectors on all M channels
$p_N^{(m)}(l)$	Probability mass function of l contending nodes on the mth channel
$g^{(m)}(z)$	Probability generating function (PGF) of the number of contending nodes on the mth channel

Source: Horn, R. A., Johnson, C. R., *Matrix Analysis*, Cambridge University Press, Cambridge, UK, 1986.

Table 2.2 Parameters Used for Performance Analysis

Parameter	*Description*
$g_c^{(m)}(z)$	PGF of the number of contending nodes with the signaling pattern c in the first t slots
$\tau_{q^{(m)}}(i)$	Probability that the contention is successfully resolved on the mth channel when i nodes select that channel
$\pi_p(n)$	Successful transmission probability given n contending nodes
$\rho_w(n)$	Throughput of the wth received packet given n contending nodes
p_{md}	Probability of miss detection
p_{fa}	Probability of false alarm
λ	Threshold of the energy detector

Source: Rbsamen, M., et al., *Signal Process.*, 90, 1338–1349, 2010.

We are now ready to formulate the problem. Let $\mathbf{p} = (p_1, p_2, \ldots, p_M)$ be the channel selection distribution, where p_m is the probability that the mth channel is selected by a sender. Assume the collision resolution algorithm uses k slots, and let $q^{(m)}$ be the probability vector of size $\sum_{i=0}^{k-1} 2^i = 2^k - 1$, used to resolve the contention on the mth channel. Figure 2.3 gives an example of how the vector space can be calculated. Therefore, the probability vectors on all M channels can be expressed with a matrix $\mathbf{q} = [q^{(1)}, q^{(2)}, \ldots, q^{(M)}]^T$. Supposing the number of senders in the contention is n, a transmission is successful on the mth channel if and only if:

- m is the first nonidle frequency channel, that is, $\prod_{t=0}^{m-1} (1 - p_t)^n$.
- There is only one node transmitting on the m th channel, that is, $\sum_{i=1}^{n} \binom{n}{i} (p_m)^i (1 - p_m)^{n-i} \tau_{q^{(m)}}(i)$.

Therefore, the success probability is derived as

$$\pi_p(n) = \sum_{m=1}^{M} \prod_{t=0}^{m-1} (1-p_t)^n \sum_{i=1}^{n} \binom{n}{i} (p_m)^i (1-p_m)^{n-i} \tau_{q^{(m)}}(i)\,, \tag{2.2}$$

where $p_0 := 0$, and $\tau_{q^{(m)}}(i)$ is the probability that the contention is successfully resolved on the mth channel when i nodes selected that channel. To further calculate $\tau_{q^{(m)}}(i)$, we need to find the probability mass function of the number of contending nodes on the mth channel. For a given vector $\mathbf{p}$, this distribution can be expressed as

$$p_N^{(m)}(l) = \sum_{n=l}^{N} \binom{n}{l} (p_m)^l (1-p_m)^{n-l} p_N(n)\,, \tag{2.3}$$

where $m \in \{1,\ldots,M\}$ and $l \in \{0,\ldots,N\}$. The probability generating function (PGF) of the number of contending nodes on the mth channel is defined as

$$g^{(m)}(z) := \mathbb{E}(z^n) = \sum_{n=0}^{N} p_N^{(m)}(n)\, z^n\,, \tag{2.4}$$

Now the $\tau_{q^{(m)}}(i)$ can be derived as $\frac{d}{dz}\sum_{c\in\mathcal{C}_k} g_c^{(m)}(z)|_{z=0}$, where $g_c^{(m)}(z)$ is the PGF of the number of contending nodes on the mth channel after the elapse of c time slot and $\mathcal{C}_k$ is the set of all binary numbers of length k from the alphabet $\{0,1\}$. It is worth noting that to avoid duplicating contents, the derivations of $g_c^{(m)}(z)$ and $\tau_{q^{(m)}}(i)$ can be directly referred from Equations 2.18, 2.19, and 2.21 with sensing errors equaling 0.

Averaging $\pi_p(n)$ over the distribution described in Equation 2.1 leads to the success probability

$$\pi_p = \mathbb{E}[\pi_p(n)] = \sum_{n=2}^{N} p_N(n)\pi_p(n)\,. \tag{2.5}$$

Now, we try to find the probability distribution $\mathbf{p}$ and matrix $\mathbf{q}$ that maximize the success probability described in Equation 2.5, i.e.,

$$\arg\max_{\mathbf{p},\mathbf{q}} \pi_p \tag{2.6}$$

Algorithm 2.1 provides the solution and describes how we can calculate the optimal vector $q^{(m)}$ for the mth channel, given the distribution of contenders on the mth channel, i.e., the value of p_m is assumed to be known. It is noted that we have used the method proposed in Ref. [70] to minimize the collision probability on the mth channel, which finds the optimum solution by approximating the collision probability with a Riemann integral.

The distribution of $\mathbf{p}$ determines the efficiency of collision resolution. For this purpose, we have chosen the truncated geometric distribution used in the design of the Sift protocol [100] to achieve fast collision resolution. It is a randomized CSMA-based protocol for wireless sensor networks, where nodes use a truncated geometric distribution for selecting their contention slots. Similarly, in our protocol, senders use this geometrically increasing probability distribution for picking their channels in the transmission cycle. Its expression for $m = 1,\ldots,M$ is given by

$$p_m = \frac{\beta^{\frac{m}{M}} - \beta^{\frac{m-1}{M}}}{\beta - 1}\,, \tag{2.7}$$

Algorithm 2.1: Maximize success probability on the *m*th frequency channel without sensing errors

Set $\hat{u}(z) := \sqrt{g''(z)}$ /* $g''(z)$ is the second derivative of $g(z) = \sum_{n=2}^{N} p_N^{(m)}(n) z^n$ with respect to z */
Initialization: Set $b := 2^k$, $B := 10\,b$, and $u_0 := 0$
for $i = 1$ to B **do**
    $u_i := u_{i-1} + \hat{u}\left(\frac{i-\frac{1}{2}}{B}\right)$
end for
Set $x(0) := 0$, and $x(b) := 1$
for $t = 1$ to $b-1$ **do**
    $x(t) = \frac{1}{B} \min\left\{i : \frac{u_i}{u_{B-1}} \geq \frac{t}{b}\right\}$
end for
Set $q^{(m)} := 1 - \frac{x\left(\frac{b}{2}\right)}{x(b)}$
for $L = 1$ to $k-1$ **do**
    Set $L := 2^{k-l-1}$
    for $j = 0$ to 2^{l-1} **do**
        Convert j into l bits binary number c
        $q_c^{(m)} := \frac{x(2L(j+1)) - x(L(2j+1))}{x(2L(j+1)) - x(2Lj)}$
    end for
end for

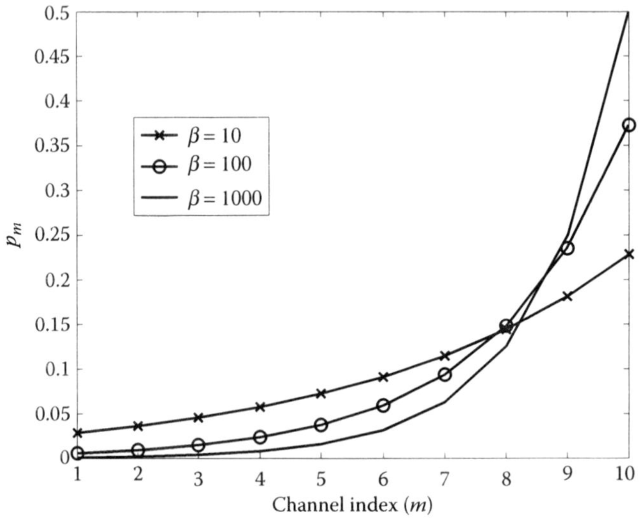

Figure 2.4 Channel selection probabilities when $M = 10$ channels are available for multiple choices of β.

where β is the parameter that needs to be carefully designed. Figure 2.4 illustrates the impact of various β values on the channel probabilities when $M = 10$. We have obtained these probabilities for three values: $\beta = 10$, $\beta = 100$, and $\beta = 1000$. It can be observed that the channel probabilities increase much faster as β increases. It is worth noting that our goal is to find the optimal probabilities (**p**, **q**) to maximize the probability of success (Equation 2.6), and thus the parameter β can be adjusted to feed the optimal requirement. Thus, in this article, we actually address the optimization by finding the optimal (**p**, **q**) via the numerical method; the specific value of β is outside the scope of this chapter.

2.4.1 Throughput evaluation

Suppose that the random variable T_1 denotes the number of transmission cycles required to successfully transmit the first packet. If there are n contenders, then

$$\mathbb{P}(T_1 = r) = \pi_p(n)(1-\pi_p(n))^{r-1}, \tag{2.8}$$

where $r \in \{1,2,\ldots\}$, and $\pi_p(n)$ are the probability of success described in Equation 2.2 when there are n contenders. Note that T_1 describes the delay that corresponds to the first packet successfully transmitted to the receiver. By a similar argument, we find the distribution of T_w, the number of transmission cycles needed to transmit w packets to the destination. Let X_i denote the number of transmission cycles required to transmit the ith packet, conditioned that the previous packets have been transmitted successfully. From Equation 2.8, it is obvious that X_i has a geometric distribution with average $\dfrac{1}{\pi_p(n-i+1)}$. We can express the random variable T_w as

$$T_w = \sum_{i=1}^{w} X_i. \tag{2.9}$$

Thus, the expected value of T_w is

$$\mathbb{E}[T_w] = \sum_{i=n-w+1}^{n} \frac{1}{\pi_p(i)}, \tag{2.10}$$

with $w \in \{1,2,\ldots,n\}$. The normalized throughput under consideration is defined as a fraction of the time the network is used to successfully transmit packets. Hence, we have the throughput that corresponds to the wth received packet as

$$\rho_w(n) = \frac{w\sigma_d}{\mathbb{E}[T_w]\sigma_{\text{cycle}}}. \tag{2.11}$$

where the transmission cycle duration is defined as $\sigma_{\text{cycle}} = k\sigma_s + M\sigma_c + \sigma_d$; σ_s represents the amount of time required by a node to determine the presence of the carrier on a frequency channel, σ_c is the time duration needed by the receiver to sample a frequency channel and switch to the next channel, and σ_d specifies the amount of time needed for transmitting a packet and receiving an ACK.

2.5 Performance Analysis under the Presence of Sensing Errors

In this section, we start by giving an overview of the sensing algorithms and discussing our design goals, then we analytically evaluate the impact of sensing errors on the performance of the proposed protocol.

2.5.1 Impulse noise filtering

The noise over DC power lines contains impulse components; thus, we need to consider signal detection schemes designed for non-Gaussian noise scenarios. There are several detection algorithms proposed for non-Gaussian noise in the literature [114]. However, these algorithms are either difficult to implement or time-consuming to compute.

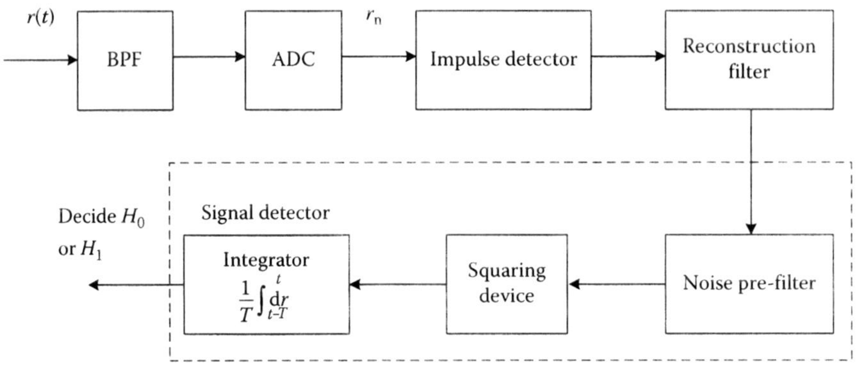

Figure 2.5 Block diagram of a robust sensing module.

Motivated by a robust prediction and whitening method in Ref. [56], we propose a detection scheme composed of a nonlinear preprocessor and a simple signal detector to reshape the designated signals into Gaussian signals. The proposed sensing module is depicted in Figure 2.5, where the received signal passes through a band-pass filter (BPF) with bandwidth W and an Analog-to-Digital Converter (ADC) with sampling rate f_s, followed by a preprocessor and a signal detector. The impulse components are removed from the received signal by using the preprocessor proposed in Ref. [56], which consists of an impulse detector, followed by a reconstruction-matched filter that chooses between input and predicted samples, i.e., Gaussian assumption. The impulse detector has a blanking nonlinearity to mitigate the effects of the impulsive noise. We use an energy detector for signal detection, where the energy of the received signal is measured over a time period and then is compared with a predetermined threshold to determine the presence or the absence of the signal [179].

The experiment is intended to demonstrate the effectiveness of the preprocessor. Assume the received signal is corrupted by a Gaussian noise with variance 1 and an impulsive noise with probability of occurrence 0.1 and variance 10. We plot the empirical cumulative distribution function (ECDF) of the received signal in Figure 2.6. As can be seen, the preprocessor removes the heavy tail of the signal and thus filters out the impulse components in the received signal. We would like to remark that any other detector of interest can also be used for signal detection.

Therefore, the signal detection problem under the output Gaussian noise can be formulated as a binary hypothesis testing problem with $\mathcal{H}_0$ (noise only) or $\mathcal{H}_1$ (signal present),

$$\begin{aligned} \mathcal{H}_0 &: r(t) = n(t), \\ \mathcal{H}_1 &: r(t) = s(t) + n(t), \end{aligned} \tag{2.12}$$

where $n(t)$ is the noise signal at the receiver and is assumed to be an additive white Gaussian noise (AWGN), and $s(t)$ is the filtered transmitted signal. It is noted that the filtered signal $s(t)$ includes the attenuation imposed by the PLC channel transfer function $h(t)$, which has been discussed in Section 2.2. In the absence of much knowledge concerning the input signal, it is appropriate to use an energy detector to determine the presence of a signal. The energy detector operates over a specific time interval by filtering, squaring, and integrating the received signal $r(t)$. For the sake of brevity, we only provide a brief discussion of the system model. More detailed derivations of these fundamental results can be referred from [232]. Thus, the probability of miss-detection (p_{md}) or a false alarm (p_{fa}) for energy detection are expressed as

$$p_{\text{md}} = 1 - Q_u(\sqrt{2\varepsilon}, \sqrt{\lambda}), \tag{2.13}$$

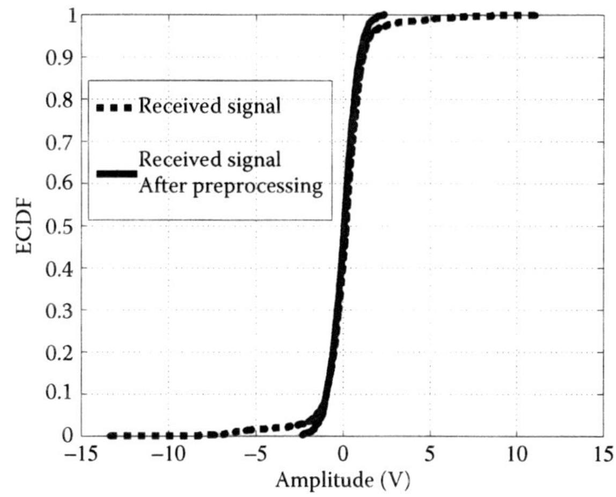

Figure 2.6 ECDF of the received signal before and after preprocessing.

$$p_{\text{fa}} = \frac{\Gamma(u, \frac{\lambda}{2})}{\Gamma(u)}, \tag{2.14}$$

where $Q_u(.,.)$ is the generalized Marcum-Q function, and $\Gamma(.,.)$ is the upper incomplete gamma function [53]. The parameter ε is the ratio of signal energy to one-sided noise spectral density at the receiver and has $\varepsilon = u \times \text{SNR}$, where u is the time-bandwidth product[2] and assumed to be a positive integer and SNR is the ratio of signal energy to the noise energy of the preprocessed signal. λ is the threshold of the energy detector. It is noted that the false alarm probability does not depend on SNR or reception schemes but directly relates to the threshold of the energy detector.

2.5.2 Impact of sensing errors

Sensing errors are inevitable in any CSMA-based MAC protocols. There are two types of sensing errors associated with any sensing algorithms: false-alarm and miss-detection. False alarms occur when idle channels are sensed to be busy, and miss detections occur when busy channels are detected idle. The performance of any detection algorithm is characterized through its receiver operating characteristic (ROC) curves [232]. ROC curves describe the trade-off between false alarm and miss-detection by plotting detection probability (p_{d}) versus false-alarm probability (p_{fa}).

We start by giving an example of the protocol operation to show how sensing errors can affect the protocol performance. Figure 2.7 shows an instance of the protocol with three slots. Assume nodes A and B have packets to transmit. In the first slot, node A selects cs operation, whereas node B emits a carrier. In the case of perfect sensing, node A, which listens to the channel, hears a carrier and loses the contention. With imperfect sensing, however, node A may not hear the carrier sent by node B with p_{md} and thus continues to the second slot. If this happens and both nodes emit a carrier in the second slot, the winners are determined in the last slot. As

[2] $u = TW$, where T is the observation time interval in seconds, and W is the one-sided bandwidth (Hz), i.e., the positive bandwidth of the low-pass (LP) signal.

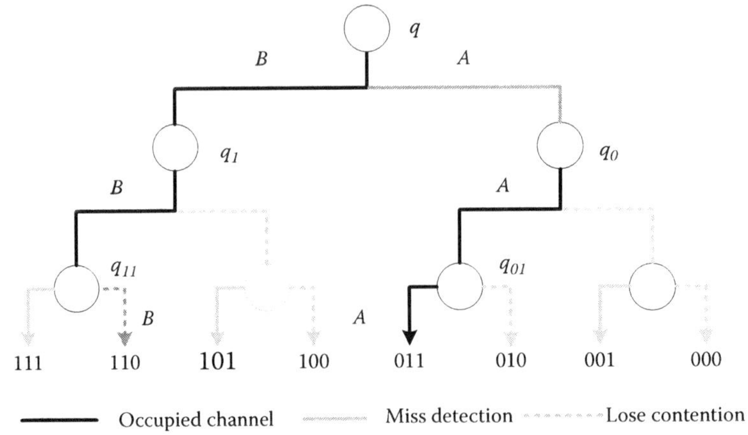

Figure 2.7 Illustration of the protocol operation with imperfect sensing.

can be seen, node A emits a carrier, while node B listens to the channel. If node B correctly identifies the carrier from A, it retires from the contention. Recall from Section 2.3 that nodes with the largest number win the contention. In this example, nodes A and B have chosen 011 and 110, respectively. Hence, node B has a larger binary value and wins the contention with the assumption of perfect sensing. However, the result is unknown with imperfect sensing, as in the above example node A wins the contention.

We next investigate the impact of sensing errors on the performance of our protocol. Suppose there are N nodes connected to the DC power line network and the number of nodes with packets follows the distribution described in Equation 2.1. Furthermore, assume that sensing is not perfect, and there are sensing errors due to the channel impairments. Let $p_{\text{fa}}^{(m)}$ and $p_{\text{md}}^{(m)}$ denote the false-alarm and miss-detection probabilities for senders on the mth channel respectively. Similarly, we define $q_{\text{fa}}^{(m)}$ and $q_{\text{md}}^{(m)}$ as the respective probabilities for false alarm and miss-detection related to the receiver on the mth channel. Next, we calculate the success probability when there are n contenders. Assume m is the channel selected by at least one node. We say that the contention is successfully resolved on the mth channel if and only if (1) only one node remains on the mth channel after the completion of the collision resolution protocol and (2) the receiver is able to correctly determine the states of the first m channels. Hence, the success probability can be expressed as

$$\pi_p(n) = \sum_{m=1}^{M}(1-q_{\text{md}}^{(m)})\prod_{t=0}^{m-1}(1-p_t)^n(1-q_{\text{fa}}^{(t)})\sum_{i=1}^{n}\binom{n}{i}(p_m)^i(1-p_m)^{n-i}\tau_{q^{(m)}}(i,p_{\text{fa}}^{(m)},p_{\text{md}}^{(m)}), \tag{2.15}$$

where $p_0 := 0$, $p_{\text{fa}}^{(0)} := 0$, and the expected value of success probability is expressed by Equation 2.5.

To derive the success probability, $\tau_{q^{(m)}}$, on the mth channel when sensing is not perfect, we need to find the PGF of the number of contenders still in the competition after the elapse of one time slot. For simplicity of computation, we assume that, by using the preprocessor suggested in Section 2.5.1, all nodes in the competition experience the same average SNR, and therefore we can use binomial distribution to denote the probability that i out of n nodes estimated the channel state correctly. Because this PGF depends on the number of nodes that selected ct operation in the previous slot, we derive its expression for the following cases:

- Case 1: If no carrier has been emitted in the first slot, then a node moves to the second slot if it senses the channel idle, which happens with probability $1 - p_{\text{fa}}$. The PGF denoted as $g_0^{(m)}$, is given as

$$g_0^{(m)}(z) = \sum_{n=0}^{N} p_N^{(m)}(n)\left(1-q^{(m)}\right)^n \sum_{i=0}^{n} \binom{n}{i} \left(1-p_{\text{fa}}^{(m)}\right)^i \left(p_{\text{fa}}^{(m)}\right)^{n-i} z^i$$
$$= \sum_{n=0}^{N} p_N^{(m)}(n)\left(1-q^{(m)}\right)^n \left(\left(1-p_{\text{fa}}^{(m)}\right)z + p_{\text{fa}}^{(m)}\right)^n$$
$$= g^{(m)}\left((1-q^{(m)})z_{\text{fa}}^{(m)}\right), \tag{2.16}$$

where $z_{\text{fa}}^{(m)} := \left(1-p_{\text{fa}}^{(m)}\right)z + p_{\text{fa}}^{(m)}$, $q^{(m)}$ is the probability of transmitting a carrier in the first time slot, and $p_{\text{fa}}^{(m)}$ is the false alarm probability on the mth channel given in Equation 2.14.

- Case 2: In this case, we consider scenarios where at least one node has emitted a carrier in the previous slot. All nodes in the contention that have emitted a carrier in the first slot survive and move to the second time slot. However, nodes that have sensed the channel busy, which happens with probability $1 - p_{\text{md}}$, will retire from contention. Others that have miss-detected the carrier on the channel, will continue the contention. Therefore, the expression for its PGF, $g_1^{(m)}$, is expressed by

$$g_1^{(m)}(z) = \sum_{n=1}^{N} p_N^{(m)}(n) \sum_{i=1}^{n} \binom{n}{i} \left(q^{(m)}\right)^i \left(1-q^{(m)}\right)^{n-i} z^i$$
$$\sum_{j=0}^{n-i} \binom{n-i}{j} \left(p_{\text{md}}^{(m)}\right)^j \left(1-p_{\text{md}}^{(m)}\right)^{n-i-j} z^j$$
$$= g^{(m)}\left(q^{(m)}z + (1-q^{(m)})z_{\text{md}}^{(m)}\right) - g^{(m)}\left((1-q^{(m)})z_{\text{md}}^{(m)}\right), \tag{2.17}$$

where $z_{\text{md}}^{(m)} := p_{\text{md}}^{(m)}z + 1 - p_{\text{md}}^{(m)}$, and $p_{\text{md}}^{(m)}$ is the miss-detection probability on the mth channel defined in Equation 2.13.

Let c be a t-bit binary number with bit notations from b_1 to b_t that shows the operations performed by a sender in each slot, where $b_i = 0$ or 1 denotes the events of choosing cs and ct in the ith slot, respectively. Following mathematical induction and using Equations 2.16 and 2.17, we are now able to find the iterative PGF of the nodes in the contention after the elapse of $t+1$ slots.

$$g_{c0}^{(m)}(z) = g_c^{(m)}\left((1-q_c^{(m)})z_{\text{fa}}^{(m)}\right), \tag{2.18}$$

where $c0$ means an additional bit $b_{t+1} = 0$ is attached after c. And

$$g_{c1}^{(m)}(z) = g_c^{(m)}\left(q_c^{(m)}z + (1-q_c^{(m)})z_{\text{md}}^{(m)}\right) - g_c^{(m)}\left(\left(1-q_c^{(m)}\right)z_{\text{md}}^{(m)}\right), \tag{2.19}$$

where $c1$ means an additional bit $b_{t+1} = 1$ is attached after c. $q_c^{(m)}$ is the probability that, in slot $t+1$, nodes emit a carrier on the mth channel, given the signaling pattern c in the first t slots, $g_c^{(m)}$ is the PGF of survivors, when the signaling pattern in the first t slots is c, and $g_\varnothing^{(m)} := g^{(m)}$.

Hence, the PGF of the number of contenders on the mth channel after k slots is $\sum_{c \in \mathcal{C}_k} g_c^{(m)}(z)$, where $\mathcal{C}_k$ denotes the set of all binary numbers of length k from the alphabet $\{0, 1\}$. So the distribution of survivors is given by

$$\Pr\{n \text{ nodes remain on the } m\text{th channel}\} = \frac{1}{n!}\frac{\mathrm{d}^n}{\mathrm{d}z^n}\sum_{c \in \mathcal{C}_k} g_c^{(m)}(z)|_{z=0}, \tag{2.20}$$

where n denotes the number of survivors at the end of the contention. The success probability, which is defined as the probability that at the end of the contention only one survivor remains, is given as

$$\tau_{q^{(m)}} = \frac{\mathrm{d}}{\mathrm{d}z}\sum_{c \in \mathcal{C}_k} g_c^{(m)}(z)|_{z=0}. \tag{2.21}$$

Based on the selection of optimal energy-detecting parameters λ and the probability distributions obtained by Algorithm 2.1, we can obtain the maximum success probability (Equation 2.5) of the sensing errors case.

2.6 Numerical Results

In this section, we present numerical results to illustrate the performance of the proposed protocol. Throughout the simulation, we assume that the number of nodes connected to the DC power line, N, is 50 and that the shape parameter of the distribution in Algorithm 2.1 is set to 0.6, unless specified otherwise. The reason behind choosing $\gamma = 0.6$ is that the system can perform well in both high and low traffic loads as shown in Figure 2.8. Here, there are $N = 50$ nodes connected to the harness, and the system uses $M = 2$ channels and $k = 6$ slots to resolve the contention between nodes. It can be noted that the system with $\gamma = 0$ (uniform distribution) performs well when the number of contenders is large, whereas the system with $\gamma = 1$ gives a

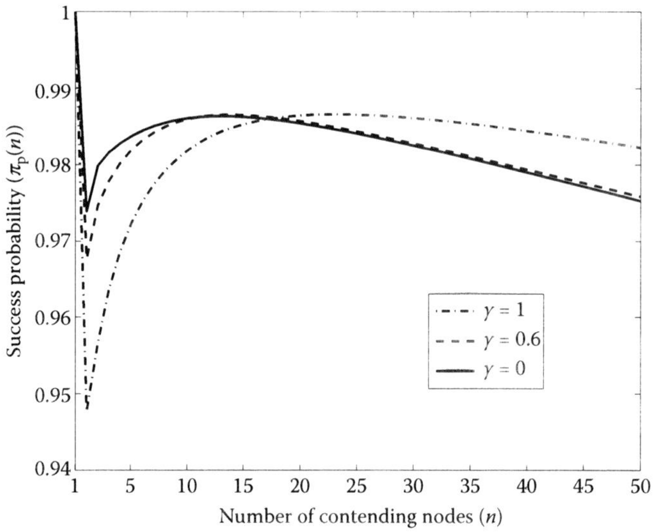

Figure 2.8 Success probability versus number of contending nodes for different values of γ, $N = 50$, k = 6, $M = 2$.

better performance when the number of contending nodes is small, and the system with $\gamma = 0.6$ provides a good performance in both cases. Our experiments confirm that $\gamma = 0.6$ provides a balanced performance for other configurations as well, i.e., different values of M and k.

Figures 2.9 and 2.10 show the success probability and the system throughput as the number of channels varies between 2 and 8 for different number of contention slots, respectively. To evaluate the effects of the number of frequency channels and contention slots on the system throughput, we define a ratio between time constants in each transmission cycle as $r := \frac{\sigma_s}{\sigma_d} = \frac{\sigma_c}{\sigma_d}$,

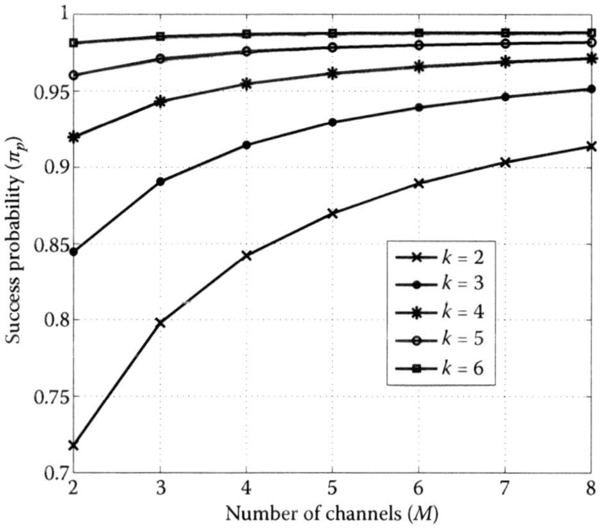

Figure 2.9 Average success probability versus number of channels for different number of time slots (k), $N = 50$, $\gamma = 0.6$.

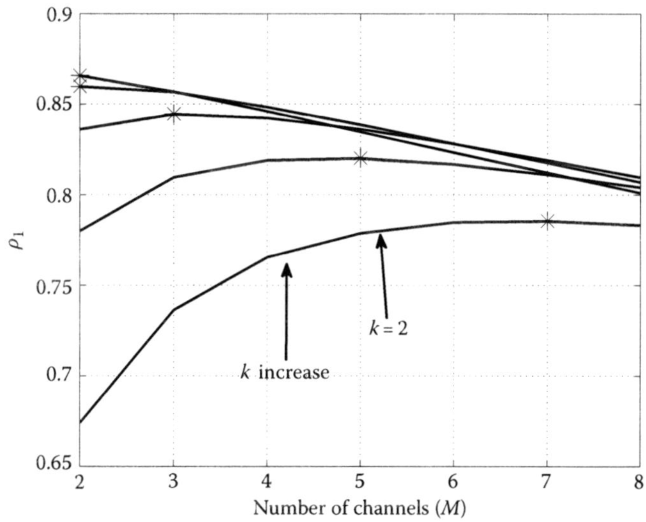

Figure 2.10 ρ_1 versus number of channels for different number of time slots (k), $N = 50$, $\gamma = 0.6$.

where σ_s, σ_c, and σ_d are defined in 2.10. Here, we considered that $\sigma_s = \sigma_c$; however, the case of $\sigma_s \neq \sigma_c$ can be easily included in the numerical evaluations as well. Expectedly, as can be observed in Figure 2.9, with an increasing number of contention slots or frequency channels, the success probability of the protocol increases. However, as the number of contention slots increases, the gain in use of multiple frequency channels decreases, since the protocol is capable of resolving the contention on each channel with a high probability. In Figure 2.10, we assume that the packet size is relatively large, and therefore $r = 1/60$. Each point describes the expected throughput computed as $\sum_{n=2}^{N} p_N(n)\rho_1(n)$, and the maximum point along each curve is specified with a marker. We can observe that, as the number of contention slots increases, the maximum point along each curve moves to the left and thus occurs at a smaller number of channels. It can also be seen that adding a channel to the system does not improve the system performance when the number of contention slots used in the system is high. The reason behind this behavior is that a system with large k can handle a wide range of traffic loads, and therefore the success probability will not improve much by separating contenders across more channels. On the other hand, adding one more channel will increase the packet overhead and thus potentially degrade the system performance.

Figure 2.11 shows the average success probability of the system as a function of the network size. The protocol operational parameters (k, M) are set to the values giving the maximum throughput as shown in Figure 2.10. It could be noticed that the system shows the least performance degradation when $k = 6$ and $M = 2$. The reason is that as we increase the number of contention slots in each channel, the collision resolution algorithm performed on each channel is more capable of resolving the contention in a wide range of traffic loads and also less sensitive to the number of contenders compared to the channel selection algorithm.

In Figures 2.12 and 2.13, we report the probability mass functions of the number of transmission cycles required to transmit the first and all packets, when there are 25 and 5 contenders, respectively. We assume there are $k = 4$ slots available on each channel to resolve the contention, and the system uses $M = 3$ channels according to Figure 2.10, this configuration

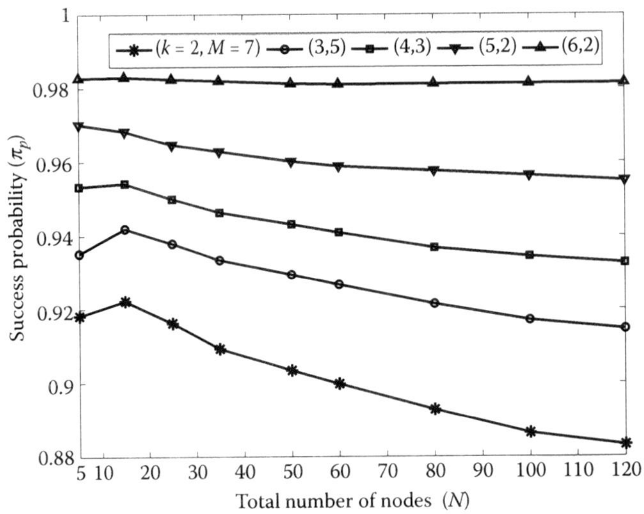

Figure 2.11 The average success probability versus total number of nodes connected to the harness (*N*), γ = 0.6.

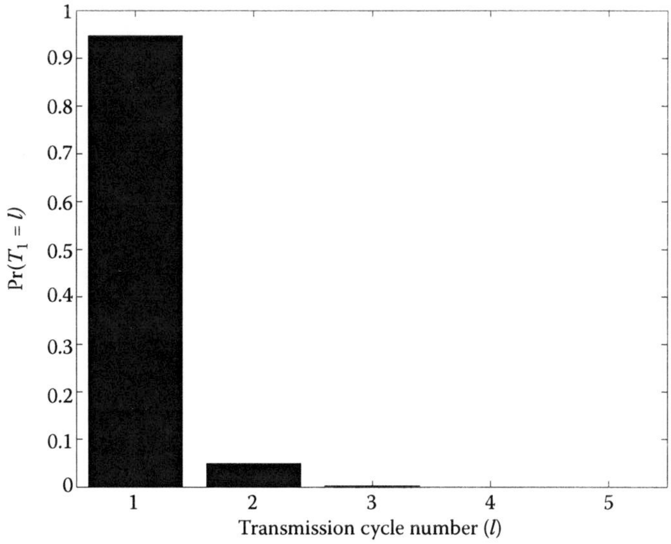

(a) First packet

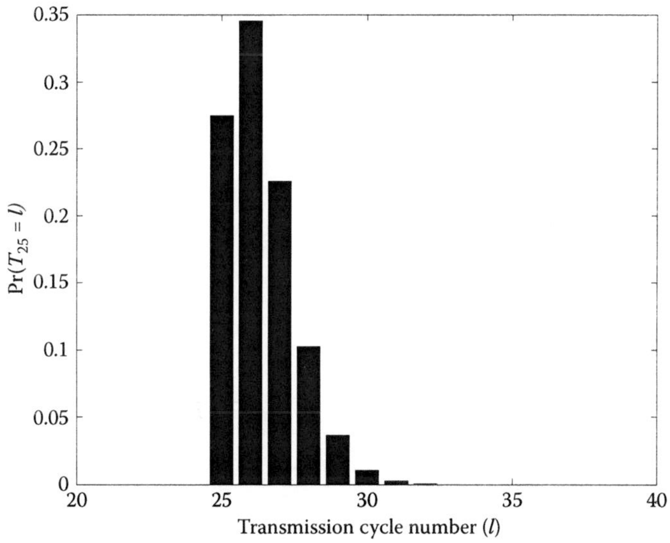

(b) All packets

Figure 2.12 Probability mass function of the number of transmission cycles required to transmit the first packet (a) all packets (b) when there are 25 contenders, $N = 50$, $\gamma = 0.6$, $k = 4$, $M = 3$.

provides the highest throughput when $k = 4$. We have plotted theses distributions by using Equation 2.9, and the results obtained by solving the optimization problem given in Equation 2.6. The figures show that the protocol delivers the packets with small latency and also scales well with respect to the number of contenders.

Figure 2.14 gives a success probability comparison between our proposed MAC protocol and CDR [13] when the number of contending nodes vary between 2 and 50. To make a fair comparison, we note that adding one more channel or slot to our system increases the packet

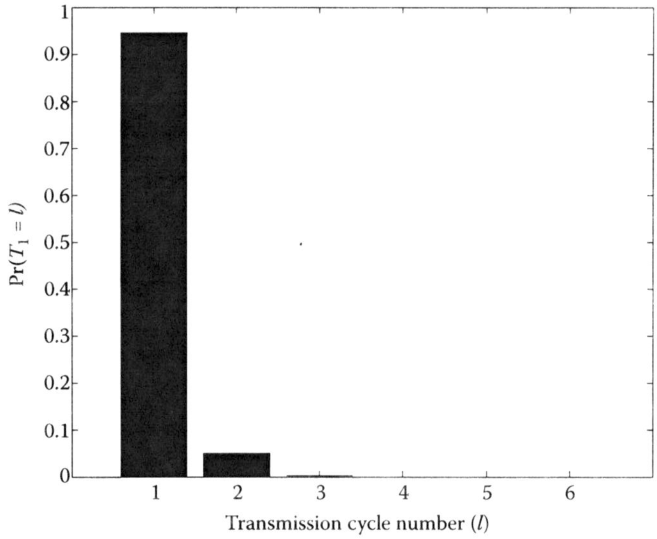

(a) First packet

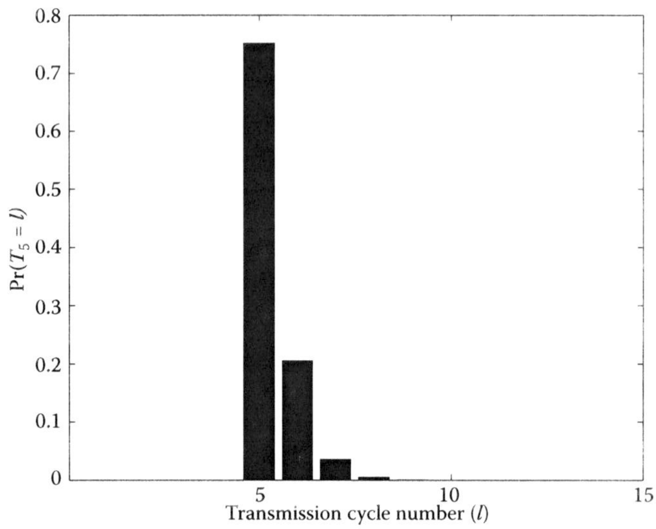

(b) All packets

Figure 2.13 Probability mass function of the number of transmission cycles needed to transmit the first packet (a) all packets (b) when there are five contenders, $N = 50$, $\gamma = 0.6$, $k = 4$, $M = 3$.

overhead by the same amount as adding one more slot to the CDR protocol. The number of channels, in our system, is fixed to 3. We would like to remark that the parameters used in our protocol are calculated by solving the optimization problem in Equation 2.6 with the distribution in Equation 2.1, when $N = 50$ and $\gamma = 0.6$, and thus the protocol does not need to know about the number of contenders. It can be observed that our protocol performs much better in all scenarios, and its performance deteriorates at a much lower rate compared to CDR as we increase the number of contending nodes. With the same assumption, Figure 2.15 further shows the throughput ρ_1 comparison of the proposed solution and CDR. We also assume that

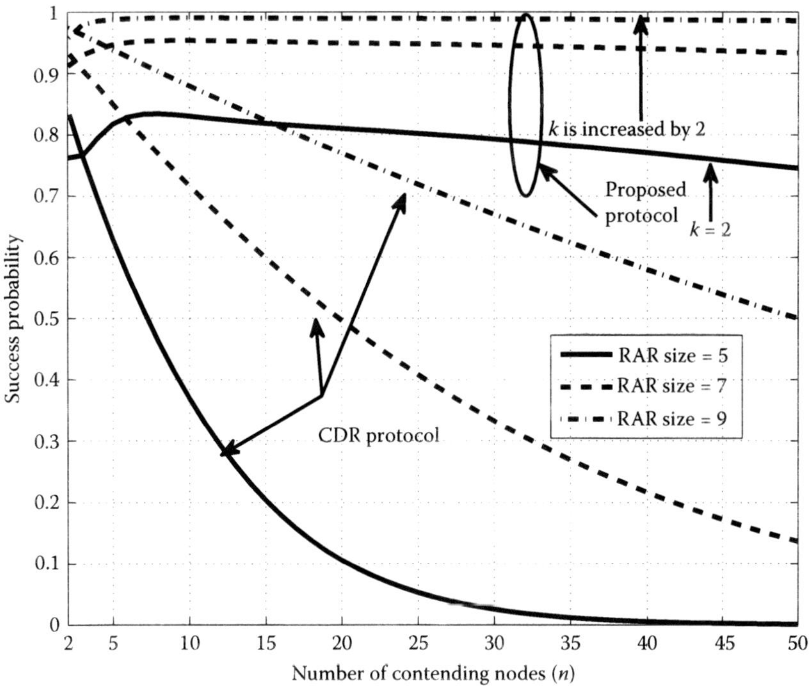

Figure 2.14 Success probability of the proposed protocol and CDR versus number of contending nodes, $M=3$, $N=50$, $\gamma=0.6$.

the packet size is relatively large and that $r = 1/60$. It is clear that the result is proportional to the probability result, which can be proved from 2.11.

According to Equations 2.13 and 2.14, the sensing errors are directly related to the threshold λ of energy detector. The effect of λ on the average success probability is depicted in Figure 2.16. We consider the scenario where $k = 6$ slots and $M = 2$ channels is used in the system as it provides a high throughput according to Figure 2.10. The values of SNRs are set between -5 to 5 dB, and are chosen randomly in each channel. Note that each point in the figure represents the average success probability as in 2.5, and is averaged over 1000 runs. We also assume that the sensing module takes 20 samples to determine the presence or absence of the carrier. It is clear that the decision of threshold λ on each channel plays a critical role in the performance of overall success probability, and there is an optimal selection of λ that can maximize the success probability. Since the main focus of this chapter is on the maximization of π_p, in the following paragraphs we will evaluate the maximum success probability 2.6 that can be achieved by optimally selecting system parameters.

We have designed a robust system by considering the carrier-sensing errors. We have plotted the results for three cases corresponding to $r = 1/60, 1/20, 3/20$. Note that in all cases the packet length is fixed, and the number of samples taken from the received signal changes instead. There are $M = 2$ channels available in the system for contention, and assume that the SNRs corresponding to these channels are 5 dB (a good channel) and 0 dB (a bad channel). Figures 2.17 and 2.18 both show the systems where the detector operating point is optimized on each channel with respect to the physical layer characteristics and correspond to the cases where the good channel is indexed as the first channel and vice versa. We can make the

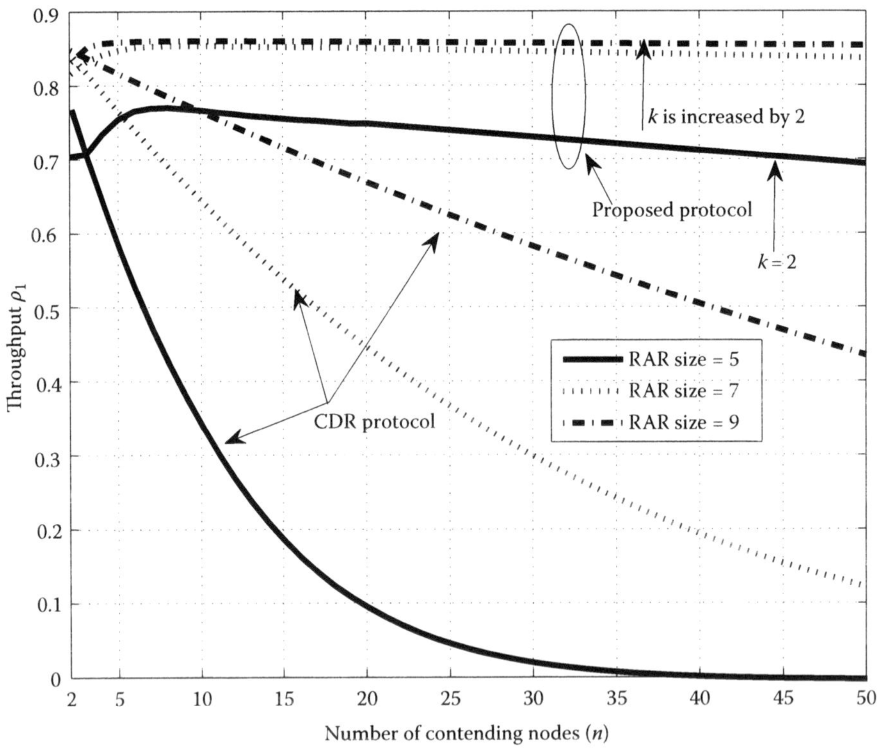

Figure 2.15 Throughput ρ_1 of the proposed protocol and CDR versus number of contending nodes, $M = 3$, $N = 50$, $\gamma = 0.6$.

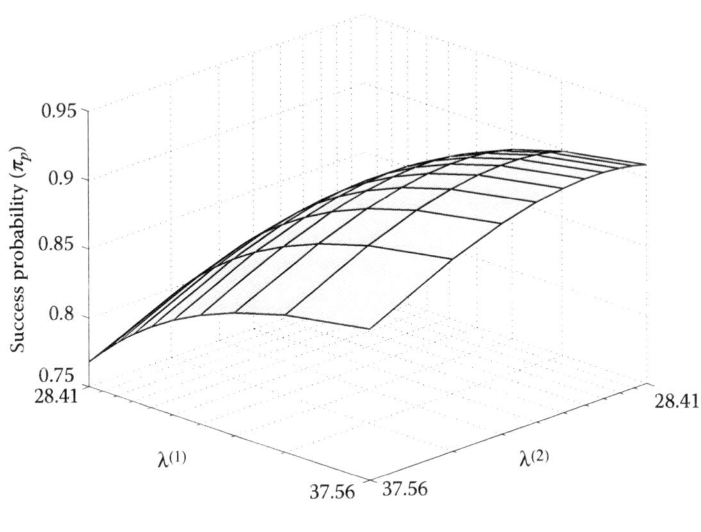

Figure 2.16 Average success probability versus threshold λ of the energy detector, $N = 50$, $\gamma = 0.6$, $k = 6$, $M = 2$, $u = 10$.

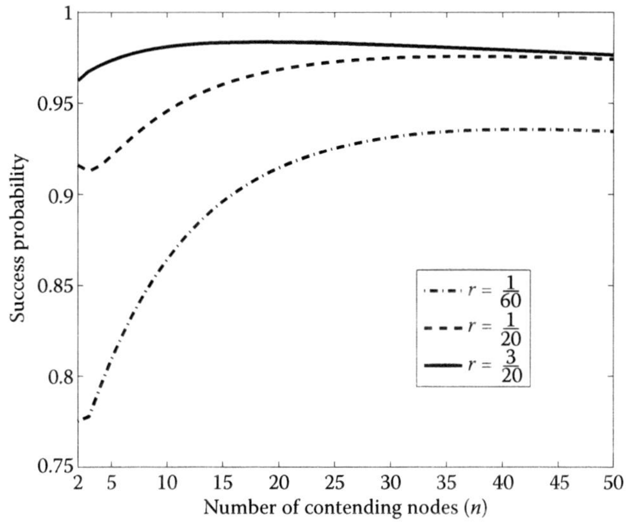

Figure 2.17 Success probability versus number of contending nodes for multiple values of r, where the good channel is indexed 1, $k=6$, $M=2$, $N=50$, $\gamma=0.6$.

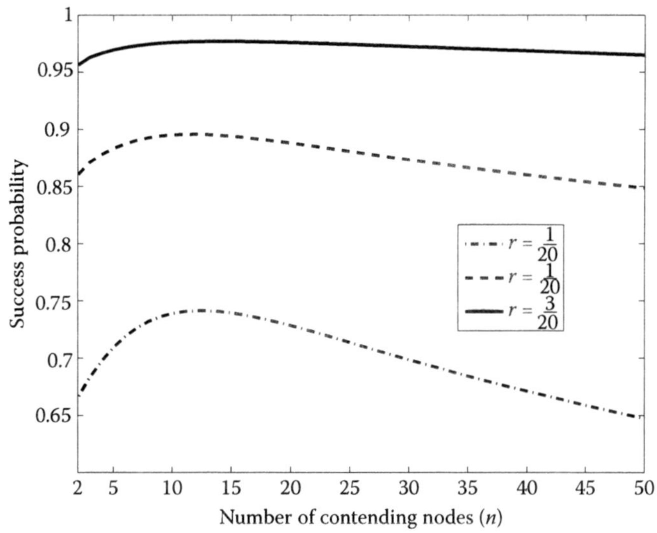

Figure 2.18 Success probability versus number of contending nodes for multiple values of r, where the bad channel is indexed 1, $k=6$, $M=2$, $N=50$, $\gamma=0.6$.

following observation: The success probability decreases when the bad channel is indexed 1. This happens since, according to Equation 2.7, the receiver will receive the packet from the low-indexed channels with high probability, and a low SNR on the selected channel can degrade the performance of the collision resolution algorithm performed on that channel. Hence, we can shuffle the order of the channels in each transmission cycle to reduce the impact of the noise and fading on the MAC protocol performance. Moreover, we obtain additional results for $M > 2$ and with equal channel quality. Figure 2.19 illustrates the success probability for

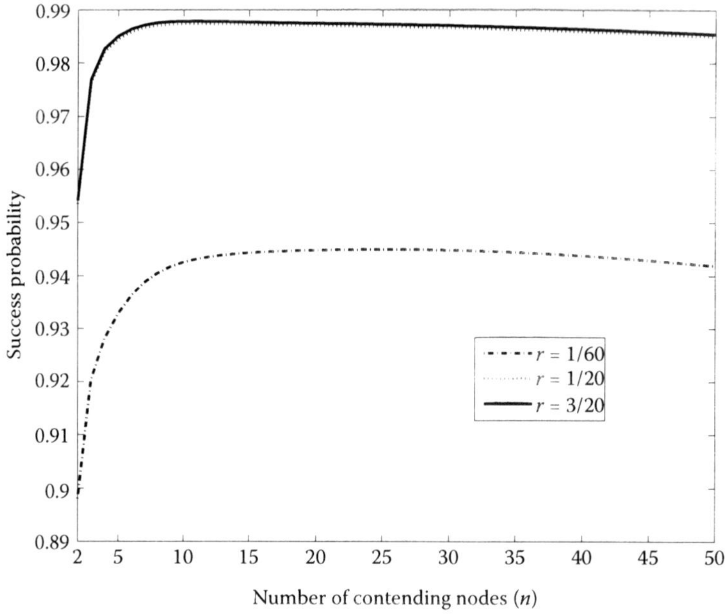

Figure 2.19 Success probability versus number of contending nodes for multiple values of *r*, where channels are equally good, $k = 6$, $M = 3$, $N = 50$, $\gamma = 0.6$.

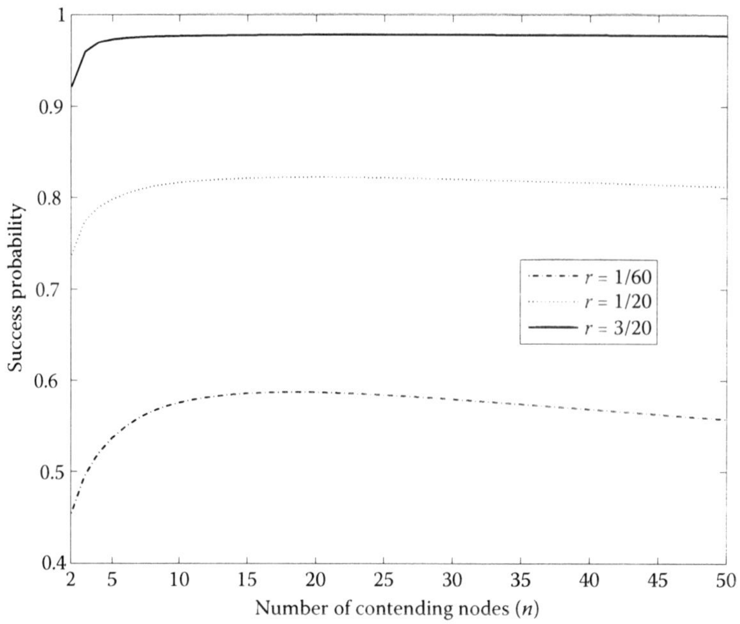

Figure 2.20 Success probability versus number of contending nodes for multiple values of *r*, where channels are equally bad, $k = 6$, $M = 5$, $N = 50$, $\gamma = 0.6$.

$M = 3$ when SNRs of all channels are equally good (5 dB), and shows that all channels can achieve decent performance when sampling rates (r) are closed. Figure 2.20 shows the result for $M = 5$ when SNRs of all channels are equally bad (0 dB). It is clear that the performance has deteriorated with poor channel conditions; however, the scheme with a larger sampling rate is robust in resisting compact carrier-sensing errors.

2.7 Conclusions

We have introduced a random access MAC protocol based on the combination of time and frequency multiplexing. Nodes in the contention randomly select a frequency channel to perform channel contention in a number of slots. After that, the receiver samples the signal level on each frequency channel and stops on the first nonidle channel to receive the packet. We mathematically analyzed the performance of the proposed MAC protocol under both perfect and imperfect sensing. With numerical evaluations, we have verified our analysis and demonstrated the effectiveness of our MAC protocol. Our results show that the system demonstrates a good performance in terms of collision probability, system throughput, and delay. In this work, we have also considered that the system is not free from carrier sensing errors, and we have observed that a great care must be taken when designing such a system.

In future, we will promote the two-dimensional MAC protocol into a more practical implementation. Specifically, a live demonstration model will be built based on a network simulator. Also, we intend to incorporate the proposed method with existing PLC solution, i.e., HomePlug, and further extend its usability and compatibility with home automation and smart grid.

Chapter 3

Worst-Delay Analysis of Intravehicle Communication Buses Using HomePlug GP

Zhengguo Sheng
University of Sussex

Roberto P. Antonioli
Federal University of Ceará

Morgan Roff
Queen's University

Yunpeng Wang
Beihang University

CONTENTS

3.1 Introduction 47
3.2 Background of HomePlug Green PHY Protocol 48
3.3 Proposed Real-Time MAC Protocol 48
3.4 Simulation Results 49
3.5 Future Research Subjects 54
3.5.1 Compatibility with standard HomePlug GP 54
3.5.2 Further improvements in delay performance 54
3.6 Conclusion 55

3.1 Introduction

In this chapter, we introduce a real-time in-vehicle media access control (MAC) protocol based on the HomePlug GP protocol that maintains the security and reliability of HomePlug while also reducing energy consumption through power-save functionality; it has a data rate of up to 9.8 Mbps [87]. Specifically, this paper analyzes the capability of the proposed protocol to enable real-time communications for safety-critical applications such as x-by-wire systems based on message delay. This is defined as the time delay from a message being created by a station, and that station receiving an acknowledgment message from the receiver for successful delivery. The performance analysis shows that with modifications and improvements of HomePlug GP, the proposed solution can better cope with stringent real-time requirements of in-vehicle environments and show advantages in delay and collision performance.

3.2 Background of HomePlug Green PHY Protocol

The HomePlug Green PHY protocol is chosen for study based on its low power use, its compatibility with existing infrastructure, and its robust error-handling procedures [87]. However, maintaining compatibility with other HomePlug versions and providing error-handling and error-checking information contribute significant overhead to messages. Considering a network comprising only HomePlug GP stations, this protocol uses 4 bytes of header information and 4 bytes of cyclic redundancy check (CRC) for 512 or 128 bytes of data. The size of the physical block (PB) depends on the modulation used: for example, the Beacon message is always sent using Mini-ROBO modulation, which is the most reliable and has the lowest data rate (3.8 Mbps); it uses PBs of 136 bytes. The other messages are transmitted using either Standard ROBO (STD-ROBO) or High-Speed ROBO (HS-ROBO) modulations, which are less reliable than Mini-ROBO, have data rates of 4.9 and 9.8 Mbps, respectively, and use PBs of 520 bytes [87]. The HomePlug GP can chain up to three PBs per message with an additional 128 bits of frame control header. If messages contain <128 or 512 bytes of data, the rest of the PB is filled with padding, creating wasted space, which will negatively impact delay performance. Compare this to the Controller Area Network (CAN), which is a commonly used in-vehicle communication protocol standard and uses up to only 8 bytes of data per message [73]; it is clear that for an in-vehicle network there would be much wasted space in using HomePlug GP messages.

To ensure timely and fair competition for the medium, bus arbitration is accomplished by utilizing a Carrier Sense Multiple Access (CSMA) approach, where stations gain access to the channel based on a 4-level priority value (2 bits), followed by a random backoff counter value. Once this backoff counter value reaches zero, the station will send its message. This can easily lead to collisions if multiple stations choose the same backoff counter value, which starts off with a range of only 0–7. If a PB is received with errors, the entire 136 or 520 bytes PB must be re-sent by the transmitting station. This leads to significant delays if errors or collisions occur.

3.3 Proposed Real-Time MAC Protocol

After reviewing the HomePlug GP protocol, a modified protocol has been devised that aims to better suit the in-vehicle network's real-time data requirements. This protocol is designed to be easily mapped back to HomePlug GP format messages for the purposes of communicating with a Home Area Network when necessary. Specifically, we propose to enhance the existing HomePlug GP by increasing the number of priority slots, reducing the length of individual messages, and removing unnecessary frame control overhead.

First of all, it is hypothesized that the addition of a single priority bit, also known as a priority resolution slot (PRS), would reduce collisions and improve quality-of-service (QoS) by allowing for eight priority levels. By spreading the messages across more priority levels, fewer messages will be competing for access using backoff counters, and therefore collisions become less likely. Higher levels can be used for infrequent but critical messages to guarantee rapid delivery. This 8-level priority message can easily be converted to a HomePlug compatible message for communication between electric vehicles (EVs) and in-home networks, since the 4-level HomePlug priority is already decided based on an 8-level user priority [87]. An adapter in the car or charging station can behave as a proxy to perform this conversion before further relaying this message between cars and grid/home networks.

Second, the length of each data message can be greatly reduced through a few simple alterations within the frame control. For in-vehicle purposes, the frame control can be reduced to 80 bits, down from 128 bits, by removing segments that are only for compatibility with other HomePlug protocols (e.g., HomePlug AV 2.0) and not needed for relatively small in-vehicle messages.

Furthermore, by altering the message length information within the Frame Control, it becomes possible for the PB, when using STD-ROBO or HS-ROBO modulations, to be set with a variable size of 0–256 bytes of data, which is justified by the fact that it is not necessary to have 512 bytes of data to transmit vehicle instructions, since most of the in-vehicle protocols, e.g., CAN and FlexRay [82], can only use up to 254 bytes to send in-vehicle commands. In case a regular HomePlug GP station transmits a 512-byte data message, the central coordinator (CCo) on a charging station is able to divide it into two segments of 256 bytes and send them in two PBs. By means of the new format of the Frame Control, it is possible to chain up to eight PBs through four new bits that were inserted in that field; one bit indicates whether the message has one or more PBs, and the other three bits indicate how many PBs comprise the message. The reduction in PB size also allows for a shorter CRC of only three bytes, giving a total PB size of 263 bytes. Another proposed change is a reduction in the CW size that limits the backoff counter range, as seen in Table 3.1. This is suggested due to the fact that the new eight-level priority values reduce the number of nodes competing for transmission after a priority check; therefore, there is no need for a range as vast as before.

The final modification is to reduce the slot time for the backoff counter and priority resolution down to 5 μs from 35.84 μs. This modification is justified by the fact that in the CAN protocol, which has a similar arbitration procedure if compared to the priority resolution slots, the signal takes 1 μs to propagate from the sending node to all nodes and return [73]; therefore, 5 μs should be enough time for the priority resolution and backoff counter slots. All time slots present in a message transmission as well as the interframe spaces can be visualized in Figure 3.1, which shows a media access control scheme in the proposed HomePlug system. It is noted that without this change, a safety-critical message may miss its deadline purely through the priority resolution and backoff process. A summary of the differences proposed in the paper can be found in Table 3.2.

Table 3.1 Original and Proposed Values of Contention Windows (CWs) as a Function of Backoff Procedure Event Counter (BPC) and Priorities

	Original Values		*Proposed Values*	
	Priorities CA3 and CA2	*Priorities CA1 and CA0*	*Priorities CA7, CA6, CA5 and CA4*	*Priorities CA3, CA2, CA1 and CA0*
BPC = 0	CW = 7	CW = 7	CW = 3	CW = 3
BPC = 1	CW = 15	CW = 15	CW = 7	CW = 7
BPC = 2	CW = 15	CW = 31	CW = 7	CW = 15
BPC > 2	CW = 31	CW = 63	CW = 15	CW = 31

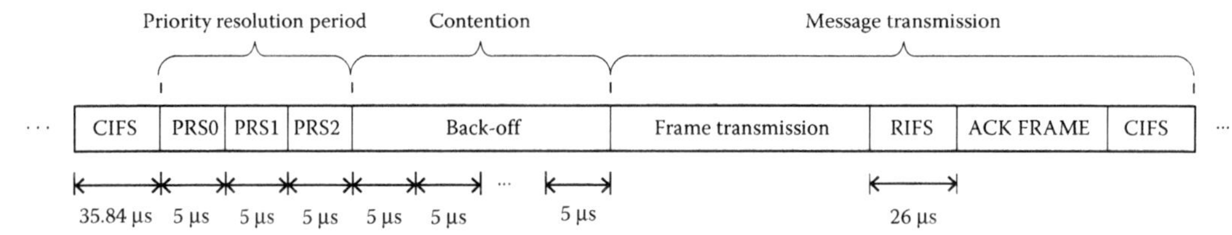

Figure 3.1 Frame structure of the proposed MAC protocol.

Table 3.2 Summary of Differences

	Regular HomePlug GP	*Proposed Real-Time HomePlug GP*
Number of PRSs	2	3
Duration of backoff slots	35.84 μs	5 μs
CW values	See Table 3.1	See Table 3.1
Frame control size (bits)	128	80
Size of PB (bytes)	512	256
Size of CRC in the PB (bytes)	4	3

3.4 Simulation Results

The simulation for the proposed HomePlug Green PHY protocol is conducted in the open-source-message-level simulator OMNeT++. The network topology is comprised of thirteen stations (nodes), which are connected via power lines, distributed among two electric/hybrid vehicles (i.e., battery, Global Positioning System (GPS), brakes, and engine), one in-home network (i.e., washing machine, wireless router, computer and temperature controller), and a charging station (i.e., CCo). The network topology is illustrated in Figure 3.2. The stations send messages at a random rate that is dependent on the selected priority value; the function that generates these random times is chosen such that messages are sent with a certain frequency, allowing collisions to happen. The values obtained are shown in Table 3.3. The bit error rate (BER) is fixed as 10^{-6}.

It is worth noting that the role of the CCo is played only by the charging station, which means that there is no transfer/handover of CCo functions to another station during the whole simulation. Moreover, for simulation purposes, all stations are considered time synchronized, though in a real physical system they use Beacon messages from the CCo to maintain synchronization [153].

Following the aforementioned modifications, some parameters need to be changed for the proposed HomePlug GP protocol: the number of PRSs is increased by one providing four additional levels of priority; consequently, the size of the CW could be reduced as shown in Table 3.1. In addition, the size of the Frame Control used when only HomePlug GP stations are present in a network is shortened from 128 bits to 80 bits, and the size of the PB sent in each message is reduced to 263 bytes from 520 bytes. Finally, a reduction in the slot time for the backoff counter and priority resolution is made and their new value is 5 μs instead of 35.84 μs.

Some of parameter values are maintained between protocols [87]. One Beacon Period, which consists of two AC powerline cycles, is 33.33 ms, based on a 60 Hz powerline frequency. The interframe space durations are unchanged: the Contention InterFrame Space (CIFS), used after a response transmission, is 35.84 μs; the Beacon to Beacon InterFrame Space (B2BIFS), used after transmission of a Beacon message by the CCo, is 90 μs; and the

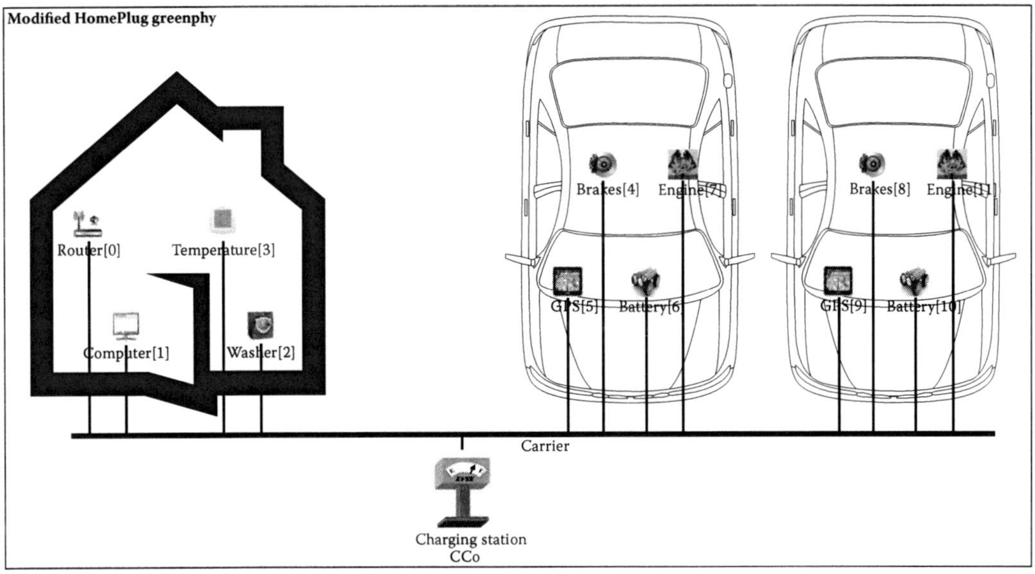

Figure 3.2 Topology of extended HomePlug Green PHY network.

Table 3.3 Minimum and Maximum Intermessage Times in the Regular and Proposed Real-Time HomePlug GP

	Proposed HomePlug GP		*Regular HomePlug GP*	
Access Priorities	*Minimum (ms)*	*Maximum (ms)*	*Minimum (ms)*	*Maximum (ms)*
0	0.408	6.12	0.408	6.12
1	0.583	8.75	0.816	12.2
2	0.758	11.4	1.22	18.4
3	0.933	14.0	1.63	24.5
4	1.11	16.6		
5	1.28	19.2		
6	1.46	21.9		
7	1.63	24.5		

Response InterFrame Space (RIFS), used before a response transmission, is 26 μs. The values for the Deferral Counter can be found in Ref. [124]. Regular messages sent by stations are transmitted using HS-ROBO (9.8 Mbps) and Beacon messages transmitted by the CCo use Mini-ROBO (3.8 Mbps).

The results obtained from the simulation show the number of message collisions and message delays for each level of priority. To obtain these results, some code is implemented using the signal mechanism provided by OMNeT++. For the message delay, the simulation records the time from a station creates a message for transmission to the time it receives acknowledgment from the receiver that the message has been successfully transmitted. These values are recorded by priority. The number of collisions is incremented every time a station detects that two or more Frame Control messages are relayed at the same time (because they have the same priority level and same backoff counter value). When this happens, a signal is emitted and the current number of collisions is recorded in a vector; also, specifically for this result, the *WATCH()* macro is used for further inspection in the graphical user interface (Tkenv) provided by OMNeT++.

Figures 3.3 and 3.4 show the mean, maximum, and minimum values for the message delays in the regular and proposed HomePlug GP Protocols, respectively. The value of the minimum delay is achieved when a station wants to send a message only at the end of the current transmission and acquires channel access on its first attempt. Analyzing Figure 3.3, for all priority levels, this value is equal to 1.2 ms, while in Figure 3.4 this value is 686.52 μs. That specific

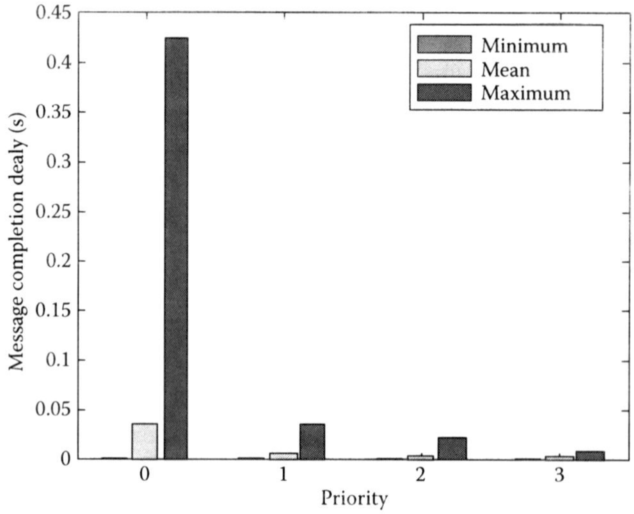

Figure 3.3 Average delay for the regular HomePlug GP messages.

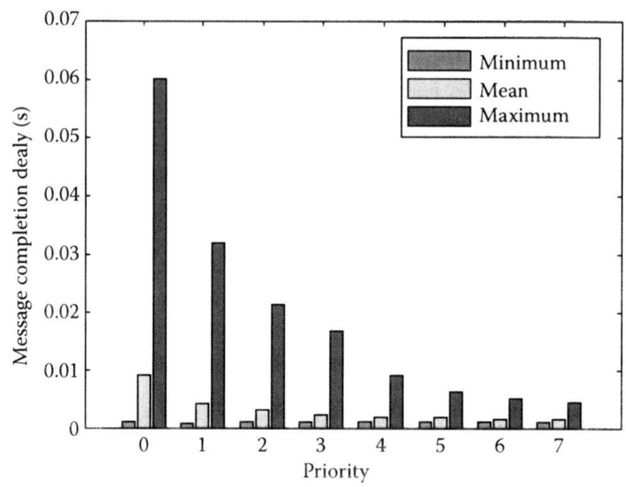

Figure 3.4 Average delay for the proposed HomePlug GP messages.

outcome shows an improvement of roughly 44% in the minimum time for a message to be delivered, which is a result of the reduction in the message size and in the length of the priority resolution slots. It should be noted that all messages in the proposed protocol are sent with a full 256 bytes of data, but the protocol allows for 0–256 bytes of data, corresponding to further reduced delay times. The largest mean value for the regular HomePlug GP protocol corresponds to 35.4 ms, and the largest one for the proposed protocol is equivalent to 9.1 ms, showing a significant decline of approximately 75%. As expected, the low-priority messages experience longer delays, since they cannot be sent until the higher priority messages cease to exist. The maximum delay time is equal to almost half a second (more precisely, 462.2 ms) in the regular version, whereas this value is reduced to only 60.2 ms in the proposed protocol, which demonstrates a notable decrease of 87%. The highest priority messages demonstrate even better results, with maximum delays of only 4.5 ms.

To examine the percentage of message collisions, the bar diagram, shown in Figures 3.5 and 3.6, is used to demonstrate the percentage of collisions that occur out of the total attempted transmissions. As can be seen just by analyzing the vertical axis, the maximum percentage of collisions is reduced by more than 50% in the proposed solution. More precisely, the maximum value for the regular HomePlug GP protocol is 28.67%, which declines by roughly 61% in the proposed HomePlug GP protocol, reaching the value 11.26%. The minimum and mean percentage of collisions are also significantly reduced from 14.72% to 20.75% in the regular protocol, and to 7.53% and 9.35% in the enhanced HomePlug GP protocol, showing a decrease of approximately 50% in both results. These improvements are reached due to the new 8-level priority that made it possible to reduce the size of the CW, since fewer stations at a time enter the backoff procedure. These results demonstrate the efficacy of the proposed improvements.

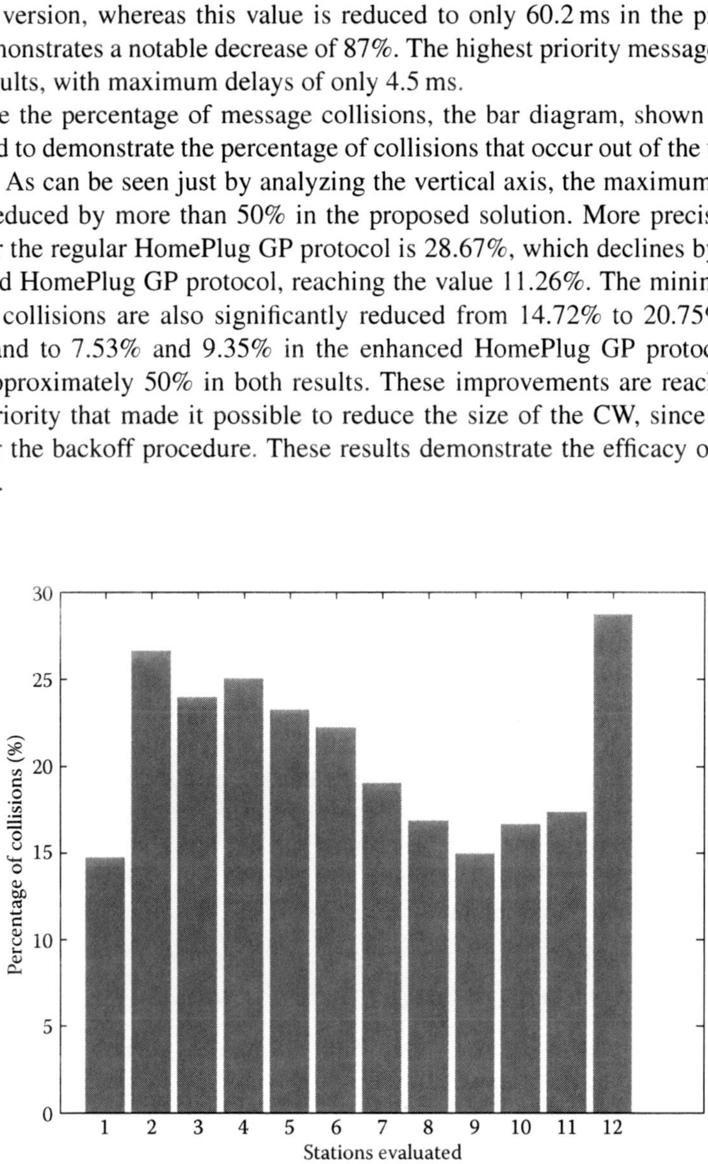

Figure 3.5 Percentage of collisions in the regular HomePlug.

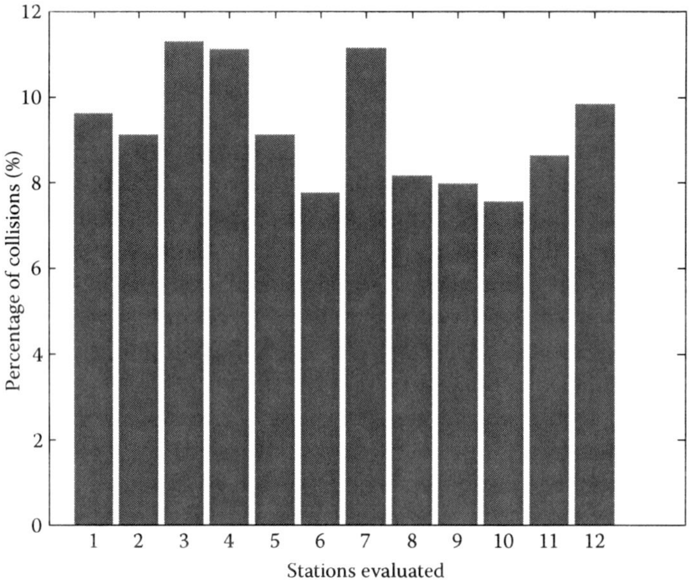

Figure 3.6 Percentage of collisions in the modified HomePlug.

3.5 Future Research Subjects

3.5.1 Compatibility with standard HomePlug GP

Recall from the previous sections that, since the regular Homeplug GP is not suitable for in-vehicle communication and there is a lack of efficient engineering design to achieve real-time features, the new method should be able to cope with new challenges of extending current standards to in-vehicle communications, both vehicle-to-grid (V2G) and Internet-of-vehicle (IoV) communications. As a practical solution, it is also important to ensure that the new solution can be compatible with existing industry standards. Since the proposed method is based on Homeplug GP with reasonable modifications and improvements to achieve real-time features, a mapping solution can be easily made to parse and encapsulate from one to another. A charging station, which is a bridge between cars and home/grid networks, could act as the CCo to perform such a translation. The same issue could also be solved by redesigning the frame control field to include more in-vehicle communication features.

3.5.2 Further improvements in delay performance

Despite demonstrated improvement in the maximum message delay, low-priority messages could still theoretically be delayed indefinitely if higher priority messages continue to be transmitted. Moreover, the minimum delay time seen by high-priority messages is still longer than the targeted 100 μs deadline for safety-critical applications such as x-by-wire systems. Some possible solutions to these ongoing issues are discussed below.

The first issue, that of an upper limit on low-priority delays, could be solved by implementing a flexible time-division multiple access (TDMA) functionality. The simulation shows a maximum delay time of only 60.2 ms, which is acceptable when compared to Lee and Park's deadline requirements of hundreds of milliseconds [128]. However, this value could soar far beyond acceptable limits as a result of an increasing number of stations, frequency of messages, or, more specifically, number of high-priority messages, which would prevent lower priority messages from being sent. A TDMA system could almost guarantee maximum delay times for lower priority messages, though it must still allow for rapid transmission of safety-critical, high-priority messages. This is where flexibility comes in.

Though HomePlug GP does not directly support TDMA functionality, other HomePlug protocols do. The implementation of TDMA in an in-vehicle protocol could follow a similar procedure, with one proposed alteration. When TDMA slots are assigned, which should only be done where necessary, they are given an override priority such that if a message with a higher priority exists, then it can be sent instead of the regular message. Essentially, one station that has frequent low-priority transmissions, which nevertheless should be received with a certain frequency, will have a TDMA slot assigned for this transmission. This station would not generally undergo any backoff procedure; however, it would send an initial empty override flag slot before transmitting. As long as it senses no transmission in this slot, it would send its message. If, however, any other stations have sufficiently high-priority, safety-critical messages that need to be sent before the end of the TDMA period, they could transmit a signal in the override slot, and these stations would then enter priority resolution and/or a backoff period before transmitting their message. In this situation, the station scheduled for that TDMA slot would not transmit. This flexible TDMA protocol has not been tested in this simulation; however, it is proposed here for possible future work.

Further reduction of the minimum delay time could be best realized through shorter interframe spaces and slot times. The total 256 bytes of data in the simulated messages is transmitted in only 26.1 μs. However, the RIFS is a 26 μs delay; the frame control transmits in 59 μs; every slot time in the backoff procedure, though reduced, is another 5 μs; and the CIFS adds another 35.84 μs. An improvement in the physical data rate would do little to reduce the minimum message delay without also reducing these preset delay values.

3.6 Conclusion

In this chapter, we have presented a MAC protocol for in-vehicle communications based on HomePlug GP. Due to the modifications proposed in this work, significant improvements were obtained, such as a reduced number of message collisions and shorter end-to-end message delay. These changes were proposed to make it possible for HomePlug GP, which comprises better power-line communication compatibility than current in-vehicle technologies such as CAN or LIN, to meet strict timing requirements for in-vehicle real-time applications. To further improve delay performance to meet safety-critical communications (100 μs), either further reduction of message overhead or a higher data rate may still be necessary for the proposed protocol to be used with safety-critical systems. These changes should be built on the proposed solution, while maintaining the reliable error-correction capabilities and energy efficiency of HomePlug GP.

INTERVEHICLE COMMUNICATIONS

II

Chapter 4

Physical Channel Modeling and Sharing

Jian Wang and Yunlei Zhang

Jilin University

CONTENTS

4.1	Introduction	59
4.2	Related Work	60
4.3	System Models	61
	4.3.1 Overlay system model of three secondary users	62
	4.3.2 Overlay system model of $n > 3$ secondary users	64
4.4	Numerical Results	66

4.1 Introduction

In wireless communication services, radio frequency (RF) spectrum is one of the most widely used resources at all times [216]. From cell phones and police scanners to TV sets and garage door openers, all the wireless devices have access to the RF wireless spectrum [216]. Over the last decades, the demand for wireless spectrum has been growing rapidly with the dramatic development of the aforementioned mobile telecommunication industry. Traditional fixed spectrum allocation is inefficient and it limits the development of radio and related industries. To use the unemployed spectrum efficiently, spectrum sharing has been proposed. The concepts of software-defined radio and cognitive radio were introduced to enhance the efficiency of frequency spectrum usage and make spectrum sharing come true [157].

Spectrum sharing can be regarded as a system that can share the available RF spectrum between different systems, with certain constraints. Spectrum sharing takes place when multiple wireless systems need to work in the same frequency band at the same time. With the development of cognitive radio technologies, dynamic spectrum access is ripe for spectrum sharing to improve the efficiency of spectrum usage. Dynamic spectrum access allows unlicensed

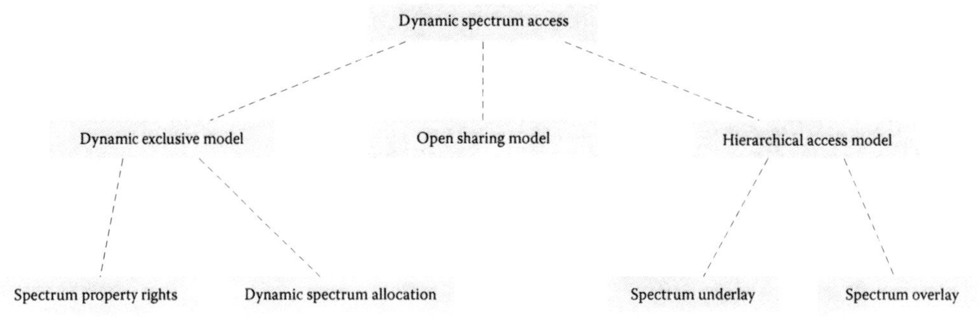

Figure 4.1 Dynamic spectrum access taxonomy.

wireless users (secondary users) to preemptively access the licensed bands when the legacy spectrum holders (primary users) leave the channel. In our work, we study spectrum sharing in Dynamic Spectrum Access Networks (DSAN). As mentioned earlier, there are two spectrum user classes of DSAN. Primary users have the highest priority to use the licensed spectrum. On the other hand, the licensed or unlicensed spectrum can be used opportunistically by the secondary users as long as it does not affect the primary users [291]. Figure 4.1 shows the taxonomy of dynamic spectrum access strategies. In this chapter, we focus on underlay and overlay spectrum-sharing schemes of continuous-time Markov chains (CTMC) in the hierarchical access model. In Ref. [165], overlay and underlay spectrum-sharing schemes have been combined as a new way to raise the system throughput by efficiently using the spectrum. In our work, we introduce multiple secondary users to the scheme and raise the access probability for secondary users to enhance the fairness of the system.

The rest of the chapter is organized as follows: in Section 4.2, we discuss the existing work in this area. In Section 4.3, we explain the multiple secondary users model and derivation process, and we also discuss the model of access data rate and channel fading. In Section 4.4, analyzing the results of simulation, we give our conclusions and propose possible future works.

4.2 Related Work

Here, we discuss some recent research results. In Ref. [38], making full use of TV White Space for spectrum sharing and using queuing theory to solve the problems in sharing have been discussed. Reference [125] has been further improved to access the channel opportunistically after spectrum detection by cognitive radio, and improve the utilization of blank spectrum.

The work in Ref. [133] raises the efficiency of spectrum sharing by competitive price game, and Ref. [134] disclosed spectrum sharing in cognitive two-way relay networks. The author proposed a two-phase spectrum-sharing protocol on account of an analog network coding strategy. A secondary relay node is enabled to share the spectrum, resulting in primary transmission via two-way relaying and meanwhile implementing secondary transmission to improve the capacity of spectrum sharing.

To promote the performance of spectrum sharing, game theory is used in Refs [39,152]. In Ref. [39], it is suggested that compared with the random access scheme, the opportunistic spectrum access scheme of the game theory can utilize the spatial temporal spectrum resources better. In Ref. [152], the game theory has been further optimized, where the spectrum is not occupied by the competitive price game when two secondary users are in a competitive relationship but rather uses a cooperative design instead of a noncooperative game. This turns rivalry into both competition and cooperation, while two secondary users contend to access the channel. The scheme not only ensures that one of the two secondary users cannot access the channel completely but also guarantees that the two secondary users utilize channel maximization.

Reference [102] discusses spectrum sharing via cognitive radio dynamic detection between two licensed wireless networks and proposes a new partial sharing policy for dynamic spectrum access. A Markov chain model has been employed to model the access process considering imperfect spectrum sensing. In Ref. [264], a continuous-time Markov model for dynamic spectrum access in open spectrum wireless networks has been presented.

The ease of modeling the dynamics of primary and secondary users encouraged us to apply CTMC. Reference [264] reveals that the Markov model matches the simulated results for the performance of agile radios exactly. Subsequent papers are all based on CTMC. Ref. [240] presents a primary Markov means to dispose of the spectrum pinch problem and ensures that it reaches higher throughput than the Carrier Sense Multiple Access (CSMA)–based approaches. Ref. [167] proposes an interference regulation scheme to be applied when primary and multiple secondary users transmit simultaneously on the same frequency band. It calculated a probability that Signal to Interference to Noise Ratio outstrips a threshold when the two types of users coexist. The discourse also contributes a cross-layer design of dynamic spectrum access. In a treatise [117] on access strategies, underlay and overlay spectrum-sharing techniques have been studied and a new strategy that combines the overlay and underlay schemes to access the spectrum has been presented. It reveals that the blended access strategy obtains a higher capacity than when the two schemes are used separately. In continuation to Ref. [117], Ref. [165] presents a spectrum-sharing scheme of mixed underlay and overlay in one primary and two secondary networks based on CTMC. It improves the spectrum-sharing efficiency further.

TV White Space has different distribution along with the map, the blank spectrum fixed relatively. It cannot utilize the spectrum dynamically. The competitive price mechanism maximizes single-user utilization of the spectrum when multiple secondary users coexist, and it does not make full use of the channel. Opportunistic access of game theory can flexibly detect the unused spectrum, but it fails to work in the real environment of the cognitive radio system. Modeling the spectrum sharing by CTMC can detect flexibly and utilize fully unused channels. Ref. [165] proposes an easy approach based on CTMC, it is obviously better than other ways. In future, the goal of spectrum sharing is to cancel the fixed allocation of spectrum. Hence, the simulation of two secondary users network cannot guarantee the complete spectrum sharing. It also cannot solve the problem of spectrum scarcity and promote the channel utilization efficiency by a big margin. Hence, we optimize Ref. [165] further by increasing the number of secondary users network. Taking an example of three secondary users network, we show how to derive the model of multiple secondary users to apply in a realistic scene. Although it has the aforementioned advantages, the fairness of the model in Ref. [165] is absent. Accordingly, we could follow the requirement of primary users to restrict secondary users. The approach will ensure transmission of primary users and improve the fairness of the system.

4.3 System Models

In this section, we present a system model of underlay and overlay schemes with one primary and three secondary users, which is derived from one primary and two secondary users, and we improve the derivation method of multiple secondary users model further. To improve the system fairness, we provide the probability of secondary users access spectrum.

We consider a dynamic spectrum access network that consists of a single primary network, P1, and three secondary networks, $S1$, $S2$, and $S3$. We expand the two secondary users model and derive three secondary users. The module and the mathematical deduction for multiple secondary users case are elaborated in this chapter. Networks $P1$, $S1$, $S2$, and $S3$ operate on and share the same spectrum band. As the primary network, $P1$ has access priority over $S1$, $S2$, and $S3$. The secondary networks keep spectrum sensing to detect the transmissions of $P1$. All of the sensing are assumed to be perfect, so the sensing will not have any false alarm or misdetection. We assume all the secondary networks have the same priority to access the spectrum and coexist in more than two secondary users network at the same time.

The mechanism adopted by these networks for coexistence is beyond the scope of this research. We consider that the secondary users employ both overlay and underlay mechanisms to access the channel. We will analyze the two schemes separately to compute the performance gains if the two schemes are used together.

Ref. [165] presents a detailed analysis and experiment of the two secondary users network. We present the simulation result of the two secondary users network model directly and compare with the simulation result of the three secondary users network model.

We assumed that all the network traffic followed the Poisson distribution, and the service times are exponentially distributed. γ_{P1} denotes the traffic arrival rate of the primary network $P1$, and β_{P1} indicates the traffic departure rate of $P1$. Analogously, the arrival rates and departure rates of secondary network $S1$ are expressed as γ_{S1} and β_{S1}. Let γ_{S2} and γ_{S3} denote the traffic arrival rate of the secondary networks $S2$ and $S3$, and β_{S2} and β_{S3} indicate the traffic departure rate of the secondary networks $S2$ and $S3$.

4.3.1 Overlay system model of three secondary users

In Ref. [165], a secondary network of the overlay access scheme accesses the medium and operates without any restrictions, and only the primary is free. The secondary network must quit the channel and search for another idle channel without primary network or simply stop the communication immediately if a primary transmission starts to use the channel. According to the characteristic of the overlay model, the value of the secondary network receiver interfered by primary transmission is reduced to zero.

The overlay system was analyzed by a continuous-time Markov model, and it is shown in Figure 4.2. The network that we assume has nine different possible states, and the channel will be one of the nine states in the overlay access scheme. The nine states are 0, $P1$, $S1$, $S2$, $S3$, $S12$, $S23$, $S13$, $S123$. The steady-state probabilities of primary and secondary users are denoted by δ_P and δ_S, and they are interpreted as the ratio of allocation time to the reference time. It gives the duration of channel occupation by the user and determines the system throughput. The state and steady-state probabilities are indicated by 0 and δ_0 when the channel is idle. The state and steady-state probabilities are denoted by P and δ_{P1} when the channel is occupied by the primary network. It has three states when the channel is occupied by one secondary network, the state and steady-state probabilities are expressed as $S1$, $S2$, and $S3$ and δ_{S1}, δ_{S2}, and δ_{S3}, respectively. The state and steady-state probabilities are denoted by $S12$, $S23$, and $S13$ and δ_{S12}, δ_{S23}, and δ_{S13} when the two secondary networks communication coexists in the channel. $S123$ indicates the coexistence of all secondary users, and δ_{S123} expresses the steady-state probability of all secondary networks concomitance. All the states transition rates are shown in Figure 4.2. The arrival and departure rates of users have been assumed as independent Poisson process. We are going to compute the steady-state probability using steady-state equations or flow equations.

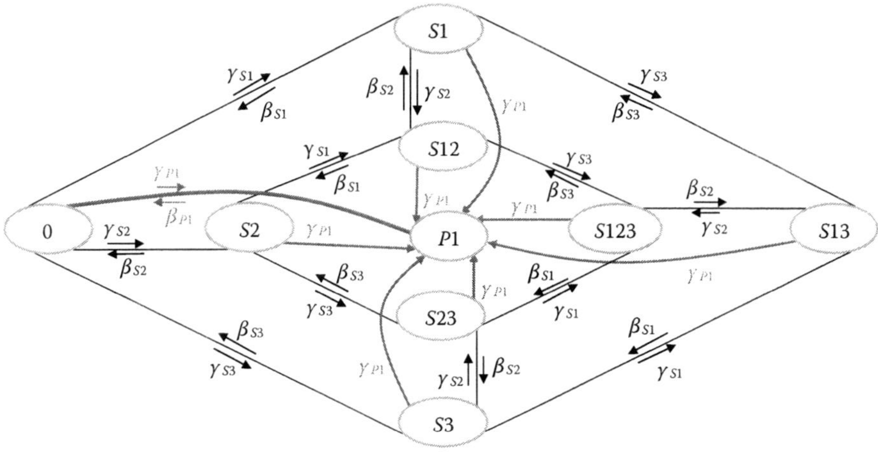

Figure 4.2 Overlay system model of three secondary users.

γ_n and β_n denote the arrival and departure rates, where $n \in S1, S2, S3, P1$. The flow equations are obtained by the CTMC for overlay as follows:

$$(\gamma_{S1}+\gamma_{S2}+\gamma_{S3}+\gamma_{P1})\delta_0 = \beta_{S1}\delta_{S1}+\beta_{S2}\delta_{S2}+\beta_{S3}\delta_{S3}+\beta_{P1}\delta_{P1}. \tag{4.1}$$

$$\beta_{P1}\delta_{P1} = \gamma_{P1}(\delta_0+\delta_{S1}+\delta_{S2}+\delta_{S3}+\delta_{S12}+\delta_{S23}+\delta_{S13}+\delta_{S123}). \tag{4.2}$$

$$\delta_0+\delta_{S1}+\delta_{S2}+\delta_{S3}+\delta_{P1}+\delta_{S12}+\delta_{S23}+\delta_{S13}+\delta_{S123}=1. \tag{4.3}$$

δ denotes the different steady-state probabilities in equations. We compute the capacity of the overlay model using conventional Shannon's capacity formula and give the secondary users' maximum data rate in the overlay scheme as follows:

Capacity of secondary user $S1$ in the overlay model

$$C^{S1}_{\text{overlay}} = W\log_2\left(1+\frac{P_0G_{S11}}{n_0}\right). \tag{4.4}$$

Capacity of secondary user $S1$ when secondary users $S1$ and $S2$ coexist in the overlay model

$$C^{S21}_{\text{overlay}} = W\log_2\left(1+\frac{P_0G_{S11}}{n_0+P_0G_{S21}}\right). \tag{4.5}$$

Capacity of secondary user $S1$ when secondary users $S1$ and $S3$ coexist in the overlay model

$$C^{S31}_{\text{overlay}} = W\log_2\left(1+\frac{P_0G_{S11}}{n_0+P_0G_{S31}}\right). \tag{4.6}$$

Capacity of secondary user $S1$ when secondary users $S1$, $S2$, and $S3$ coexist in the overlay model

$$C^{S321}_{\text{overlay}} = W\log_2\left(1+\frac{P_0G_{S11}}{n_0+P_0G_{S21}+P_0G_{S31}}\right), \tag{4.7}$$

where W and P_0 denote the system bandwidth and the secondary users power, respectively. G_{XY} represents the power gain from user X to Y. n_0 is the noise power.

Capacity of secondary users $S2$ and $S3$ in the earlier situations is computed the same as $S1$.

Every steady-state probability is solved by the use of the CTMC for the overlay model. The average throughput of each user for overlay is shown as follows:

Throughput of S1

$$\begin{aligned} Th_{\text{overlay}}^{S1} = & \ C_{\text{overlay}}^{S1} \times \delta_{S1} + C_{\text{overlay}}^{S21} \times \delta_{S12} + C_{\text{overlay}}^{S31} \times \delta_{S13} \\ & + C_{\text{overlay}}^{S321} \times \delta_{S123}. \end{aligned} \quad (4.8)$$

Throughputs of $S2$ and $S3$ are the same as $S1$. The overlay system throughput consists of all the secondary users throughput, which is given as:

$$C_{\text{overlay}}^{\text{system}} = Th_{\text{overlay}}^{S1} + Th_{\text{overlay}}^{S2} + Th_{\text{overlay}}^{S3} \quad (4.9)$$

4.3.2 Overlay system model of $n > 3$ secondary users

There are five possible states that can be given by 0, $S1$, $S2$, $P1$, $S12$ in the overlay model of one primary user network and two secondary users network. In all the overlay states, state 0 indicates that the channel is idle. Each primary and secondary users network has one state in the model and they are denoted by $P1$, $S1$, and $S2$. The characteristic of overlay model is exclusive, so the primary and secondary users cannot coexist in the overlay model simultaneously. Because the secondary users network could coexist in the overlay scheme, the state that the two secondary users coexist is denoted by $S12$. In the overlay model, state 0 could change to $P1$, $S1$, and $S2$ and vice versa. Primary user $P1$ has higher priority than secondary users, so the secondary users' transmissions have to stop in the channel if the primary user transmission starts. In other words, if state $P1$ exists then states $S1$, $S2$, and $S12$ vanish, so states $S1$, $S2$, and $S12$ can convert into state $P1$, but the process cannot happen contrarily. States $S1$ and $S2$ cannot transform each other, and transformation can occur between states $S1$ and $S12$ and states $S2$ and $S12$, but states 0 and $S12$ cannot transform reciprocally.

There are nine possible states that can be given by 0, $S1$, $S2$, $S3$, $P1$, $S12$, $S23$, $S13$, and $S123$ in the overlay model of one primary user network and three secondary users network. In all the overlay states, state 0 indicates that the channel is idle. Each primary and secondary users network has one state in the model, and they are denoted by $P1$, $S1$, $S2$, and $S3$. The characteristic of the overlay model is primary, and secondary users cannot coexist in the overlay model simultaneously, but the secondary users network could coexist in the overlay scheme, so the states that two of the secondary users coexist are denoted by $S12$, $S23$, and $S13$, and three of the secondary users coexist is denoted by $S123$ only. In the overlay model, state 0 can change to $P1$, $S1$, $S2$, and $S3$ and vice versa. Primary user $P1$ has higher priority than secondary users, so the secondary users transmissions have to stop in the channel if the primary user transmission starts. In other words, if state $P1$ exists then all the states vanish except states 0 and $P1$; so each state combination of the three secondary users can convert into state $P1$, but the process cannot happen contrarily. State $P1$ can transform into state 0 only and vice versa. States $S1$, $S2$, and $S3$ cannot transform each other, that is to say, all the single secondary user states cannot transform reciprocally. Transformation can occur between state S1 and the state of two secondary users coexist, which includes $S1$; it is the same as states $S2$ and $S3$. State 0 cannot transform into any states without the single secondary user state and $P1$. State $S123$ can convert into $S12$, $S23$, and $S13$ and vice versa, but it cannot change to $S1$, $S2$, $S3$, and 0.

In the overlay model of one primary user network and n secondary users network, state 0 indicates that the channel is idle. Each primary and secondary users network has one state in the model, so there are $C_n^1 + 1 = n + 1$ states. The characteristic of the overlay model is that primary and secondary users cannot coexist in the overlay model simultaneously, but the secondary users network could coexist in the overlay scheme, so the states of two secondary users coexist and three secondaries coexist are C_n^2 and C_n^3, respectively. According to the method mentioned earlier, we can derive that there is $C_n^n = 1$ state when n secondaries coexist in the channel simultaneously. In the overlay model, state 0 can change to $P1$ and vice versa. State 0 cannot transform into any states without the $C_n^1 = n$ single secondary user state. Primary user $P1$ has higher priority than all the secondary users, so all the secondary users transmissions have to stop in the channel if the primary user transmission starts. In other words, if state $P1$ exists then all the states vanish except states 0 and $P1$, so each state combination of the n secondary users can convert into state $P1$ but the process cannot happen contrarily. State $P1$ can transform into state 0 only and vice versa. All the single secondary user state cannot transform reciprocally. Transformation can occur between state X and the state of two secondary users coexist, which includes X; it also happens between X and 0. We deduce that the transformation can occur between the state of m secondary users coexist that includes state XY and the state of $m-1$ secondary users coexist that includes state X but not Y in the n secondary users network. It is the same between the state of m secondary users coexist that includes state XY and the state of $m+1$ secondary users coexist that includes state XY. The state of n secondaries coexist can convert into the state in which $n-1$ secondaries coexist and vice versa, but it cannot change to any other states except the state of $n-1$ secondaries coexist.

The flow equations of the overlay system in n secondary users network are given as follows:

$$\begin{aligned}\beta_{P1}\delta_{P1} &= \gamma_{P1}(\delta_0 + \delta_{S1} + \delta_{S2} + \cdots + \delta_{Sn} + \delta_{S12} + \delta_{13} \\ &\quad + \cdots + \delta_{S1n} + \cdots + \delta_{S123} + \cdots + \delta_{S1,2,3,\cdots,n}).\end{aligned} \tag{4.10}$$

$$\begin{aligned}(\gamma_{S1} + \gamma_{S2} + \gamma_{S3} + \cdots + \gamma_{Sn} + \gamma_{P1})\delta_0 &= \beta_{S1}\delta_{S1} + \beta_{S2}\delta_{S2} \\ &\quad + \beta_{S3}\delta_{S3} + \cdots + \beta_{Sn}\delta_{Sn} + \beta_{P1}\delta_{P1}.\end{aligned} \tag{4.11}$$

$$\begin{aligned}(\beta_{S1} + \gamma_{S2} + \gamma_{S3} + \cdots + \gamma_{Sn} + \gamma_{P1})\delta_{S1} &= \gamma_{S1}\delta_0 + \beta_{S2}\delta_{S12} \\ &\quad + \beta_{S3}\delta_{S13} + \cdots + \beta_{Sn}\delta_{S1n}.\end{aligned} \tag{4.12}$$

$$\begin{aligned}(\gamma_{S1} + \beta_{S2} + \gamma_{S3} + \cdots + \gamma_{Sn} + \gamma_{P1})\delta_{S2} &= \gamma_{S2}\delta_0 + \beta_{S1}\delta_{S12} \\ &\quad + \beta_{S3}\delta_{S23} + \beta_{S4}\delta_{S24} + \cdots + \beta_{Sn}\delta_{S2n}.\end{aligned} \tag{4.13}$$

$$\begin{aligned}&(\beta_{S1} + \beta_{S2} + \beta_{S3} + \cdots + \beta_{Sn} + \gamma_{P1})\delta_{S1,2,3,\cdots,n} \\ &= \gamma_{S1}\delta_{S2,3,\cdots,n} + \gamma_{S2}\delta_{S1,3,4,\cdots,n} + \gamma_{S3}\delta_{S1,2,4,\cdots,n} + \cdots + \gamma_{Sn}\delta_{S1,2,3,\cdots,n-1}.\end{aligned} \tag{4.14}$$

$$\begin{aligned}&\delta_0 + \delta_{P1} + \delta_{S1} + \delta_{S2} + \delta_{S3} + \cdots + \delta_{Sn} + \delta_{S12} + \delta_{S13} + \cdots + \delta_{S1n} \\ &+ \delta_{S23} + \cdots + \delta_{Sn-1,n} + \delta_{S123} + \delta_{S124} + \cdots + + \delta_{S1,2,3,\cdots,n} = 1.\end{aligned} \tag{4.15}$$

We compute the capacity of the n secondary users network overlay model, and the secondary users' maximum data rate in the overlay scheme is the same as the three secondary users network aforementioned.

Capacity of secondary user $S1$ in the overlay model

$$C_{\text{overlay}}^{S1} = W\log_2(1+\frac{P_0 G_{S11}}{n_0}). \tag{4.16}$$

Capacity of secondary user $S1$ when secondary users $S1, S2,\cdots, Sn$ coexist in the overlay model

$$\begin{aligned} C_{\text{overlay}}^{Sn,n-1,\cdots,3,2,1} = & \; W\log_2(1+P_0G_{S11}/(n_0+P_0G_{Sn,1} \\ & +P_0G_{Sn-1,1}+\cdots+P_0G_{S31}+P_0G_{S21})). \end{aligned} \tag{4.17}$$

Capacity of secondary user Sn when secondary users $S1, S2,\cdots, Sn$ coexist in the overlay model

$$\begin{aligned} C_{\text{overlay}}^{Sn-1,\cdots,3,2,1,n} = & \; W\log_2(1+P_0G_{Snn}/(n_0+P_0G_{Sn-1,n} \\ & +\cdots+P_0G_{S3n}+P_0G_{S2n}+P_0G_{S1n})). \end{aligned} \tag{4.18}$$

Throughput of A

$$\begin{aligned} Th_{\text{overlay}}^{S1} = & \; C_{\text{overlay}}^{S1}\times\delta_{S1}+C_{\text{overlay}}^{S21}\times\delta_{S12}+C_{\text{overlay}}^{S31}\times\delta_{S13} \\ & +\cdots+C_{\text{overlay}}^{Sn,1}\times\delta_{S1,n}+C_{\text{overlay}}^{S321}\times\delta_{S123} \\ & +C_{\text{overlay}}^{S421}\times\delta_{S124}+\cdots+C_{\text{overlay}}^{Sn21}\times\delta_{S12n} \\ & +\cdots+C_{\text{overlay}}^{Sn,n-1,\cdots,3,2,1}\times\delta_{1,2,3,\cdots,n-1,n}. \end{aligned} \tag{4.19}$$

The throughput derivation of $S2$ to Sn is the same as $S1$, the overlay system throughput is the sum throughput of all the secondary users, which is given as:

$$C_{\text{overlay}}^{system} = Th_{\text{overlay}}^{S1}+Th_{\text{overlay}}^{S2}+Th_{\text{overlay}}^{S3}+\cdots+Th_{\text{overlay}}^{Sn} \tag{4.20}$$

4.4 Numerical Results

In this section, we integrate the proposed scheme and validate it using MATLAB®. All the meanings and values of the parameters are listed in Table 4.1.

Table 4.1 Parameters

Parameters	*Meaning*	*Default Value*
$\gamma_{S2}, \gamma_{S3}, \gamma_{P1}$	The arrival rates of secondary networks $S2$, $S3$, and $P1$	100
γ_{S1}	The arrival rates of secondary networks $S1$	(100,180)
$\beta_{S1}, \beta_{S2}, \beta_{S3}$, and β_{P1}	The departure rates of secondary networks $S1$, $S2$, $S3$, and $P1$	100
γ_W	Wavelength	0.0508 m
W	Bandwidth	10 MHz
n_0	Noise	10^{-15} W
P_0	The transmission power of primary and secondary users in overlay model	10 mW
G_{XY}	Power gain from user X to Y	All the power gains are 1

We assume the same bandwidth for all users. The transmitter of user $P1$ is at (0 m, 0 m) and the receiver of user $P1$ is at (300, 0 m). The transmitter and the receiver of user $S1$ are at (600, 0 m) and (700, 0 m), respectively. Similarly, the transmitter and the receiver of user $S2$ are at (400, 0 m) and (450, 0 m), respectively. The transmitter and the receiver of user $S3$ are at (500, 0 m) and (550, 0 m), respectively.

The state probabilities of channel occupancy in Figure 4.3 are computed using the flow Equations 4.1 through 4.3. We can gain the throughput of each secondary user and the system from Figure 4.4, and then compare the difference between two and three secondary users in

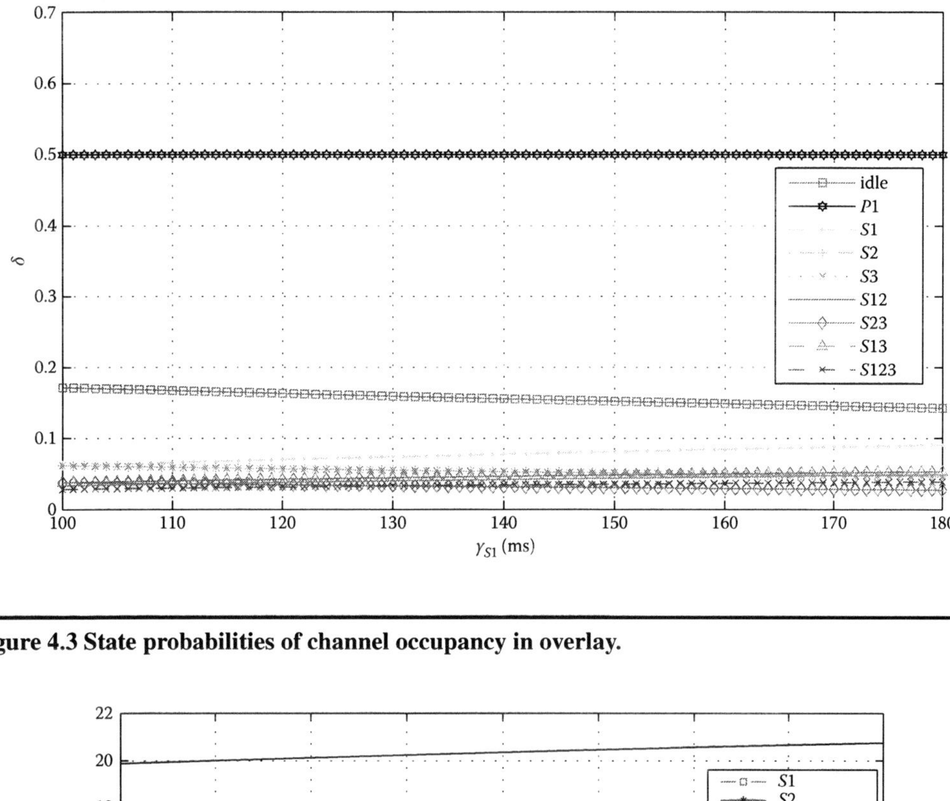

Figure 4.3 State probabilities of channel occupancy in overlay.

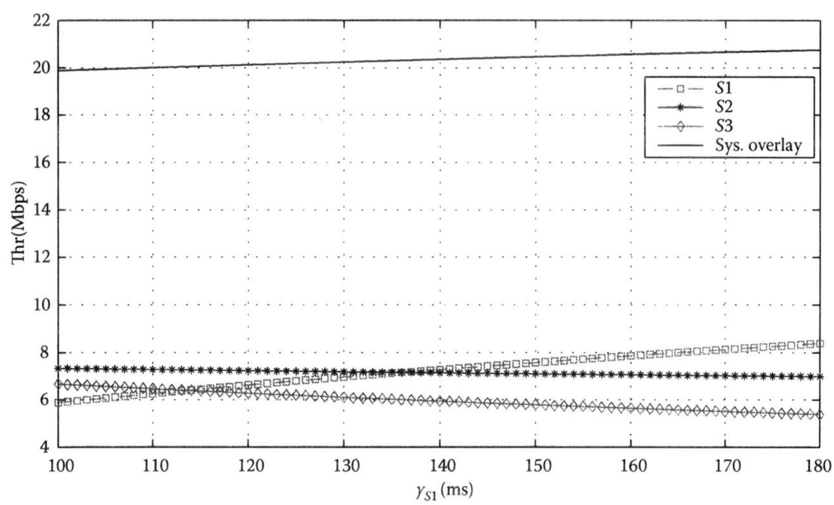

Figure 4.4 Secondary system throughput comparison between two and three secondary users.

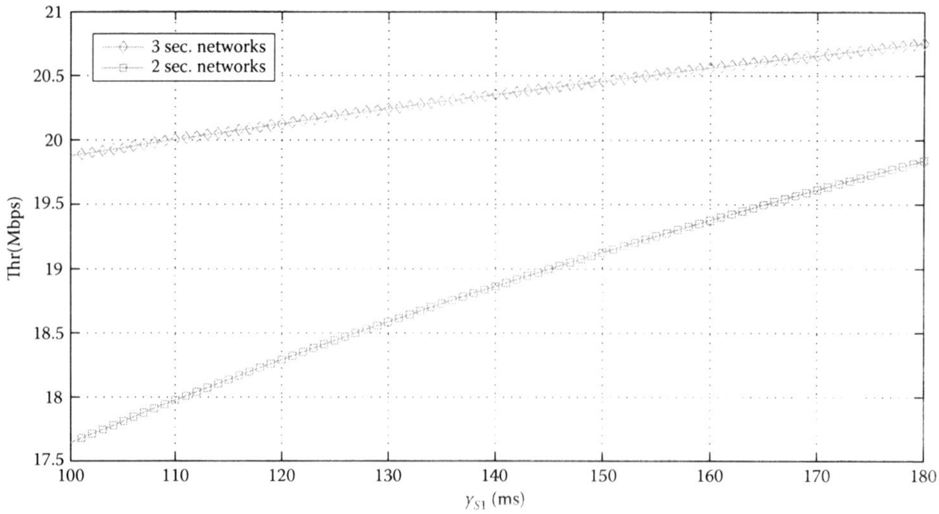

Figure 4.5 Throughput comparison between two and three secondary users in overlay scheme.

overlay model. We can observe from Figure 4.5 that the throughput of three secondary users is higher than two secondary users in overlay, and it will improve the performance of the spectrum-sharing scheme further.

Finally, the throughput of three secondary users is higher than two secondary users in simulation results; it also improves the secondary users number of spectrum sharing. We can also improve the throughput by raising the power gain, then the noise–signal ratio improves.

Chapter 5

Short-Range Communication Technology in the Internet of Vehicles

Daxin Tian, Jianshan Zhou, Yingrong Lu, and Yunpeng Wang
Beihang University

Zhengguo Sheng
University of Sussex

CONTENTS

5.1	Introduction	70
5.2	Related Work	71
	5.2.1 Dedicated short-range communication	71
	5.2.2 Bio-inspired model for making the handover decision	72
5.3	A Self-Adaptive V2V Communication System with DSRC	73
	5.3.1 V2V communication system with DSRC	73
	5.3.2 The embedded self-adaptive configuration scheme	75
5.4	Bio-inspired QoS-Oriented Handover Model in Heterogeneous Wireless Networks	76
	5.4.1 System model	76
	5.4.1.1 Problem formulation	76
	5.4.1.2 Basic attractor selection model	76
	5.4.1.3 Extended attractor selection model	77
	5.4.1.4 Model validation and discussion	78
	5.4.2 QoS-oriented handover decision	81
	5.4.2.1 Handover decision framework based on the extended attractor selection model	82
	5.4.2.2 Quantification of terminal's QoS satisfaction	83
5.5	Experiment and Analysis	85

5.5.1 Self-adaptive V2V communication system 85
5.5.2 Bio-inspired QoS-oriented handover model 86
5.5.2.1 Simulation scenario 87
5.5.2.2 Simulation settings 87
5.5.2.3 Numerical evaluation 89
5.6 Conclusion 93

5.1 Introduction

In recent years, as an instance of a vehicular ad hoc network, the telematics integrating the utility of telecommunications and informatics has dramatically promoted the development of wireless communications in vehicular environments, intelligent transportation systems, and the automotive electronics industry. Nowadays, IEEE 802.11p and IEEE 1609 family standards are called wireless access in vehicular environments (WAVE) standards, since they are developed to facilitate the provision of wireless access in vehicular environments. The WAVE technology is classified under dedicated short-range communications (DSRC). Several ongoing research projects supported by car manufacturers, electronic industries, governments, and academia have been underway to accelerate the deployment of short-range wireless networks that exploit vehicle-to-vehicle (V2V) and vehicle-to-roadside (V2R) infrastructure communications based on WAVE. These networks are characterized by rapidly changing topologies and short connection lifetimes, and one of the most important goals of the emerging DSRC-based V2V communication standards is to enable road safety applications that could save thousands of lives.

Safety-related applications are geared primarily toward avoiding the risk of car accidents, by using cooperative collision warning, precrash sensing, or lane change and traffic violation warnings. These applications all have real-time constraints, which always rely on one-hop broadcasting or multihop V2V and vehicle-to-infrastructure (V2I) communications. Moreover, the quality of vehicle safety applications degrades with the increase in packet loss and delay in vehicular wireless communications. Timely warning messages transmitted by a braking or slowly moving vehicle enable approaching traffic to take appropriate action such as slowing down or changing lanes earlier than they could have without the warning messages, and thereby can reduce the chance of crashes or chain collisions.

Wireless communication technologies such as the wireless LANs (WLAN IEEE 802.11a/b/g/n/p standards), WiMAX (IEEE 802.16 a/e standards), and the third- and fourth-generation cellular wireless (3G or 4G) and satellite communications have been developed. The substantial significant infrastructure of these wireless networks has been deployed to support the dramatically increasing demand for mobile terminals' access to network services anywhere and anytime. However, no single wireless access technology can be efficient enough to satisfy the demands of various users for reliable connection and quality of service (QoS) in all situations. Consequently, the next-generation wireless communication system is evolving, which depends on those heterogeneous wireless networks.

In this chapter, we will propose two kinds of short-range communication technologies in the Internet of Vehicles—"a self-adaptive V2V communication system with DSRC" and "a bio-inspired QoS-oriented handover model in heterogeneous wireless networks"; we will use experiments to highlight their usefulness and efficiency.

5.2 Related Work

5.2.1 Dedicated short-range communication

The V2V communication system was developed based on an 802.11p protocol. When compared to traditional wireless networks, it is more efficient and more secure. This is due to its protocol stack architecture. To build this communication system and to collect its packet loss rate and delay data, we have reviewed many sources and numerous works on IEEE 802.11p and some other standard drafts. Furthermore, a novel V2V communication system software has been proposed and realized in this work; it has been used to set up some testing experiments in a real V2V environment for validating the ability of IEEE 802.11p protocol with the real data collected.

The IEEE 1609 working group has cooperated with the IEEE 802.11p for developing specifications of the additional layers in the protocol suite known as the WAVE stack [78], which is illustrated in Figure 5.1. IEEE 802.11p is known as an approved amendment added to the IEEE 802.11 standard family, which defines wireless access in vehicular environments (WAVE as given in Figure 5.1). The goal of IEEE 802.11p is to support a wide range of ITS applications, especially the safety-realted applications based on reliable real-time data exchange between high-mobility vehicles and between vehicles and roadside infrastructure, such as intersection collision warning, road obstacle warning, cooperative chain collision avoidance, etc. This amendment is sometimes considered for dedicated short-range communications (DSRC) technologies, proposed by a U.S. Department of Transportation project, i.e., the ITS America VII (Vehicle-Infrastructure Integration). As shown in Figure 5.1, WAVE physical layer (WAVE PHY) and MAC layer (WAVE MAC) can provide a basis for other upper layers, including

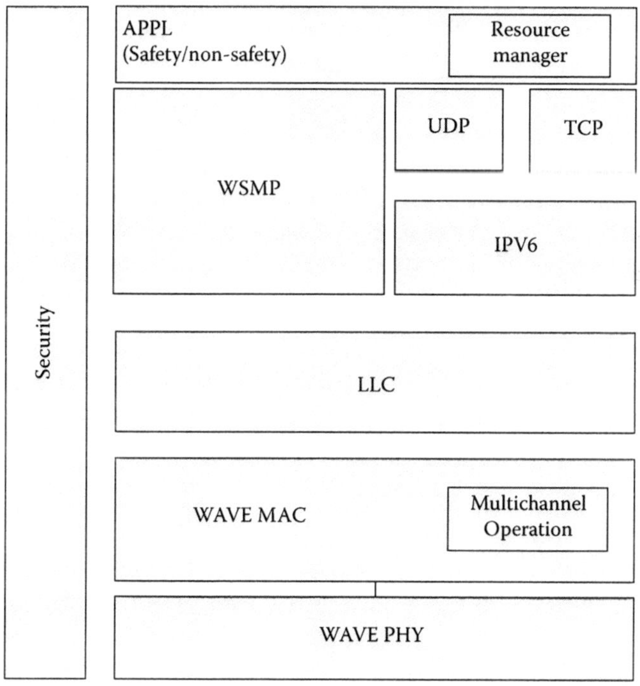

Figure 5.1 WAVE stack.

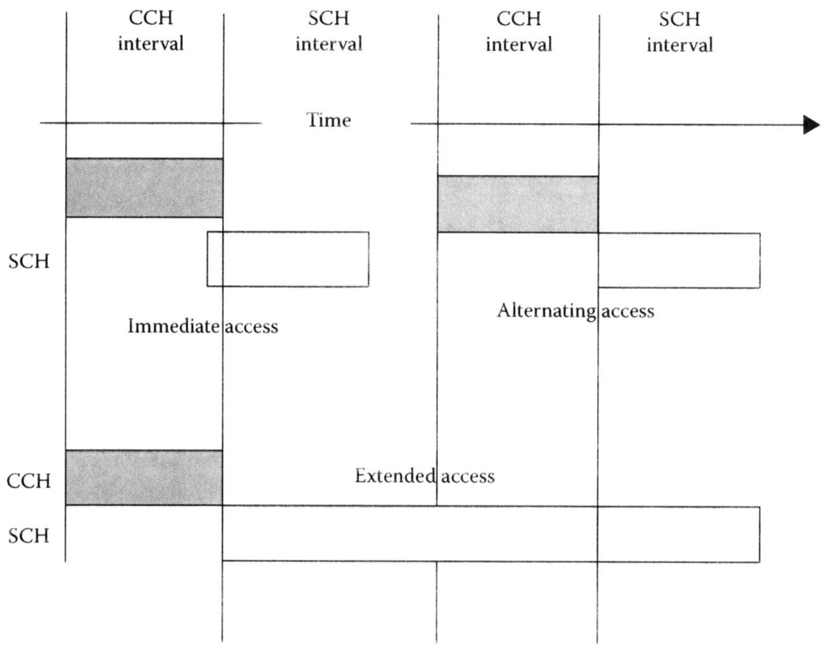

Figure 5.2 Channel access mechanisms. CCH, control channel; SCH, service channel.

the logic link control (LLC) layer, the WAVE short message protocol (WSMP) layer or the Internet protocol version 6 (IPV6) layer as well as the related user datagram protocol (UDP), and transmission control protocol (TCP) sub-layers. On the top layer of the WAVE stack is the application layer (APPL), which can support a variety of safety and non-safety applications and those relevant to resource management [22,42,86,93,215].

IEEE 1609.3 and IEEE 1609.4 provide a lot of support for WAVE/DSRC standards. They define a control channel and several other transmission channels. Different types of information can be transmitted on different channels. A synchronization interval is divided into a control channel (CCH) interval and a service channel (SCH) interval; WAVE/DSRC device control channel spacing is CCH. These are specified in IEEE1609.4, as shown in Figure 5.2.

5.2.2 *Bio-inspired model for making the handover decision*

In existing heterogeneous wireless networks, multiple wireless access technologies as well as multiple radios have to interwork and be used in a cooperative manner to realize the always best connected (ABC) concept in terms of high level of QoS satisfaction and fairness of resource allocation [79].

The handover decision is one of the most important issues (including handover management, resource allocation, mobility support, etc.) related to the heterogeneous wireless networks. It should be efficiently addressed in order to realize the envisioned next-generation communication system [285]. However, there exist some significant challenges in developing the essential functional components and in designing the corresponding algorithm for handover decision, such as impracticality of centralized control, dynamic nature, resources constraint, and heterogeneity in the heterogeneous wireless environment [273].

However, the bio-inspired paradigm provides a novel approach to designing a new, powerful solution for many engineering problems [54,246]. Similar challenges met by the heterogeneous wireless communication system arising from dynamic nature, system complexity, heterogeneous architectures, and absence of centralized control have been well addressed by the biological system [155]. Many biological mechanisms such as adaptability to environmental changes, inherent robustness to external perturbation, and self-optimization are appealing to be introduced in the handover decision solution to deal with those aforementioned significant challenges. As some interdisciplinary studies have argued, many biological mechanisms resulting from the evolution of nature over millions of years always go far away beyond the traditional technologies so that they are promising to be used to settle some complex engineering problems [120,121,292].

Attractor selection is one type of bio-inspired mechanism that induces cellular gene expression to adaptively respond to the dynamically changing environment. Its related model, the attractor selection model [113], has drawn much attention and has consequently been extended to be implemented in many engineering domains. For example, it has been applied to robust robot control [66], error-tolerant wireless sensor networks control [94], adaptive virtual network topology control [123], adaptive routing protocol in mobile ad hoc networks or overlay networks [129], and to resource allocation among multiple users and multiple applications in the heterogeneous wireless environment [108].

Motivated by the attractor selection of cellular gene network, we adopt this bio-inspired mechanism for modeling the vertical handover decision in heterogeneous wireless networks. The goal of this chapter is to deal with the varying heterogeneous wireless environmental conditions, and at the same time to guarantee user satisfaction level and to ensure the fairness of network resource allocation among multiple mobile terminals. We extend the basic attractor selection model that has been proposed to a novel one with a higher dimension for multiple-attributes decision making. The upper and lower bounds of QoS requirements for multiple applications are combined with the dynamic wireless network conditions, including bandwidth, end-to-end delay, and packet loss ratio to formulate a function for evaluating the terminal's QoS satisfaction. This utility function is used to assist the bio-inspired model in performing the attractor selection mechanism. The handover decision induced by the attractor selection mechanism allows us to capture the dynamic nature of the heterogeneous wireless environment and to evaluate the goodness of accessing wireless networks, so as to enable the handover decision adaptive to the wireless environmental changes.

5.3 A Self-Adaptive V2V Communication System with DSRC

5.3.1 *V2V communication system with DSRC*

In this section, the proposed V2V communication system with DSRC is presented. A Linux-based platform was built for our system, owing to its characteristics such as that it is free, portable, scalable, secure, and versatile. Figure 5.3 depicts the software architecture of the V2V communication system devices. It shows that the architecture consists of IEEE 802.11p, IEEE 1609.4, IEEE 1609.3, and user application. The implementation of both the control plane and data plane on each component is detailed in the subsequent paragraphs.

1. IEEE 802.11p: The ath5k driver can be achieved through IEEE 802.11p [258]. This is considered to achieve the physical components of the device chip. We can meet all the requirements of the standard IEEE 802.11p by modifying the functional components of the device chip. DSRC device can also modify the MAC layer filtering.

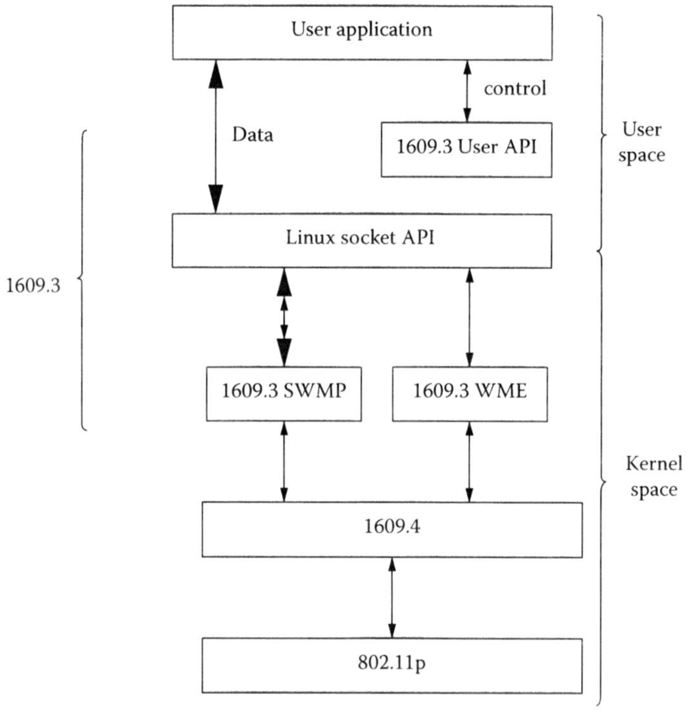

Figure 5.3 V2V communication system architecture with DSRC devices.

2. IEEE 1609.4: According to the specifications of the IEEE 1609.4, WSA frame should broadcast regularly. To improve performance, the IEEE1609.4 module is a portable Linux kernel space, using the Linux kernel timer to achieve this functionality. A set of application programming interface (API) functions are also developed according to IEEE 1609.3.

3. IEEE 1609.3: IEEE 1609.3 mainly handles the data delivery between DSRC devices and manages network services. The core of the IEEE 1906.3 module is divided into two parts: data plane and control plane. It is similar to the implementation of the IEEE 1609.4; both the parts are implemented on the kernel space of the Linux to improve system performance. In the data plane part, the well-known Linux socket is utilized to develop a socket-like API and an extra PF WSMP of the protocol family. As for the control plane, we exploit a socket-like mechanism to be an interprocess communication (IPC) between the kernel space and the user space. Moreover, an API of services registration on user space is provided for user applications. The related information will be stored on kernel space in order to speed up data processing, because referring to channel information and provider service identifier (PSID) is necessary for data delivery in DSRC systems.

To provide the APIs for other up-layer applications, we furthermore design the implementation model of the V2V communication system based on the aforementioned software architecture [92,159,181,241] as illustrated in Figure 5.3. The V2V communication system model is shown in Figure 5.4. Our system can be divided into two parts [256,278]: the host and the router.

1. Host: The host side not only runs the Java implementation based on the transmission control protocol/Internet protocol (TCP/IP), which can be used on almost any operating

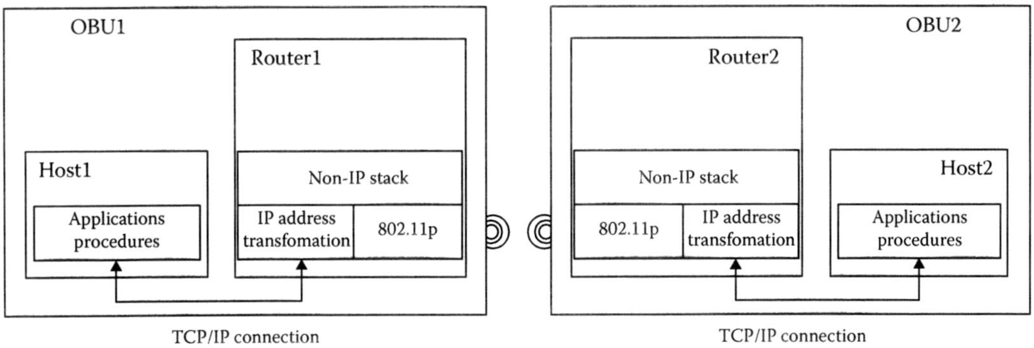

Figure 5.4 The V2V communication system model.

system, because of Java's good compatibility and portability, but also executes many applications based on Satellite Navigation System (SNS) and Controller Area Network (CAN). The Java implementation itself utilizes the TCP/IP protocol to communicate on thc router side.

2. Router: On the router side, an 802.11p driver and application process has been deployed. The 802.11p layer of the V2V communication software is embedded in the router.

5.3.2 The embedded self-adaptive configuration scheme

In this section, we mainly focus on developing the self-adaptive scheme embedded in the V2V communication system for transmission frequency configurations in order to reduce the probability of a rear-end collision between two cars [159,181,241]. First, let t and t_c be the V2V delivery delay and the driver's reaction time, respectively. We can obtain the maximum acceptable delivery delay through certain experiments. We consider the performance of the V2V communication system in the presence of packet loss. To achieve a reliable and adaptive transmission, the periodical rebroadcasting of messages is used in DSRC-based devices. A collision between vehicles can thus be avoided, provided that at least one of the retransmitted packets reaches the destination vehicle within the time window $0 \leq t \leq t_{\mathrm{c}}^{\max}$. In the next step we derive the probability of collision avoidance as a function of the packet loss rate, p, under the assumption that the source vehicle retransmits the warning packet every T seconds. The probability that the first warning packet successfully reaches the other vehicle after $n,(n = 1,2,3\ldots)$ retransmissions is given by: $(1-p)p^n$. The probability distribution function for the arrival time of an intact packet at this vehicle is then given by:

$$\text{probability}^* = \sum_{n=0}^{\infty}(1-p)p^n\delta(t-(t_{\mathrm{c}}+nT)). \tag{5.1}$$

Vehicle collision can be avoided if $t < t_{\mathrm{c}}^{\max}$, from which the probability of collision avoidance, probability*, is derived by:

$$\begin{aligned} \text{probability}^* &= \sum_{n=0}^{\infty}(1-p)p^n\Theta(t_{\mathrm{c}}^{\max}-(t_{\mathrm{c}}+nt)) \\ &= 1-p^{\frac{\lfloor t_{\mathrm{c}}^{\max}-t_{\mathrm{c}}\rfloor}{T}}. \end{aligned} \tag{5.2}$$

In Equations 5.1 and 5.2, $\delta(\cdot)$ and $\Theta(\cdot)$ are the delta and the step functions, respectively. According to Equation 5.2, the minimum message retransmission frequency, f, can be obtained from Equation 5.3, which is required to reduce the probability of collision below a system-defined value, $\varepsilon = 1 -$ probability*:

$$f = \left(\frac{1}{[t_c^{\max} - t_c]}\right)\left(\frac{\log \varepsilon}{\log p}\right). \tag{5.3}$$

The Equation 5.3 should satisfy the inequality $0 \le t \le t_{\mathrm{c}}^{\max}$.

Because the model Equation 5.3 as expressed here can be used to adjust the transmission frequency parameter of the V2V communication system, it is very useful. Once we have detected the real-time transmission time parameter t_{c} and the loss rate parameter p and the proposed collision avoidance probability is set up, we can dynamically achieve better transmission frequency and optimize the frequency configuration of the V2V communication system proposed according to Equation 5.3.

5.4 Bio-inspired QoS-Oriented Handover Model in Heterogeneous Wireless Networks

5.4.1 System model

In this section, the core problem to be solved in the handover decision is described first, then the basic attractor selection model as well as its corresponding bio-mechanism is presented. Following the mathematical form of the basic attractor selection model, we extend it to a novel form and apply this extended model for multiple attributes' decision making.

5.4.1.1 Problem formulation

We assume that each mobile terminal moving in a given heterogeneous wireless environment is equipped with a multimode communication device. These terminals with multiple wireless interfaces are able to access different wireless networks simultaneously. Namely, under this assumption, a mobile user is allowed to assign different wireless links to its different applications that are running in the terminal device. We consider a heterogeneous wireless environment composed of a network set of multiple heterogeneous wireless networks. We denote this network set as $NetSet = \{net_1, net_2, \ldots, net_M\}$. The parameter M here denotes the total number of those considered wireless networks. We then consider that there are in total Q vehicular terminals moving in this given heterogeneous wireless environment. All of these mobile terminals compose a set that is denoted by $User = \{u_1, u_2, \ldots, u_Q\}$, and each one $u_k \in User$ has a certain number of applications running in its terminal device. For instance, we denote those applications simultaneously running in u_k as a set $S_{u_k} = \{s_1, s_2, \ldots, s_N\}$. Thus, each u_k is required to make a decision to select the most suitable wireless network for each of its applications. Each application $s_i \in S_{u_k}$ may connect to the same network or may use different wireless links. Each u_k performs the handover decision process during every discrete period $\triangle t$. When the network selection is done, the wireless interface of each application in u_k is switched from the previous one to the new determined network.

5.4.1.2 Basic attractor selection model

The basic attractor selection model is inspired by cell biology. It is used to describe the adaptive response of the gene expression in an *Escherichia coli* (*E. coli*) cell to the changes in its available nutrients, especially when there is not enough molecular machinery for signal transduction from the environment to the DNA expression [113]. The basic mathematical model of attractor selection can be expressed by two nonlinear differential equations with stochastic noise as follows:

$$\begin{cases} \frac{dm_1}{dt} = \frac{S(A)}{1+(m_2)^2} - D(A) \times m_1 + \eta_1 \\ \frac{dm_2}{dt} = \frac{S(A)}{1+(m_1)^2} - D(A) \times m_2 + \eta_2 \end{cases}, \tag{5.4}$$

where m_1 and m_2 represent two different mRNA concentrations respectively corresponding to two different nutrients. $S(A)$ and $D(A)$ are two different rate coefficients of nutrient synthesis and degradation. They are defined as the monotonously increasing function of cellular activity that is represented by the parameter A. The $S(A) = 6A/(A+2)$ and $D(A) = A$. In Ref. [113], η_1 and η_2 represent independent white Gaussian noise that is inherent in cellular gene expression.

From the viewpoint of a dynamics system, the variable pair $[m_1, m_2]^T$ can be treated as a state of cellular metabolic phenotype, and Equation 5.4 essentially represents a cell dynamics system. An attractor is a stable state of the dynamics system where the phase space trajectory of the system will converge, no matter what the initial conditions are. In fact, fluctuation inherently exists in the actual biological system, so the gene expression or other behaviors of a biological entity is not purely deterministic. Even though the state of the cellular dynamics system is perturbed by stochastic noise arising from the external environmental fluctuations, the system is able to gradually come to stability over time and finally stays at a new growth rate as well as at a well living state. The adaption of a cell to changes in its survival environment is analogous to the attractor selection of the dynamics system given in Equation 5.4, in which the system will select and switch to a new stable and suitable state when the environmental conditions have been changed or perturbed to be unsuitable for previous state.

In addition, the cellular activity A is an important parameter that lumps the fitness of the environmental conditions for the cells survival to a single real value, which ranges from 0 to 1. This parameter is used to comprehensively reflect the information about the cells external environment, control the influence of the noise on the behavior of the dynamics system, and capture the phenotypic consequence that enables cellular adaptation.

5.4.1.3 Extended attractor selection model

Following the basic attractor selection model, we would like to introduce the appealing bio-inspired attractor selection mechanism to decision making under varying conditions in terms of improving the robustness and adaptability of the decision solution. To select the most appropriate wireless access network $net_j \in NetSet$ for any one application $s_j \in S_{uk}$ that is running in the terminal device of a user $u_k \in User$, we first define a decision vector as $X_{S_i}(t) = [x^{S_i}_{net_j}(t)]_{net_j \in NetSet} = [x^{S_i}_{net_1}(t), x^{S_i}_{net_2}(t), \ldots, x^{S_i}_{net_M}(t)]$ for each application $s_j \in S_{uk}$. Each state value $x^{S_i}_{net_j}(t)$ in this decision vector refers to the score or the goodness of the network net_j relevant to the application s_i at time t. Therefore, any $u_k \in User$ should maintain a set of $|\ S_{uk}\ |$ decision vectors, since it has $|\ S_{uk}\ |$ applications (it should be noted that the notation $|\ S_{uk}\ |$ represents the number of elements in the set S_{uk}). We then use these decision vectors to construct a matrix as follows:

$$[X^{s_1}(t), X^{s_2}(t), \ldots, X^{s_N}(t)] = \begin{bmatrix} x^{S1}_{net_1} & x^{S2}_{net_1} & \cdots & x^{SN}_{net_1} \\ x^{S1}_{net_2} & x^{S2}_{net_2} & \cdots & x^{SN}_{net_2} \\ \vdots & \vdots & \ddots & \vdots \\ x^{S1}_{net_M} & x^{S2}_{net_M} & \cdots & x^{SN}_{net_M} \end{bmatrix}. \tag{5.5}$$

The matrix shown in Equation 5.5 is called a "decision matrix" or a "score matrix," whose rows index different wireless networks and whose columns index different applications. Each component value $x_{net_j}^{s_i}$ in this decision matrix indicates the proportion of selecting the wireless network net_j as the target network for supporting the application s_i at time t. Furthermore, once the decision matrix can be obtained, we can use the following Equation 5.6 to determine the target network for the application s_i:

$$net^*(s_i) = \underset{net_j \in NetSet}{argmax} \{x_{net_j}^{s_i}(t)\}. \tag{5.6}$$

Aiming to update the state of the decision matrix, we here extend the basic attractor selection model to the new form with a higher dimension. We propose the novel extended model as follows:

$$\frac{\mathrm{d}x_{net_j}^{s_i}(t)}{\mathrm{d}t} = \frac{syn(\alpha)}{1+[x_{net^*(s_i)}^{s_i}(t) - x_{net_j}^{s_i}(t)]^2} - deg(\alpha) \times x_{net_j}^{s_i}(t) + \eta_{net_j}^{s_i}(\mu,\sigma), \tag{5.7}$$

where $x_i \in S_{uk}$, $net_j \in NetSet$, and $x_{net^*(s_i)}^{s_i}(t)$ is the maximum state value in the decision vector $X^{s_i}(t)$ corresponding to the application s_i. $\eta_{net_j}^{s_i}(\mu,\sigma)$ denotes the white Gaussian noise whose mean value is μ and whose standard deviation is σ. According to the basic attractor selection model, $syn(\alpha)$ and $deg(\alpha)$ should be designed as the monotonously increasing functions of activity α. Similar to Ref. [129], we adopt the polynomial form to formulate $syn(\alpha)$ and directly set $deg(\alpha)$ identical to α as follows:

$$\begin{cases} syn(\alpha) = \alpha \times (\beta \times \alpha^n + m) \\ deg(\alpha) = \alpha \end{cases}, \tag{5.8}$$

where β and m are both the positive real values, and n is a positive integer. On the basis of Equation 5.7, we can use the model defined in Equation 5.8 to dynamically and self-adaptively update the decision matrix defined in Equation 5.5. Thus, based on the decision matrix, the handover decision can be made in terms of guaranteeing the terminal QoS satisfaction and self-adaptability.

5.4.1.4 Model validation and discussion

To analyze the proposed model given by Equation 5.7, we split the stochastic nonlinear differential equation into two parts, the deterministic term and the stochastic term. We define the notation $\phi_{net_j}^{s_i}(\alpha,t)$ as representing the deterministic term

$$\phi_{net_j}^{s_i}(\alpha,t) = \frac{syn(\alpha)}{1+[x_{net^*(s_i)}^{s_i}(t) - x_{net_j}^{s_i}(t)]^2} - deg(\alpha) \times x_{net_j}^{s_i}(t). \tag{5.9}$$

$\phi_{net_j}^{s_i}(\alpha,t)$ is a complex multivariate function of the deterministic parameters α and t. Following the notation above, we then reshape the extended attractor selection model defined in Equation 5.4 into a simpler formulation that is composed of the deterministic term $\phi_{net_j}^{s_i}(\alpha,t)$ and the stochastic term $\eta_{net_j}^{s_i}(\mu,\sigma)$:

$$\frac{\mathrm{d}x_{net_j}^{s_i}(t)}{\mathrm{d}t} = \phi_{net_j}^{s_i}(\alpha,t) + \eta_{net_j}^{s_i}(\mu,\sigma). \tag{5.10}$$

From Equation 5.10, it can be obviously observed that if the value of the activity α decreases due to some changes in the external conditions of this dynamics system, the magnitude of the deterministic term $\phi^{s_i}_{net_j}(\alpha,t)$ will become smaller and it may decreasingly approach the magnitude of the stochastic term $\eta^{s_i}_{net_j}(\mu,\sigma)$, which means that $|\phi^{s_i}_{net_j}(\alpha,t)| \approx |\eta^{s_i}_{net_j}(\mu,\sigma)|$. In this situation, Equation 5.9 tells us that the influence of the random noise becomes relatively enhanced and the behavior of the dynamics system is expected to be mostly dominated by the randomness of $\eta^{s_i}_{net_j}(\mu,\sigma)$. On the other hand, when activity α increases to make the magnitude of the deterministic term $\phi^{s_i}_{net_j}(\alpha,t)$ much larger than that of the stochastic term $\eta^{s_i}_{net_j}(\mu,\sigma)$, the deterministic term will govern this system so that its phase trajectory can asymptotically approach to a more stable state against the fluctuation resulting from stochastic noise. At this point, the process of tending to a stable state can be analogous to adaptive attractor selection in the gene expression of cells, since the dynamics of gene expression switches between different patterns, which is as well influenced by a combination of the deterministic and the stochastic behavior of cellular system.

Additionally, when the stochastic term $\eta^{s_i}_{net_j}(\mu,\sigma)$ is assumed to be zero, we can easily obtain the deterministic maximum value of the system state variable by setting $\phi^{s_i}_{net_j}(\alpha,t) = 0$ and $net_j = net^*(s_i)$ as follows:

$$\begin{aligned}\phi^{s_i}_{net_j}(\alpha,t) &= \frac{syn(\alpha)}{1+[x^{s_i}_{net^*(s_i)}(t) - x^{s_i}_{net_j}(t)]^2} - deg(\alpha)\times x^{s_i}_{net_j} \\ &= syn(\alpha) - deg(\alpha)x^{s_i}_{net^*_j}(t) = 0. \qquad (5.11)\end{aligned}$$

Therefore, the deterministic state of maximum value at time t derived from Equation 5.11 is

$$x^{s_i}_{net^*_j}(t) = \frac{syn(\alpha)}{deg(\alpha)}. \qquad (5.12)$$

In the simple case of considering the specific formulations of $syn(\alpha)$ and $deg(\alpha)$ given by Equation 5.12, we can get the closed-form expression for $x^{s_i}_{net^*_j}(t)$ as

$$x^{s_i}_{\max} = \frac{\alpha(\beta\times\alpha^n + m)}{\alpha} = \beta\times\alpha^n + m, \qquad (5.13)$$

where we use the notation $x^{s_i}_{\max}$ to represent $x^{s_i}_{net^*_j}(t)$ for simplicity.

As mentioned before, we consider that these parameters β, n, and m are positive real values. Thus, based on Equation 5.13, $x^{s_i}_{\max}$ is an increasing function of the activity α. This means that when α is increased, the maximum system state is also increased along with this increasing α. From the biological perspective, if cells successfully express their genes that are able to make them well survive and optimally grow in an uncertain and highly dynamic environment, their activity is consequently expected to approach the best level. It can be said that those cells select the adaptive attractor that corresponds to their suitable gene expression pattern. From this point, the level of the maximum system state given by Equation 5.13 can capture the fitness degree of the dynamics system in the dynamic environment, once activity α is correlated with the varying environmental conditions. This also explains why it is appropriate to adopt the form of increasing function to represent $syn(\alpha)$ and $deg(\alpha)$.

Next, we come to consider the influence of random noise on the dynamics system. In fact, it is impossible to achieve the aforementioned selection of different attractors when only considering the deterministic term $\phi^{s_i}_{net_j}(\alpha,t)$ in the dynamics system defined in Equation 5.10. The biological system is always evolving along with inherent randomness. As is discussed in Refs. [58,85], the stochastic fluctuation in cell systems is the one of the most significant factors that drives the process of gene expression switching between attractor states. Some numerical simulations are done to validate how the deterministic and the stochastic terms affect the behavior of the dynamics system represented by the extended attractor selection model.

Following the basic attractor selection model defined in Equation 5.4, we adopt the assumption of white Gaussian noise for the stochastic term $\eta^{S_i}_{net_j}(\mu,\sigma)$, and then without loss of generality we set $\mu = 0$ in the following experiments. In the simulation, the standard deviation σ is respectively set to be 0.5, 1, and 2.5, and the parameters β, n, and m are fixed at 5, 3, and 5, respectively. These simulations allow us to perform comparative analyses and investigate the properties of the bio-inspired model in the dynamic environment of different stochastic perturbation magnitudes.

Furthermore, because our goal here is to investigate the properties of the proposed model in terms of different random noises, without loss of generality we consider the specific form of the model with only three state variables, respectively denoted by x_1, x_2, and x_3:

$$\frac{\mathrm{d}x_i}{\mathrm{d}t} = \frac{syn(\alpha)}{1+(x_{\max}-x_i)^2} - deg(\alpha)\times x_i + \eta_i(0,\sigma), \tag{5.14}$$

where i = 1,2,3 and $x_{\max} = \max_{i=1,2,3}\{x_i\}$. Also, it is worth pointing out that the state variables are all restricted in the positive real number domain, i.e., $x_i \geq 0$ for all $i = 1,2,3$, since these state variables have actual meanings when they are related with the level of cells' gene expression, we will reset it to 0 when any one state variable changes to be lower than zero.

Now, we set the simulation time from 0 to 3600 s. And then we employ a changing environment in the simulations where a square wave with period 900 seconds is adopted to simulate the time series for the varying activity α. The maximum and the minimum peaks of the square wave are set at 1 and 0, respectively, and the wave's duty cycle (i.e., the percent of the period in which the signal is larger than 0) is set at 60%. The simulated varying activity α is given in Figure 5.5.

In Figure 5.6, the subgraphs (a) through (c), show the variation of the dynamics system. In the first case, with the standard deviation of the random noise $\sigma = 0.5$, the dynamics system composed of x_i ($i = 1,2,3$) stably stays at the attractor whose maximum state is x_2 and whose lower states are x_1 and x_3. This is because the magnitude of the stochastic term in the model is very small, while the deterministic term always dominates the behavior of the dynamics system, even when the activity α decreases to 0 (see Figure 5.5). This means that the system is trapped into one attractor and will not switch to another. When the magnitude of the stochastic term

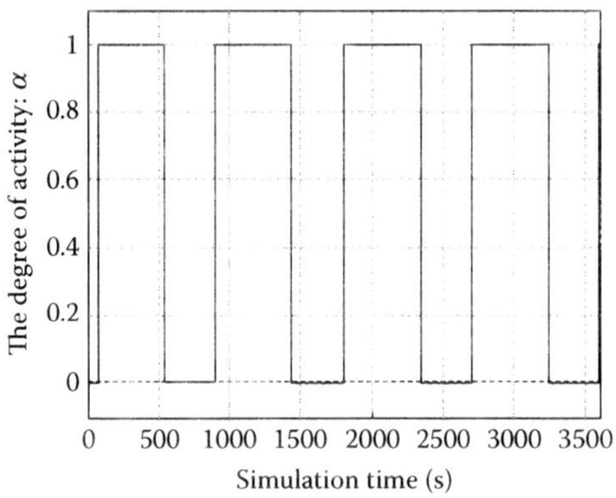

Figure 5.5 Varying activity α in simulations.

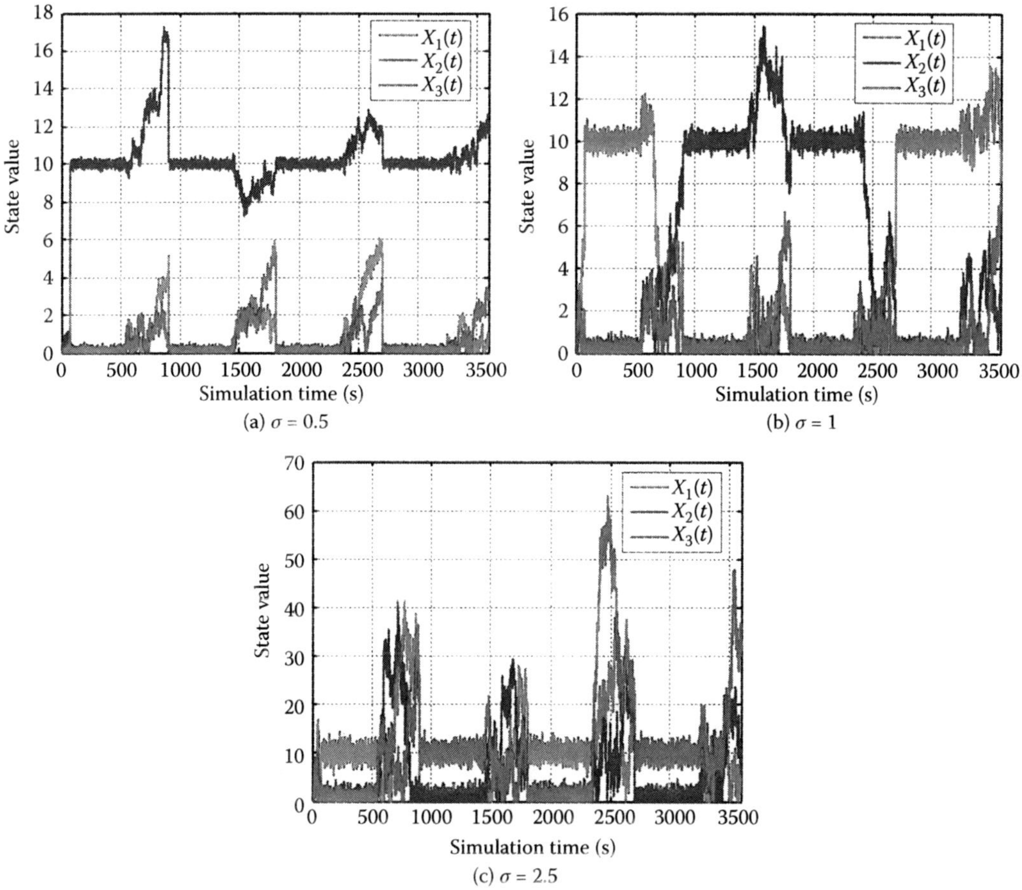

Figure 5.6 Evaluation of the impact of random noise variability on the dynamics system.

increases (namely, when the standard deviation of the random noise is set to be larger as shown in the cases of (b) and (c)), the state of the dynamics system fluctuates fiercer with the low value of activity α. However, when activity α jumps from the minimum to the maximum peak, the dynamics system always evolves to a stable state, i.e., one attractor. For instance, in the subgraph (c), the system stays at the attractor whose maximum state value is x_1 and whose lower states are x_2 and x_3 during the simulation time interval, which is from 70.5 to 535 s (see Figure 5.5), whereas this system switches to another attractor whose maximum state value changes to be x_3 after a relatively shorter time interval [535, 908.3 s]. From the experimental results, it is confirmed that the proposed bio-inspired model well inherits the mechanism of attractor selection and is able to dynamically capture the variation of the environmental conditions.

5.4.2 QoS-oriented handover decision

This section gives the handover framework used in this work, as well as the formulation for evaluating the terminal QoS satisfaction. Following this, we also present the detailed QoS-based handover decision scheme based on the extended attractor selection model.

5.4.2.1 Handover decision framework based on the extended attractor selection model

On the basis of the extended attractor selection model, we propose a distributed handover decision-making scheme framework. This framework is outlined in Figure 5.7. At each time period $\triangle t$, each individual mobile terminal $u_k \in User$ can perform the handover decision process independently. Because the extended attractor selection model only needs the information on the QoS requirements of the applications that belong to an individual terminal and the information on the current wireless network conditions, multiple terminals do not need to exchange their decision information with each other. In our scheme, the handover decision is made at the terminal side instead of the network side. Therefore, it is not necessary to deploy a centralized control entity for managing the handover process.

Under the distributed handover decision framework shown in Figure 5.7, each application of a mobile terminal first provides its QoS requirements to its terminal during each time period. Meanwhile, the mobile terminal needs to sense the current network conditions by interacting with the networks some signaling messages through air interface. Then the utility of each application is evaluated by using the proposed utility function (the utility function is developed in the following subsection.), i.e., quantifying the degree at which the QoS requirements of each application are satisfied by its current wireless link. On the basis of the QoS satisfaction degree of each application, we evaluate the terminal QoS satisfaction so as to map the degree of the terminal's QoS satisfaction to activity α. Furthermore, activity α is inputted to the extended attractor selection model and is used to drive the model to update the handover decision matrix given in Equation 5.5, as well as to determine the target networks by using Equation 5.6. Finally, the handover can be done according to the determined target networks.

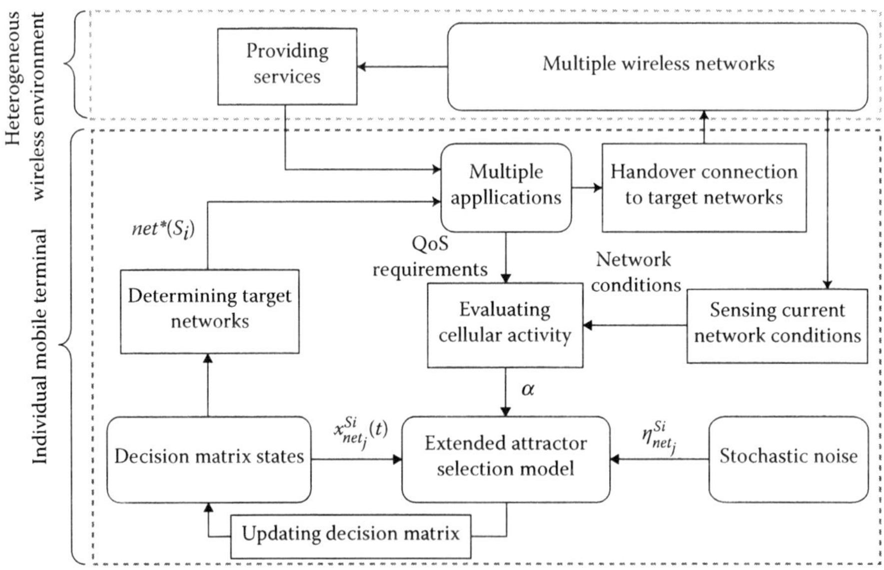

Figure 5.7 The handover decision framework based on the extended attractor selection model.

5.4.2.2 *Quantification of terminal's QoS satisfaction*

In this work, we take the upper and lower bounds of QoS requirements of each application into consideration. The QoS-related attributes considered include bandwidth, end-to-end delay, and packet loss ratio. The bandwidth, delay, and packet loss ratio are, respectively, indexed by the notations b, d, and p. On the basis of this, we denote the upper bound of bandwidth, delay, and packet loss ratio required by the application $s_i \in S_{uk}$ as $U(S_i,x)$ $(x = b,d,p)$, and the lower bound of those requirements as $L(S_i,x)$ $(x = b,d,p)$. We assume that the application s_i is currently connected to the network $net_j \in NetSet$. Additionally, the current conditions (including available bandwidth, transmission delay, and packet loss ratio) of the network net_j at time t are denoted as $C(net_j,x,t)$ $(x = b,d,p)$. Then we evaluate the utility of each attribute perceived by the application s_i with the linear normalization strategy as follows:

$$\begin{cases} R(s_i,x,t) = \frac{\min\{U(s_i,x),C(net_j,x,t)\}-L(s_i,x)}{U(s_i,x)-L(s_i,x)} & x = b \\ R(s_i,x,t) = \frac{U(s_i,x)-\max\{L(s_i,x),C(net_j,x,t)\}}{U(s_i,x)-L(s_i,x)} & x = d,p \end{cases} \quad (5.15)$$

In addition, we adopt the weighted sum method to lump different $R(s_i,x,t)$ $(x = b,d,p)$ to one variable as follows:

$$G(s_i,t) = \sum_{x=b,d,p} \omega(s_i,x) \times f(R(s_i,x,t)), \quad (5.16)$$

where $\omega(s_i,x)$ is the positive weight corresponding to the QoS attribute x $(x = b,d,p)$,and they must satisfy the constraint $\sum_{x=b,d,p} \omega(s_i,x) = 1$. $f(\bullet)$ is a monotonously increasing function that can map $R(s_i,x,t)$ to [0,1]. In this work, we formulate the function as the sigmoid form:

$$f(v) = \frac{1}{1+\exp(-c_1 \times v + c_2)}, \quad (5.17)$$

where c_1 and c_2 are both positive real parameters.

To investigate the influence of c_1 and c_2 on $f(v)$ given in Equation 5.17, we vary these parameters and plot the variation of the function with respect to different parameter settings. For simplicity but without loss of generality, we set $c_1 = 2c_2$ and obtain the results shown in Figure 5.8. The figure shows the slope of $f(v)$ with considering different values of c_1 and c_2. As we can see, the larger the parameters c_1 and c_2 are, the steeper the slope of $f(v)$ becomes. And the function value can stably stay at a relatively low (or high) level when the variable v approaches to 0 (or 1). Because $R(s_i,x,t)$ represents the degree of the wireless network satisfying the application s_i in terms of the requirements on bandwidth, transmission delay, and packet loss ratio, its value may change abruptly due to the dynamic nature of the wireless environment. On the basic of the results of Figure 5.8, we adopt $f(v)$ to map $R(s_i,x,t)$ to the interval [0,1] and fix the parameters c_1 and c_2 at 16 and 8, respectively, so that the weighted sum term $G(s_i,t)$ given in Equation 5.20 is limited in [0,1] and its sensitivity is inhibited when $R(s_i,x,t)$ becomes either too small or large.

On the other hand, the function value $G(s_i,t)$ comprehensively represents the degree of the QoS satisfaction of the application s_i. To evaluate the QoS satisfaction of the terminal u_k, we combine all of the $G(s_i,t)$ $(s_1 \in S_{uk})$ with the cumulative product strategy and yield

$$F(u_k,t) = \prod_{s_1 \in S_{uk}} G(s_i,t). \quad (5.18)$$

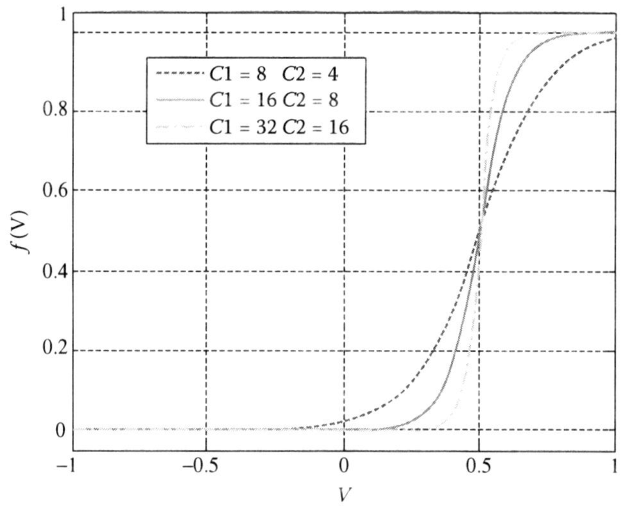

Figure 5.8 The variation of the adopted sigmoid function with different parameter settings.

Since the terminal QoS satisfaction $F(u_k,t)$ may change suddenly at the instant time t along with the varying wireless environment, it is not feasible to directly set the activity α equal to $F(u_k,t)$ as the input of the extended attractor selection model. We further adopt the weighted moving averaging method to map $F(u_k,t)$ to α as follows

$$\alpha = \frac{\int_{t-T}^{t} \lambda(\tau) \times F(u_k, \tau)\mathrm{d}\tau}{T}, \tag{5.19}$$

where T is the fixed time window $(0 < T < t)$. $\lambda(\tau)$ $(\tau \in [t-T,t])$ is the time-dependent positive weight that is used to reflect the significance of $F(u_k,\tau)$ and needs to satisfy the constraint of $\int_{t-T}^{t} \lambda(\tau)\mathrm{d}\tau/T = 1$. $\lambda(\tau)$ should be an increasing function of the time variable τ. That is, the closer τ is to the current time t, the larger $\lambda(\tau)$ is, implying that $F(u_k,\tau)$ at time τ reflects more significant information. Therefore, for simplicity, we design $\lambda(\tau)$ as follows

$$\lambda(\tau) = \frac{2\tau}{2t-T}. \tag{5.20}$$

It is easy to validate that Equation 5.20 satisfies the constraint $\int_{t-T}^{t} \lambda(\tau)d\tau/T = 1$ and $\lambda(\tau) \geq 0$.

On the basis of Equation 5.19, we can suppress the sensitivity of the activity α to the fierce changes in the dynamic wireless environment. Once α is obtained by Equation 5.19, we treat α as the external input of the extended attractor selection model. Consequently, the model is driven to update the decision matrix in Equation 5.5 so as to make the handover decision adaptively and automatically. Since each individual mobile terminal can independently process the extended attractor selection model and can perform the handover decision according to the aforementioned distributed framework, each individual terminal is essentially analogous to a cellular system.

5.5 Experiment and Analysis

5.5.1 Self-adaptive V2V communication system

To verify our proposed V2V system, the assessment of performances for the proposed frequency-adaptive V2V system with DSRC scheme when referring to both vehicles mobility, real road environment has been carried out through real experiments. We first set up the testing scenario in the Beihang University Campus as shown in Figure 5.9.

In the real testing scenario in Figure 5.9, two cars are deployed in a straight road and their relative distance is changing dynamically in terms of their changing velocities. Two OBU with our system is located onboard, which are also embedded with GPS system for recording the real vehicle velocity and its real longitude and latitude data. By using vehicle position data, the dynamic relative distance can be calculated, and both frequency-adaptive V2V system as described in previous session and nonfrequency-adaptive V2V communication system are considered for performance comparison when referring to two important metrics transmission packet loss rate and delay.

Figures 5.10 and 5.11 show the performances of the two metrics transmission delay and packet loss rate. Figure 5.10 describes the transmission delay of frequency-adaptive V2V system and nonfrequency-adaptive V2V system as the relative distance increases. The two V2V systems transmission delay curves exhibit a slightly increasing trend, mainly due to the growing wireless communication distance; however, as shown in Figure 5.11, our developed system ensures lower delay when compared to the nonfrequency-adaptive V2V system. On the other hand, as regards the packet loss rate in Figure 5.7, our V2V system guarantees a lower packet loss rate especially when the communication distance between the two OBUs is larger than 400 m; this, thanks to the adaptive transmission frequency configuration mechanism enforced by our developing V2V system. From the results of the performance comparison, we can observe that both V2V communication systems with DSRC are efficient in the transmission distance ranging from 300 to 400 m, and with the relative distance more than 400 m, the packet loss rate and the delay of both V2V schemes are increasing dramatically while our system obtains a better performance.

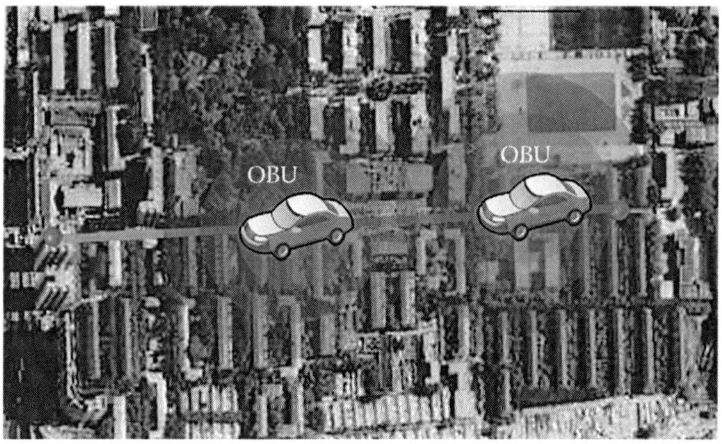

Figure 5.9 The real testing scenario.

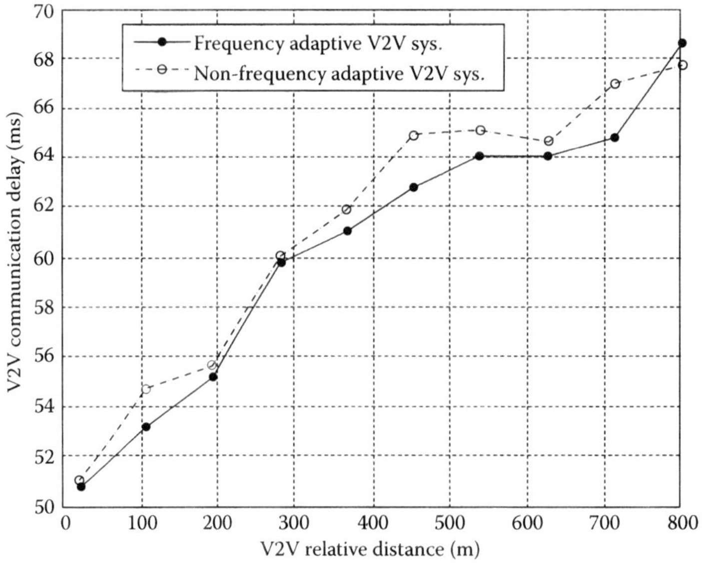

Figure 5.10 Performance comparison: Transmission delay versus relative distance.

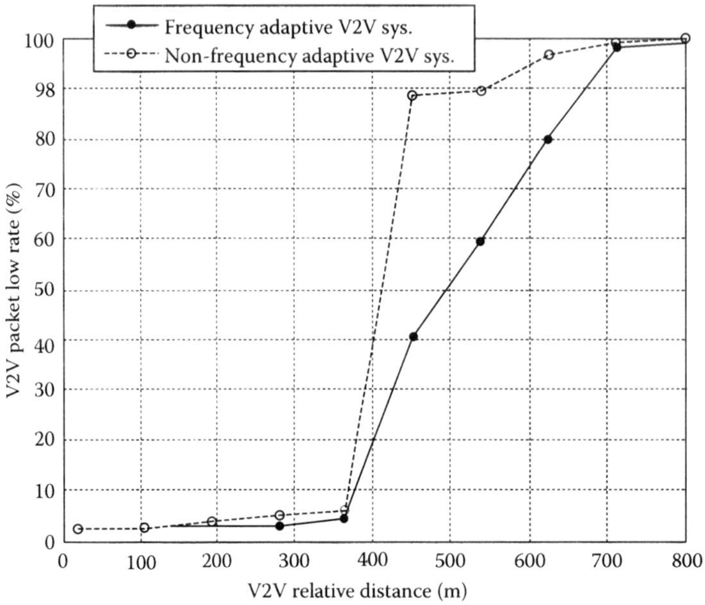

Figure 5.11 Performance comparison: Packet loss rate versus relative distance.

5.5.2 Bio-inspired QoS-oriented handover model

We perform some comparative simulations to evaluate the proposed bio-inspired handover decision scheme in this section through a discrete event simulator that we have developed in MATLAB® with the object-oriented programming. First, we present a typical heterogeneous wireless environment as the simulation scenario. And then we present the performance evaluation of our bio-inspired scheme and compare our scheme with the typical utility-function-based scheme with the simple additive weighting (SAW) [208].

5.5.2.1 Simulation scenario

We consider a vehicular heterogeneous wireless environment where there exist three types of wireless networks: 3G cellular network (wideband code division multiple access (WCDMA)), Wi-Fi (IEEE 802.11n), and DSRC (IEEE 802.11p). This simulation scenario is given in Figure 5.12. The cellular network and DSRC are both assumed to be able to cover the expressway denoted by the segment AE, whose length equals 1000 m. There are three Wi-Fi access points (APs) in total, deployed at the location of Points B, C, and D, which are along the expressway AE. The coverage of Wi-Fi APs is set to be 200 m as illustrated in Figure 5.5. In the simulation, multiple vehicular terminals are stochastically generated and uniformly distributed on the expressway at the beginning of the simulation. The direction of these vehicles is from A to E, and these vehicles are moving at a constant velocity during the simulation. In addition, each of these vehicular terminals runs three types of networking applications, including the voice application, the video application, and the data stream. In addition, we have set the time interval $\triangle t = 0.5$ s as the discrete period during which the decision matrix is updated at a time through processing the extended attractor selection model.

5.5.2.2 Simulation settings

In this simulation, we use a code division multiple access (CDMA)-based cellular network, and IEEE 802.11-based wireless local area network (WLAN) access networks including Wi-Fi and DSRC. The network conditions–related settings are referred to in the work [109] and are given in Table 5.1.

It is worth pointing out that the network conditions are varying during simulation. Since the number of the connections to a network changes all the time and has a significant influence on the network resource, we assume that the number of connections to one network is the main factor to change these networks' conditions over time. To simulate the dynamic nature of the heterogeneous wireless networks, we vary those network QoS attributes during simulation. We denote the number of applications that are currently connected to a network

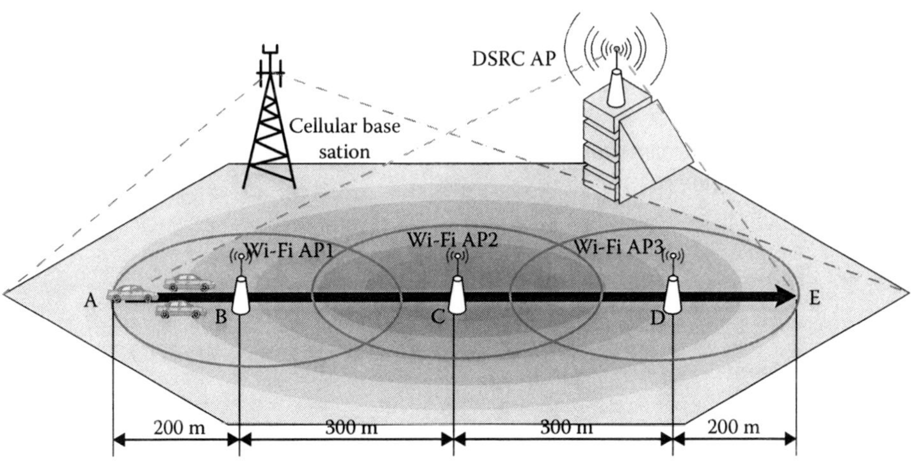

Figure 5.12 A simulation scenario.

Table 5.1 Network Conditions Settings

Network	*Capacity (Mbyte/s)*	*Delay (ms)*	*Packet Loss Ratio (%)*
Cellular	1.3	25	0.08
Wi-Fi(AP1)	25	8	0.04
Wi-Fi(AP2)	25	25	0.04
Wi-Fi(AP3)	25	45	0.04
DSRC	27	50	0.03

$net_j \in NetSet$ at time t as $num(net_j,t)$. For simplicity but without loss generality, we simulate the time-dependent QoS attributes by using the following formulations.

$$\begin{cases} C(net_j,b,t) = [\frac{Capacity(net_j)}{num(net_j,t)}] \\ C(net_j,d,t) = Delay(net_j) \times [1+0.5\times f(\frac{num(net_j,t)}{\sum_{net_j \in NetSet} num(net_j,t)})] \\ C(net_j,p,t) = Packet\_loss\_ratio(net_j) \in [f(\frac{num(net_j,t)}{\sum_{net_j \in NetSet} num(net_j,t)})] \end{cases}, \tag{5.21}$$

where the notations $Capacity(net_j)$, $Delay(net_j)$, and $Packet\_loss\_ratio(net_j)$, respectively, represent the capacity, delay, and packet loss ratio of the network net_j, whose values are given in Table 5.1. $f(\bullet)$ is also the sigmoid function as shown by Equation 5.17; $\lfloor \bullet \rfloor$ is the floor function. From Equation 5.21, it is obvious that the more the amount of the users connected to a network is, the worse the performance of this network will become. Thus we are allowed to simulate the dynamic nature of the wireless network in the experiments.

Furthermore, Table 5.2 gives three typical applications and the upper and lower bounds of their QoS requirements. Their parameters settings are referred to Ref. [110]. In our simulations, the total number of the applications run by each terminal is limited at three, and the applications of each terminal are generated from Table 5.2 at random. For example, a vehicular terminal may run all the three types of the applications including voice, video, and data stream, while one other terminal may run two voice applications and a video application.

Additionally, we adopt the model settings illustrated in Table 5.3 for our extended attractor selection model. Since different applications are sensitive to different QoS attributes, the weights of the QoS attributes required by different applications are also different from each other. For instance, the voice application may require lower end-to-end delay and the video application may need more bandwidth for transmission. We use the detailed settings on $\omega(s_i,x)$ $(x=b,d,p)$ given in Table 5.4 for our experiments.

Table 5.2 Applications Settings

QoS Attributes	*Requirements*	*Voice*	*Video*	*Data Stream*
Bandwidth (Kbyte/s)	$U(s_i,b)$	64	128	500
	$L(s_i,b)$	9	30	128
Delay (ms)	$U(s_i,d)$	150	150	120
	$L(s_i,d)$	0	0	0
Packet loss ratio (%)	$U(s_i,p)$	0.08	0.03	0.08
	$L(s_i,p)$	0	0	0

Table 5.3 Model Settings

Parameter	*Value*
μ	0
σ	1
β	5
m	5
n	3
c_1	16
c_2	18
T	25S

Table 5.4 Settings on the Weights

		x		
	$\omega(s_i, x)$	*b (bandwidth)*	*d (delay)*	*p (packet loss ratio)*
	Voice	0.3	0.5	0.2
s_i	Video	0.5	0.3	0.2
	Data stream	0.4	0.3	0.3

5.5.2.3 Numerical evaluation

To analyze the process of the handover decision driven by the attractor selection mechanism from the extended attractor selection model, we initially set the number of the total vehicular terminals equal to 90 and the initial velocity of those terminals is set to be 45 km/h. We randomly chose one of those terminals and illustrate its relevant simulation results in Figure 5.13. This vehicular terminal has two types of applications, one of which is the voice-related application and the other two are the video-related applications. The subgraphs (a through c) in Figure 5.13, respectively, show the variation of the decision vector state corresponding to each application against the simulation time. For example, in the subgraph (a), the dark dashed line plots the variation of $x_{\text{Celluar}}^{\text{voice}}(t)$ that represents the variation of the fitness of the cellular network for the voice application against the simulation time.

From these results, it is observed that the applications of this terminal are able to adapt its wireless access link according to the varying network conditions. According to the subgraph (a), the voice application stably accesses the cellular network during the whole simulation time, while the other two video applications switch their connections among different wireless networks in order to guarantee their QoS requirements. Even though these two applications are the same type, their decision behaviors differ. As shown in the subgraphs (b) and (c), the accessing link of the first video application switches to the cellular network from the first Wi-Fi network when this terminal moves out of coverage of the first access point (marked as Wi-Fi AP1 in Figure 5.12), while the second video application switches its connection to the second Wi-Fi network and maintains this wireless connection for a while, after which it also selects and stably connects to the cellular network. Since the heterogeneous network conditions are changing all the time, this terminal selects different appropriate networks for each of its applications according to the real-time network conditions so that it can ensure QoS satisfaction. The average activity of this terminal reaches 0.98962. According to the definition of the activity illustrated by Equation 5.19, the larger value of the activity indicates the higher degree of terminal QoS satisfaction. Thus, this result implies that this terminal achieves a good QoS satisfaction during the process of handover decision making.

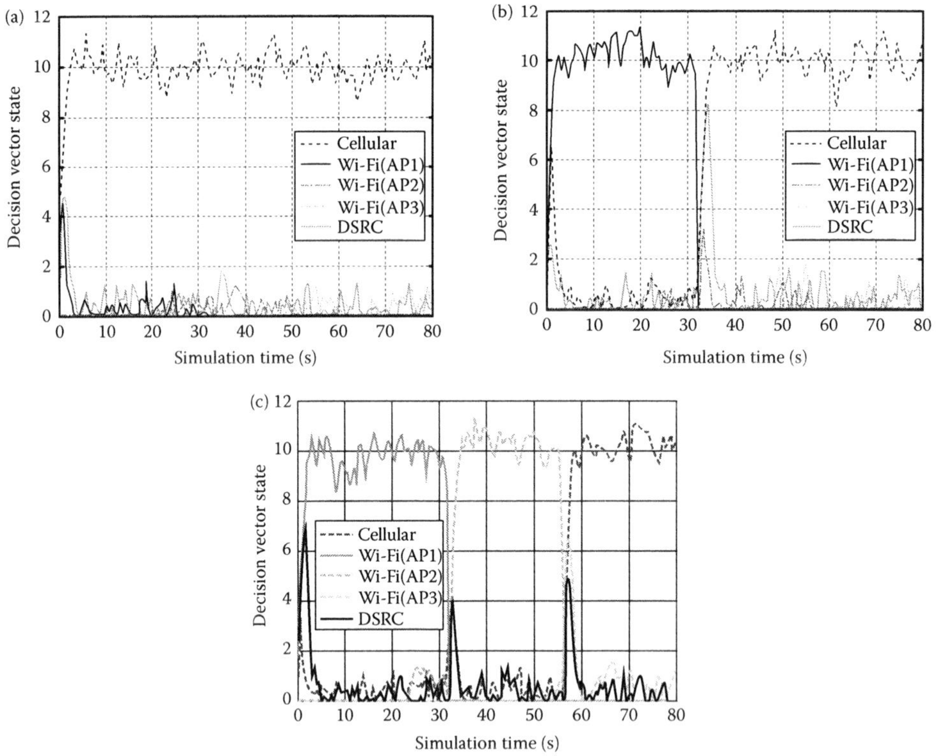

Figure 5.13 The variation of the decision vector states of the different applications: (a) the voice application, (b) one video application, and (c) another video application.

Next, we compare the simulation results obtained by our proposed bio-inspired handover decision scheme with those obtained by the utility-function-based scheme with SAW. Additionally, to perform the comparative evaluation in terms of the fairness of network resource allocation, we refer to the concept of the fairness index of resource allocation among multiple entities in [98], and then define the following equation for calculating the fairness metric:

$$\mathit{The\ Allocation\ of\ Fairness} = \frac{[\sum_{u_k \in User} \overline{\alpha}(u_k)]^2}{Q \times [\sum_{u_k \in User} (\overline{\alpha}(u_k))^2]}, \qquad (5.22)$$

where Q is the number of the total vehicular terminals and $\overline{\alpha}(u_k)$ is the average activity corresponding to u_k that can be calculated by averaging all the values of the activity obtained at every time period $\triangle t$.

First, we comparatively analyze the results of the two schemes under the specific simulation condition, where the initial velocity is set to be 45km/h and the number of total vehicular terminals ranges discretely from 10 to 160 so as to simulate the specific scenarios of different traffic densities. We calculate the average value and standard deviation of activity α per individual terminal against different total numbers of vehicular terminals and the fairness of overall network resource allocation according to Equation 5.22. These results are illustrated in Figure 5.14. Since the amount of total vehicular terminals increases and the overall network resources are limited, the competition among multiple terminals becomes much fiercer, so that the terminal QoS satisfaction and the fairness metric obtained by both schemes decreases along with

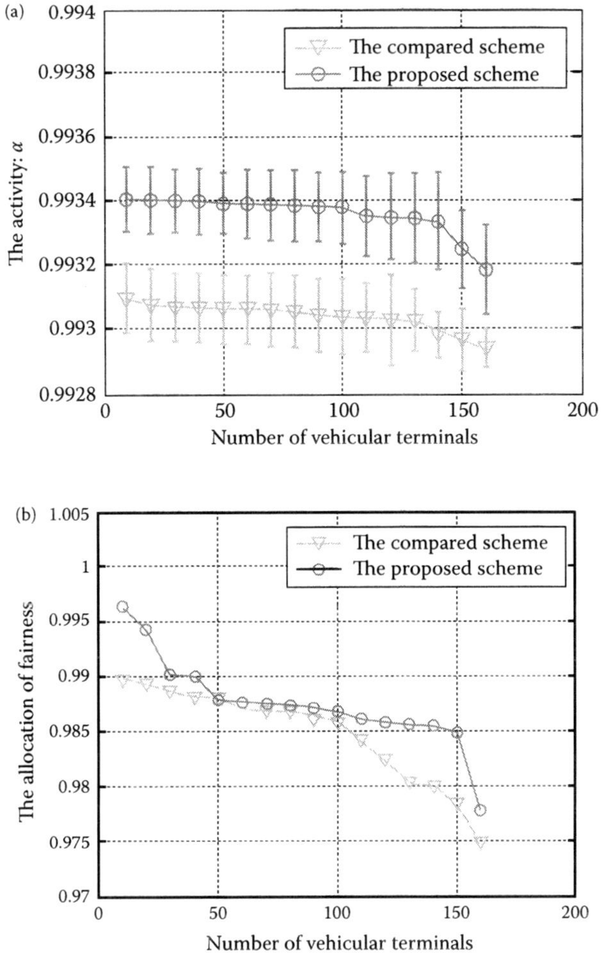

Figure 5.14 The simulation results under different vehicular terminals: (a) the results of the activity α and (b) the results of the fairness.

increasing the amount of terminals. However, from the subgraphs (a) and (b) in Figure 5.14, it can be found that the proposed scheme achieves a better activity and a better fairness of network resource allocation on average when compared with the results obtained by the compared scheme.

Furthermore, we perform the comparative simulations under different initial terminal velocities. We fix the number of vehicular terminals at 100, and vary the terminal velocities. The initial velocity is discretely set at 15, 45, 75, and 100 km/h so as to simulate different mobility scenarios. We also calculate the average value and standard deviation of the activity α per individual terminal against different initial terminal velocity as well as the average value of the fairness metric. Figure 5.15 demonstrates those numerical results. Because the faster the vehicular terminal moves, the shorter the duration when the applications of each terminal maintain their wireless links will last; a relatively high mobility may increase the times of switching wireless connection. This will reduce the efficiency of network resource allocation. Thus, the performance of both handover decision schemes degrades along with increasing the velocity.

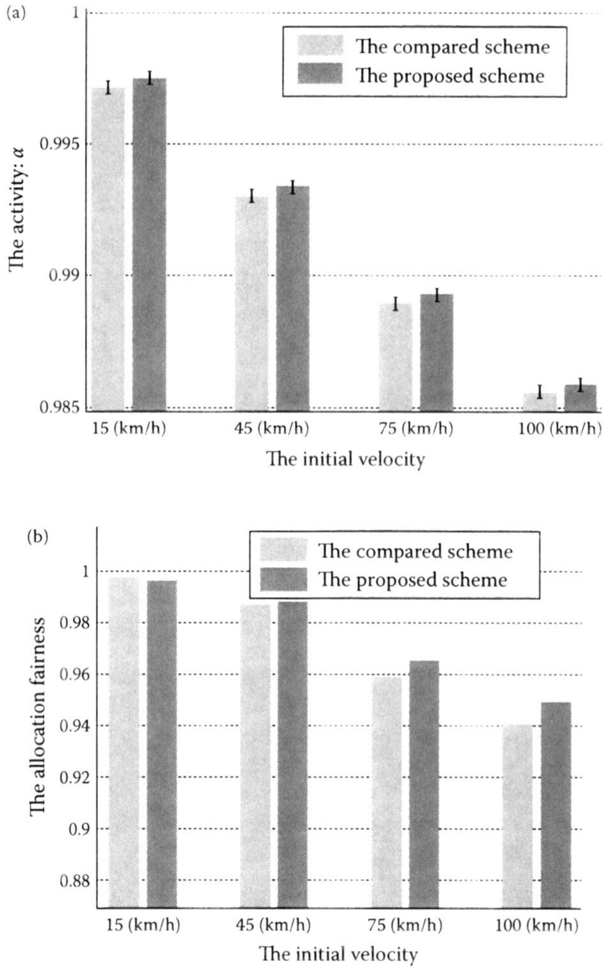

Figure 5.15 The simulation results under different initial velocities: (a) the results of the activity α and (b) the results of the fairness.

On the other hand, as shown in Figure 5.15, the activity on average obtained by our proposed scheme is greater than the compared scheme. Furthermore, the average degree of fairness of the network resource allocation obtained by our scheme is larger than that of the compared scheme.

In fact, the utility-function-based handover decision with SAW selects the access link for each application in a deterministic manner. It ranks the networks based on terminal utility and attempts to maximize every terminal's QoS satisfaction. Under the distributed framework, each application of these terminals tends to access the best network. Then, when a large number of applications access to the same best network at the same time, this network will become congested and its performance will deteriorate rapidly. Consequently, this network becomes nonoptimal and its users will again switch to another network simultaneously. Thus, multiple terminals may switch their connection between the optimal and the sub-optimal networks more frequently. Therefore, the compared scheme potentially increases the handover times and reduces the efficiency of network resource allocation as well as the overall level of multiple terminals' QoS satisfaction. Unlike the deterministic optimization that behaves in a greedy manner, the attractor selection mechanism searches the attractor with stochastic optimization, i.e., searching the optimal or sub-optimal state with certain noises. Although the attractor selection mechanism cannot guarantee that an application is connected to the best network all the time, it drives the multiple applications to select their appropriate access wireless links in a collaborative manner and makes these applications' connections adapt to the varying wireless network conditions, meanwhile meeting their QoS requirements to some extent. The mechanism drives multiple terminals to make handover decisions in a way that is similar to the coexistence and self-adaptability of multiple cells behaving in a dynamic environment. Therefore, the bio-inspired attractor selection model is able to achieve a better performance from a global perspective.

5.6 Conclusion

In this chapter, we put forward two kinds of short-range communication technologies in the Internet of Vehicles—"a self-adaptive V2V communication system with DSRC" and "a bio-inspired QoS-oriented handover model in heterogeneous wireless networks." On the one hand, we have investigated the feasibility of V2V communications and constraints on multichannel operations envisioned by IEEE 802.11p/WAVE and realistic settings for signal propagation and vehicle mobility in real environments. To increase the number of vehicles that are able to make the best of a short-lived connectivity to other vehicles and to reduce the probability of collision, we proposed a scheme of self-adaptive transmission frequency on figuration that exploits real wireless transmission information such as delay and package loss rate. On the other hand, we propose a bio-inspired model for making handover decisions in dynamic, heterogeneous, wireless environments. Our scheme provides a QoS-oriented handover solution for selecting an appropriate wireless network that can well satisfy the QoS requirements of each of the individual terminal applications in a dynamic context. For supporting adaptive and automatic decision making, we have introduced the attractor selection mechanism and proposed the distributed handover decision framework based on this bio-inspired model. We then modeled the activity parameter in terms of the individual terminal QoS satisfaction by a novel utility function, treat each individual terminal as a cellular system by analogy, and use the activity as the input of the extended attractor selection model to drive the process of updating the state value of the decision matrix as well as to make handover decision.

In the first experiment, the analysis results show that our designed system with DSRC and capability of self-adaptive frequency configuration achieves good wireless communication in vehicular environment in terms of dramatically low delay and low packet loss rate, and it out performs the traditional TCP/IP-type wireless communication Wi-Fi, 3G in terms of the adaptability of mobility and topology variation in vehicular network. In the second one, the experimental results prove that the QoS-oriented handover decision scheme induced by the bio-inspired attractor selection model achieves better adaptation to the varying heterogeneous wireless environment and has better performance in terms of guaranteeing better QoS satisfaction and ensuring better fairness of network resource allocation when compared with the traditional utility-function-based scheme. In future, we will extend our scheme by taking more decision factors into account such as the user profile–related and terminal-related decision factors, and based on these experiments, we will validate the proposed scheme under some more complex heterogeneous environments where more dynamic characteristics are considered.

Chapter 6

Cross-Layer Multihop Broadcasting

Jian Wang and Fangqi Liu

Jilin University

CONTENTS

6.1	Introduction	95
6.2	Related Work	96
	6.2.1 Analytic model for IEEE 802.11 broadcast service	96
	6.2.2 Multicast methods for VANET	97
6.3	System Model	97
	6.3.1 Analytic model	97
	6.3.2 Decision-making process	100
6.4	Numerical Results	103

6.1 Introduction

In recent years, with the rapid development of in-vehicle technology and the general improvement in people's living standards, vehicle penetration has increased greatly in urban areas. Vehicles have brought us convenience, but at the same time, they have caused many negative effects such as the frequent occurrence of road accidents, traffic congestion, parking difficulties, and more. In addition, as people spend more time in their cars, the need for more convenient and efficient services is urgent. Therefore, how to use high-tech means to prevent accidents and provide passengers with convenient network services has become the main focus of research. Vehicular ad hoc network (VANET) has emerged in this context.

VANET is a wireless mobile ad hoc network that includes intervehicle communications (IVC), vehicle-to-vehicle (V2V) communications, and vehicle-to-roadside infrastructure (V2R) communications. Through real-time information exchanges between vehicles and vehicle sensors or roadside systems, we can decrease traffic accidents and improve road safety. At the same time, VANET provides multimedia services and entertainment services to users.

The evolution of vehicle networks dates back to 1987, when the world's first car navigation system came into being in the Toyota Crown sedan, which was the symbol of the arrival of the vehicle network electronic age. Since the 1990s, vehicle network systems have been in the stage of upgrading, from simple global positioning system (GPS) navigation to the Intelligent Transport System (ITS). In recent years, vehicle networks have received broad attention worldwide. In 2003, VANET was proposed at the World Telecommunication Standardizing Assembly (WTSA) established by the International Telecommunication Union Telecommunication Standardization Sector (ITU-T) study group. In 2005, the Car2Car communication consortium was set up by the European Union. Furthermore, the concept of ITS was formally proposed by general motors (GM) in the United States in 2010.

As an integral part of the ITS, VANETs can be completely self-organized. Unlike ordinary networks, they are characterized by nodes with high-speed mobility, fast-changing network topologies, and the absence of roadside infrastructure and harsh communications environments; thus, traditional methods of data transmission cannot adapt to the needs of vehicle networks well. Therefore, designing communication protocols that suit VANET to solve the problems of network radio resource allocation has become a main focus of research.

In 2012, ISO/ETSI jointly raised the protocol for cooperative awareness messages (CAM), which is the heartbeat of communication protocols for vehicles in VANETs. In this protocol, vehicle stations periodically transmit information about themselves, such as location, speed, and network status, to their neighbor stations that are present within the reception range of their direct signal. This agreement enables drivers to learn the real-time information about surrounding road conditions and the driving direction of approaching vehicles, so as to help drivers predict risk and take timely measures.

In this chapter, we have designed a multicast CAM method for VANET. This measure modifies the traditional single-hop transmission measures into a method based on multihop multicast. In addition, by using the multicast decision function, vehicles can decide whether to re-multicast the CAM packet that they have received autonomously and dynamically. Based on the characteristics of the vehicle network and the CAM application, we use a comprehensive evaluation method to evaluate the cost of the next re-multicast. We have defined a series of evaluation indexes, such as the price of CAM packets, the benefit of the next multicast, the cost of a re-multicast, and so on. Eventually, based on a cost/benefit trade-off, vehicle nodes can make the optimal multicast decision. Compared to the traditional single-hop method, decision-making process allow vehicle nodes in the network to obtain information from later vehicles. In addition, because of the consideration of the cost of the multicast, we can control the packets transmission delay in an acceptable range. Furthermore, the fact that vehicles make multicast decisions autonomously in our method makes nodes better adapted to the mobility of the vehicle network and makes the data multicast method more flexible; as a result, all of these advantages let this method meet the requirements of an efficient communication protocol.

6.2 Related Work

6.2.1 Analytic model for IEEE 802.11 broadcast service

To evaluate the performance of IEEE 802.11 ad hoc broadcast networks, an analytic model has been proposed in Ref. [149]. The authors used a discrete time $M/G/1$ queue to model occasional occurrences of safety-related messages in each vehicle. By means of a probability generating function (PGF) and a recursive algorithm, they obtained the analytic model for the performance indices for IEEE 802.11 broadcast service. Based on the solutions from Ref. [149], we are building an analytic model for multicast CAM messages in VANET with an assumption of a saturation condition to acquire the important performances in data transmissions, such as packet delay, packet delivery ratio, and service time distribution for the price evaluation discussed later.

6.2.2 Multicast methods for VANET

Many broadcast protocols have been proposed to address the challenge of the frequent topology changes in mobile ad hoc networks (MANETs). Ni et al. [230] discussed many methods for broadcasting in MANETs. The first one is the probabilistic scheme, which is similar to flooding, except that nodes only rebroadcast with a predetermined probability. The second one is the counter-based scheme, which is an inverse relationship between the number of times a packet is received at a node and the probability of that node being able to reach additional area on a rebroadcast. The third one is the location-based scheme, which uses a more precise estimation of expected additional coverage area in the decision to rebroadcast. Wu and Dai [259] proposed a broadcast protocol in MANETs based on self-pruning, which is a neighbor knowledge method. Their approach is based on selecting a small subset of nodes to form a forward node set to carry out a broadcast process.

There has also been some recent works that consider broadcasting in delay tolerant networks (DTNs). Goundan et al. [74] discussed a mechanism for energy-efficient broadcasting. This is a k-neighbor broadcast scheme, where nodes do not broadcast all of the time but wait for an opportunity to reach multiple nodes with one transmission, thereby reducing the number of transmissions overall. Karlsson et al. [112] proposed a design for an open, receiver-driven broadcasting system that relies on delay-tolerant forwarding of data chunks through the mobility of wireless nodes. The system provides public broadcast channels, which can be openly used for both transmission and reception.

This chapter proposes a broadcast method for VANET, which uses a price evaluation method, comparing the cost and profit of media access control (MAC) message multicasting. Following the method in Ref. [230], we also use the method of estimation of the anticipated additional coverage area of the next-packets multicast. In addition, in our multicast method, we also include the estimation of the delay time of a re-multicast. In this chapter, to evaluate the price of a re-multicast, we abstract the coverage of message transmission as the price of the message's influencing, at the same time use the price of the reliability of the message to express the effect of the time duration during packet transmission. In addition, we also consider the delay produced by a re-multicast by defining the cost of the next multicast. By trading off the coverage and the delay of the CAM message multihop multicast, the nodes in VANET can make the best decisions as to whether to re-multicast messages or not.

6.3 System Model

6.3.1 Analytic model

The MAC protocol for vehicular ad hoc wireless networks is carrier sense multiple access with collision avoidance (CSMA/CA) with binary exponential backoff deployed by IEEE 802.11. The saturation performance of the distributed coordination function (DCF) mode of IEEE 802.11 MAC protocol is extensively studied [29,171]. As we know, the data transmission method in CAM application is multicast, with reference to the analytic model for analysis of broadcast service in Ref. [149], we establish an analytic model for the performance of VANET.

The DCF basic access method for multicast in a vehicle network is shortly summarized as follows. A node in a network with CAM packets to transmit monitors channel activities until an idle period equal to a DCF interframe space (DIFS) is detected. After this DIFS medium idle time, the node shall then generate a random initial backoff time for an additional deferral time before transmitting. The initial backoff time T_{ib} is generated using

$$T_{\text{ib}} = \text{Uniform}(0, W_0 - 1) \times \sigma, \tag{6.1}$$

where W_0 is the initial minimum backoff window size, and σ is the length of one backoff slot.The backoff time counter is decremented in terms of slot as long as the channel is sensed idle.The counter is stopped when a transmission is detected on the channel and reactivated when the channel is sensed idle again for more than a DIFS. Eventually, the node transmits its packet when the backoff time counter reaches zero. We use the discrete-time Markov chain model to simulate the process. We make some assumptions to give a simplified yet reasonable model. All the vehicles in the CAM application multicast CAM packets in a certain transmission range.

$$k = \pi r^2 \times q, \tag{6.2}$$

where q is the density of vehicles within the range r. k is the average number of neighbor vehicle nodes within the range r.

The one-dimensional random process $b(t)$ is a discrete-time Markov chain. We use ω to describe the state of each vehicle, and W stands for the backoff counter value and in value $(0, 1, \ldots, W_{0-1})$. According to Ref. [149], we can acquire b_0, which is the probability of the counter value being equal to zero when any transmission occurs in steady state; i.e., b_0 is the probability of packet transmission.

$$b_0 = \frac{2}{W_0 + 1}. \tag{6.3}$$

In the backoff process, when detecting an ongoing successful transmission, the backoff timer will be suspended and deferred a time period of T, which is expressed as:

$$T = \frac{L_{\text{H}}}{R_{\text{d}}} + \frac{E[L_{\text{P}}]}{R_{\text{d}}} + \text{DIFS} + \delta, \tag{6.4}$$

where the L_{H} stands for the length of the CAM packet header, and we assume that a packet holds size L_{P} with average packet length $E[L_{\text{P}}]$. Let R_{d} be the system transmission data rate, δ be the propagation delay, and DIFS be the time period for a DIFS.

We define p_{b} as the probability that the transmission channel for a vehicle node is busy. Knowing that the channel is busy if there is at least one vehicle transmitting, we can get:

$$p_{\text{b}} = 1 - (1 - p_0 \times b_0)^k, \tag{6.5}$$

where p_0 is the probability that there is a packet ready to transmit at the MAC layer in each vehicle node in current time.

Unlike the wired networks, VANET suffers from channel fading, which causes packet errors. Here, we suppose that when the signal-to-noise ratio (SNR) of the channel is S_{i}, the bit error rate of CAM packet transmission is p_{ber}. Assuming that the modulation mechanism is binary phase shift keying (BPSK), we can get the p_{ber} through:

$$p_{\text{ber}} = \frac{1}{2} erfc(\sqrt{S_i}). \tag{6.6}$$

So, we can acquire the packet error probability p_e as

$$p_e = 1 - (1 - p_{ber})^{E[L_P]+L_H}. \tag{6.7}$$

where L_H is the length of the CAM packet header, and $E[L_P]$ is the average packet length of CAM packets.

From the definitions mentioned here, we can obtain the packet delivery ratio PDR as:

$$PDR = (1 - p_e)(1 - p_0 \times b_0)^{k-1} = (1 - \frac{1}{2}erfc(\sqrt{S_i}))^{E[L_P]+L_H}(1 - p_0 \times b_0)^{k-1}. \tag{6.8}$$

We define σ as the length of a time slot. Here, we approach service time distribution through PGF. We understand that the backoff counter in each vehicle node will be decremented by a slot once an idle channel is sensed. For a node in multicast communication, the transition time (the amount of time slots) for a backoff counter to be decremented by one slot can be expressed as [149]:

$$H_d(z) = (1 - p_b) \times z + p_b \times z^{\lfloor \frac{T+\sigma}{\sigma} \rfloor}. \tag{6.9}$$

Here we can acquire the generalized state transition process. Considering that successful transmission and transmission with collision take the same amount of time in multicast, we can derive the PGF of these two transmission times (the number of time slots) as follows:

$$SC(z) = z^{\lfloor \frac{E[L_P]+L_H}{R_d \times \sigma} \rfloor}. \tag{6.10}$$

Denote q_i as the steady-state probability that the packet service time is $i\sigma$. Let $Q(z)$ be the PGF of q_i, so we have [149]:

$$Q(z) = \sum_i q_i z^i = \frac{SC(z)}{W_0} \sum_{i=0}^{W_0-1} H_d^i(z) = \frac{z^{\lfloor \frac{L_P+L_H}{R_d \times \sigma} \rfloor}}{W_0} \sum_{i=0}^{W_0-1} H_d^i(z). \tag{6.11}$$

Therefore, we can acquire the average service time T_{ave} by:

$$T_{ave} = \sigma \times \sum_i q_i(i\sigma) = \sigma \times \frac{dQ(z)}{dz}|_{z=1} = \sigma \times Q'(1). \tag{6.12}$$

To derive the average service time distribution, first of all, we define v_a as the average number of CAM packets arriving at each vehicle node per second (v_a can be known from nodes' dynamic sense). Thus, p_0 calculation depends on duration of service time (T_{ave}); here we define v_s as the service rate, so we get it through $v_s = 1/T_{ave}$.

Now we calculate the probability p_0. We apply the iterative algorithm to get p_0 as follows:

Step 1. Initialize $p_0 = 1$, which is the saturated condition.

Step 2. With p_0 calculate p_b using Equation 6.5.

Step 3. Calculate service time distribution through PGF, then acquire v_s.

Step 4. The current node dynamically gets v_a.

Step 5. If $v_a < v_s$, $p_0 = v_a/v_s$, otherwise $p_0 = 1$.

Step 6. If p_0 converges with the previous values, then stop the algorithm; otherwise, go to Step 2 with updated p_0.

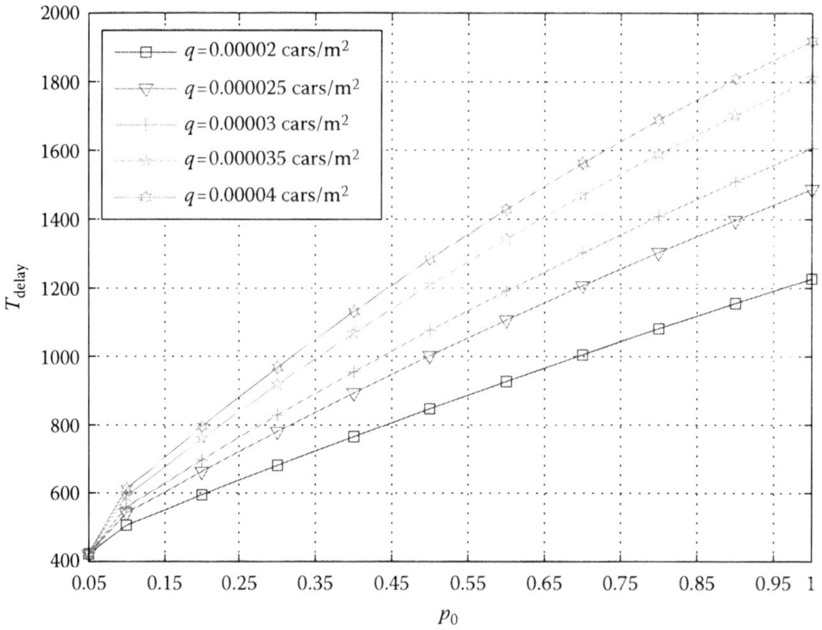

Figure 6.1 Relation between the probability of sending packet p_0 and packet transmission delay $T_{\mathbf{delay}}$.

Packet transmission delay is the average delay a packet experiences between the time at which the packet is generated and the time at which the packet is successfully received. It includes the medium service time (due to backoff and busy channels, interframe spaces, transmission delay, and propagation delay, etc.). In this chapter, we ignore the impact from queuing processes to the transmission delay. Using all of derivations mentioned earlier, we can obtain the average CAM packet transmission delay (T_{delay}) [149].

$$T_{\text{delay}} = T_{\text{ave}} + DIFS + \sigma + \delta. \tag{6.13}$$

Here, we discuss the impact generated by probability p_0 of sending a packet to the the average CAM packet transmission delay T_{delay} under different vehicle densities in Figure 6.1.

6.3.2 Decision-making process

As we have modified the traditional single-hop multicast to a multihop multicast in order to avoid a duplicate multicast, in our multicast method, when the vehicle node receives a CAM packet, the node should determine whether the CAM packet has already been received. The method of determining the duplicate is the same as that in a flooding broadcast. The node maintains a linked list that records the already multicasted information; in addition, each piece of recorded information has its corresponding expiration time; the linked list updates the recording information according to their expiration time.

After confirming that the CAM packets have not been multicasted, we begin to collect the information. First, we acquire the information about the initial vehicle node, where the CAM packets are generated originally. Define D_s as the distance between the initial vehicle node and the current node that receives the CAM packet, and T_S is the duration from the initial time of the CAM packet transmission to current time. According to the information mentioned here, we can get an important parameter D, which is the expected distance of the next multicast. D is calculated by

$$D = D_s + r, \tag{6.14}$$

where r is the expected range of CAM packet multicast in this time.

Also, we can define T_{all} as the expected total transmission delay of the current CAM packet. We can calculate T_{all} using

$$T_{all} = T_s + T_{delay}, \tag{6.15}$$

where T_{delay} is the expected transmission delay of the current CAM packet.

For the decision-making process, i.e., whether to multicast the packet received in the current node, the vehicle node should get the running speed of the initial vehicle (v). Furthermore, the current node should collect the real-time network environment information, such as the density of vehicles (q), the system transmission data rate (R_d), the average amount of CAM packets arriving at each vehicle node per second (v_a), the SNR of the channel (S_i), and so on, to provide necessary parameters for the calculation of the packet delivery ratio (PDR) and T_{delay} using the analytic model for performances evaluation mentioned earlier.

The multicast method designed here is mainly for the CAM application in VANET. We are evaluating for the profit of the currently received CAM packet multicast using a comprehensive evaluation method. Based on the purposes of CAM application, here we define the value of the next multicast that contains two aspects of evaluation criteria—the influence and the accuracy of the CAM packet. Either of these evaluation criteria occupies a part weight in the profit evaluation.

The influence of the current CAM packet that decides whether or not to multicast is the correlation degree between the next hop vehicle node and the originating vehicle. The size of the correlation degree is decided by D, which is the expected distance of next multicast. We assume that the range of packet multicast is r; thus, we can acquire D by $D = n \times r$, where n is the time the CAM packet has been multicasted then plus one.

According to the actual-1pc situation, the purpose of CAM packets is to announce the source vehicle node's information; thus we can learn that the closer the intermediate nodes away to the initial node, the larger the influence of the packet.

In the model, nonlinear function is used to change the evaluation criteria D to the evaluation index F_{value} that falls into the range of $[0, 1]$ through

$$F_{value} = e^{-\alpha \times D^2} = e^{-\alpha \times (n \times r)^2} \quad \alpha > 0, \tag{6.16}$$

where the value of parameter α should be taken according to the demand of practical applications. For instance, if the application requires long-range traffic information, the value taken by α will be relatively small. From Equation 6.16, we can derive that when $D = 1/\sqrt{\alpha}$, $F_{value} = 1/e \approx 0.3679$.

In Figure 6.2, we assume that $\alpha = 5 \times 10^{-7}$. We change the value of the multicast range r to discuss the impact of the multicast hops n on the influence of the packet in different multicast coverage ranges.

Figure 6.2 shows that the bigger the multicast range r, as the amount of packets being multicasted increase, the quicker the rate of descent of the F_{value}'s value. In addition, when the value of n is constant, the wider the multicast range, the smaller the value of F_{value}.

The data in CAM packets is the information related to the initial vehicle node of the current CAM packet, such as location information and, traveling state. After the CAM packet is sent, there will be a time interval T_{all} before the packet is received by the destination node. During the

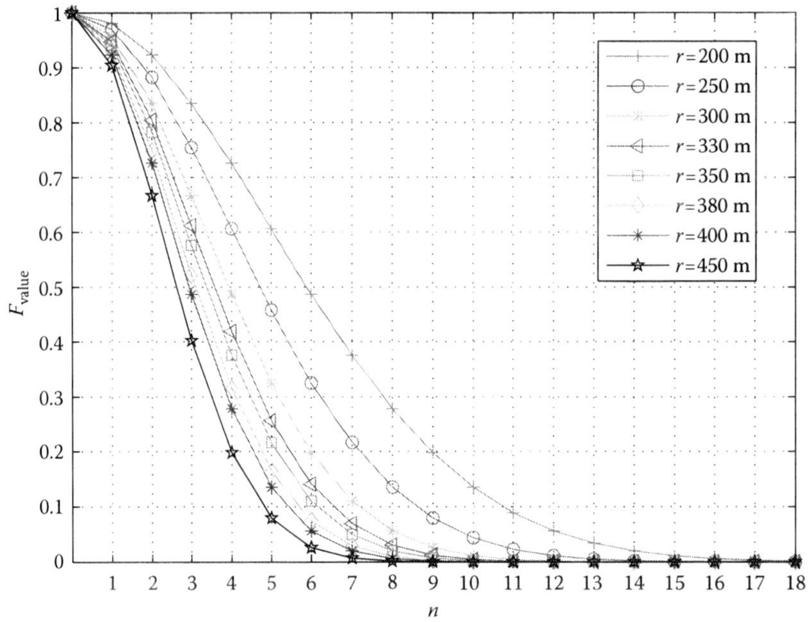

Figure 6.2 Relation between F_{value} and n.

interval, the initial vehicle will generate a certain state deviation; thus the information coming from the data received by destination node will have some errors when compared to the current actual state of the initial vehicle node. In this chapter, we use the accuracy as the evaluation criterion to evaluate the impact from state deviation to CAM packets. In the proposed method, we mainly use the position deviation that is generated by the initial node during T_{all}—E to evaluate the accuracy of the CAM packet. Here we also need the running speed of the initial node v, acquired during the preparing stage. We calculate the value of E by:

$$E = v \times T_{all}. \tag{6.17}$$

To evaluate the accuracy of the CAM packet, we change the evaluation criteria E to the evaluation index R_{value}, which is within the range of $[0,1]$ through

$$R_{value} = e^{-\beta \times (T_{all} \times v)^2} \quad \beta > 0, \tag{6.18}$$

where the value of parameter β should be taken according to the demand in the precision of the CAM packet in the current network environment. For instance, the smaller tolerance of the position deviation or the higher demand of the precision of the packets in the current application, the value of β will be relatively small. According to the definition of R_{value}, we can learn that when $T_{all} \times v = 1/\sqrt{\beta}$, $R_{value} = 1/e \approx 0.3679$.

Using the two evaluation indexes for the value of CAM packets F_{value} and R_{value}, we can obtain V of the CAM packet eventually by:

$$V = F_{value} \times w + R_{value} \times (1 - w), \tag{6.19}$$

where w is the weight occupied by F_{value} in the CAM packet's value evaluation, and the value of w is taken according to the demand of actual application.

On the basis of the practical demands of the network, we can state that there are two evaluation indexes to evaluate the profit of a CAM packet multicast: the one evaluation index is the value of the packet expected to be multicasted in current time, and the other evaluation index is the packet delivery ratio *PDR* acquired by Equation 6.8. The profit of the next multicast—*S* is co-determined by these two evaluation indexes as follows:

$$S = V \times u + PDR \times (1-u) \quad 0 < u < 1, \tag{6.20}$$

where u is the weight occupied by V in the profit of the next multicast evaluation, and the value of u is taken according to the demands of the actual application.

From Equation 6.20, we can know that the higher the value of the packet (V) or the *PDR*, the profit generated by the next multicast (S) is larger.

When a CAM packet is being multicasted, because the wireless channel is being shared, all vehicle nodes within the multicast range will wait for a deferral time to avoid transmission collision, which will have a negative impact on the performance of the network and will be taken as the cost of the next multicast. In this chapter, we use the expected average CAM packet transmission delay T_{delay}, which has to be calculated using the analytic model for performance evaluation mentioned earlier as the evaluation criterion, to evaluate the profit of the next multicast.

The profit of next multicast and the expected average CAM packet transmission delay T_{delay} have different units of measurement; therefore, we should change the evaluation criteria T_{delay} to the evaluation index C, which is within the range of $[0,1]$ through

$$C = 1 - e^{(-\varepsilon \times T_{\text{delay}})} \quad \varepsilon > 0, \tag{6.21}$$

where parameter ε should be set according to the requirement of the CAM application. When the tolerance of the transmission delay is relatively little, the value of ε can be taken larger, otherwise the ε's value should be taken small.

To make the decision whether to multicast the current CAM packet, we define the decision function:

$$E_{\text{decide}} = \frac{S}{C} \tag{6.22}$$

The physical significance of decision value E_{decide} is the ratio of the profit in comparison to the cost of next the multicast. When the value of E_{decide} is equal to 1, it means that there is no benefit to the next multicast. When the E_{decide} is greater than 1, it means that the next multicast will produce some profit. When a multicast decision is made, we compare the decision value E_{decide} with a threshold value θ to finally decide whether to multicast the current CAM packet. If the E_{decide} is greater than θ, the node will re-multicast the current CAM packet; however, the node will drop the multicast and discard the CAM packet.

The process of the multicast is shown in Figure 6.3.

6.4 Numerical Results

We use a simulation tool—MATLAB®—to simulate the network environment and to analyze the impact from certain network parameters on the result of the decision-making process. Each vehicle node in the network is equipped with the wireless ad hoc network capability with communication parameters shown in Table 6.1.

To discuss the effect of vehicle density on the decision-making process, in the process of simulation, we let the value of network environment parameters be constant as shown in

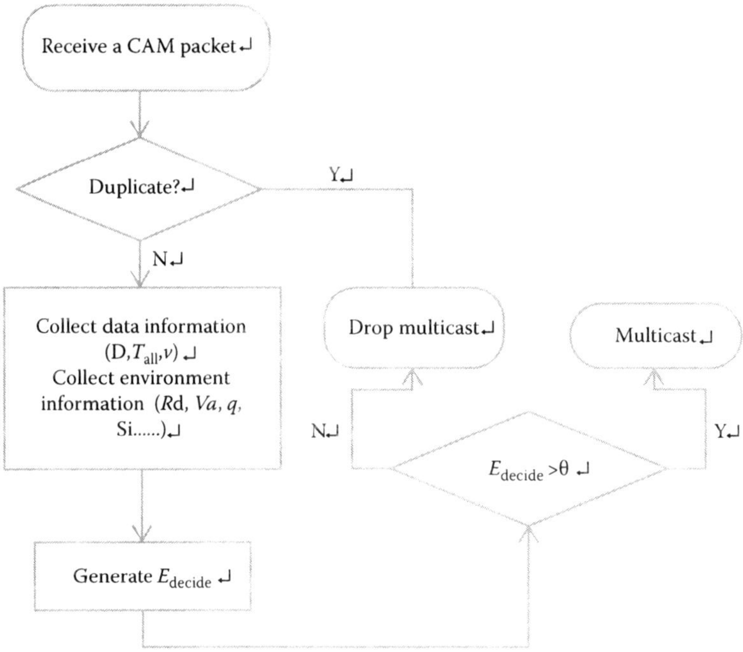

Figure 6.3 Flow diagram of the decision-making process.

Table 6.1 Parameters

Parameters	*Meaning*	*Default Value*
DIFS	A DCF interframe space	58 μs
L_P	CAM packet length	96 bytes
L_H	The length of the CAM packet header	96 bits
R_d	System transmission data rate	6 Mbps
σ	The length of a time slot	13 μs
W_0	The initial minimum backoff window size	32
δ	Propagation delay	1 μs
p_{ber}	The bit error rate of CAM packets transmission	0.0001
p_0	The probability that there is a packet ready to transmit	0.8
r	The radius of packets multicast range	300 m
α	The parameter for influence evaluation index	5×10^{-7}
β	The parameter for accuracy evaluation index	5
ε	The parameter for cost evaluation index	0.0003
w	The weight of F_{value} in calculating V	0.6
u	The weight of V in calculating S	0.7
v	The vehicle speed in network	120 km/h
q	The vehicle density	0.00003 cars/m^2

Table 6.1, except the value of vehicle density q. In Figures 6.4 and 6.5, by changing the value of vehicle density q, we can research the influence of q on the decision-making process.

According to the given definitions, we can know that the rise of vehicle density leads to an increase in the number of vehicle nodes that compete for the resource of channel with the current node; as a result, the transmission delay will be longer. The increase of T_{delay} results

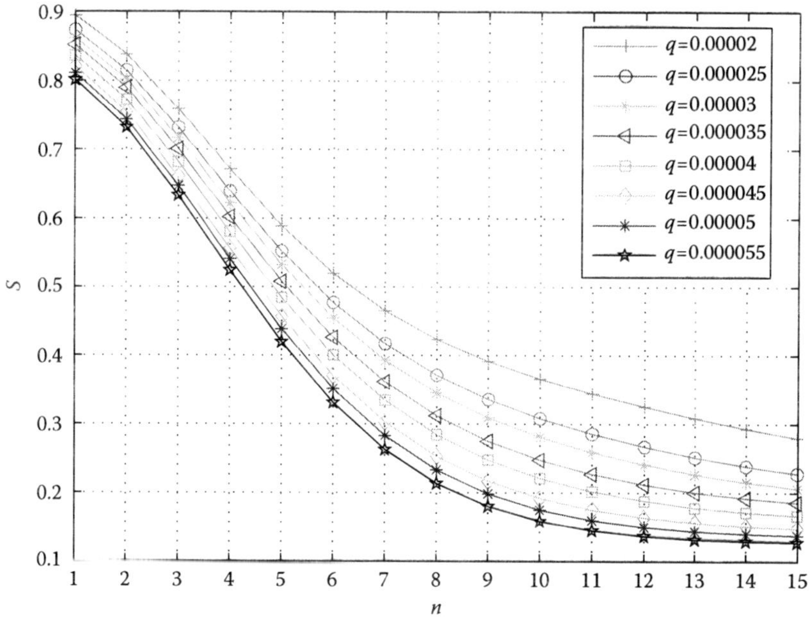

Figure 6.4 Relation between the number of multicast *n* and *S* under different vehicle densities.

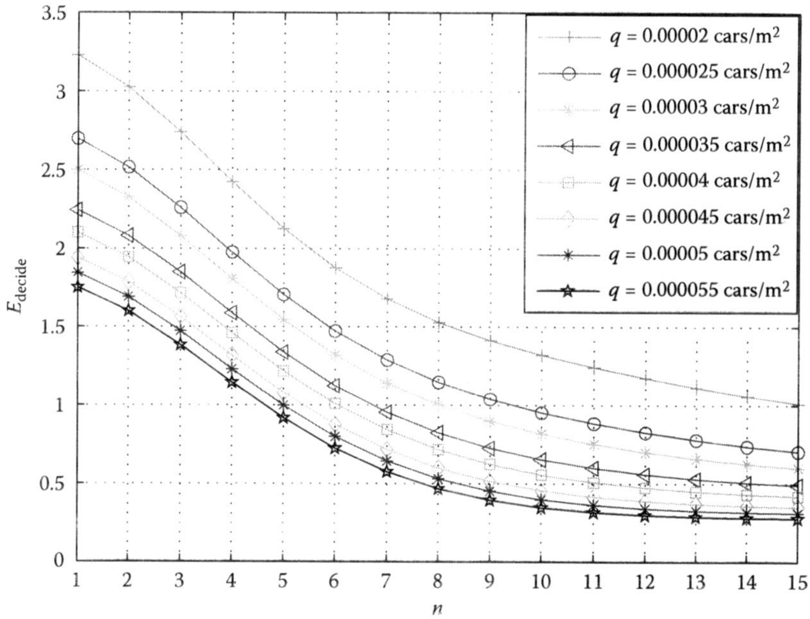

Figure 6.5 Relation between the number of multicast *n* and E_{decide} under different vehicle densities.

in the decrease of the value of the CAM packet and also makes the cost of a multicast packet greater. The decline of the profit of next multicast S and the rise of the cost C result in the multicast time being constant when the value of E_{decide} is small.

As shown in Figure 6.5, we can see that with the increase in the time of multicast, the E_{decide} declines gradually; when the n is constant, the vehicle density q is greater, and the value of E_{decide} is smaller. If we set the threshold value θ to 1.5 and the value of q between 0.00002 and 0.00045 cars/m^2, the total time of multicast is 8 times, 6 times, 5 times, ..., 2 times, respectively. From this, we can see that when the vehicle density q is smaller, the time taken for a packet to be multicasted is more, and the data transmission range will be larger. At the same time, if the value of θ decreases, the time packet being multicasted in in the decision-making process will be smaller.

To study the relation between the multicast range r and the result of the decision-making process, we focus on the parameter r to observe the result of multicast in different multicast ranges.

We can conclude from the definitions mentioned in this chapter that when the radius of packets multicast range rises, the number of nodes that conflict with the current node in the time of sending packet will increase, which will cause the cost of multicast larger. In addition, the increase in multicast range will result in the decline of the value of the CAM packet and the packet delivery ratio *PDR*. All of these lead to a result that when the time of multicast n is constant, the wider the multicast is, the smaller the value of E_{decide} is.

From Figures 6.6 and 6.7 we can see that, if we set the threshold value $\theta = 1.5$, when the radius of packets multicast range is within 250–550 m, the time of decision-making process is 13 times, 11 times, 8 times, 5 times, ... 4 times. Thus it can be seen that the wider the multicast range is the shorter the time of multicast is.

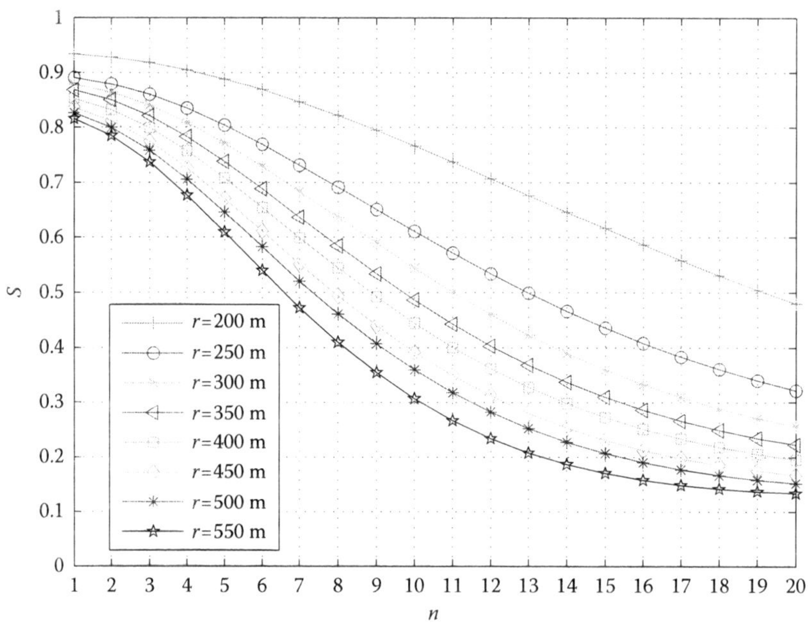

Figure 6.6 Relation between the number of multicast n and S under different multicast ranges r.

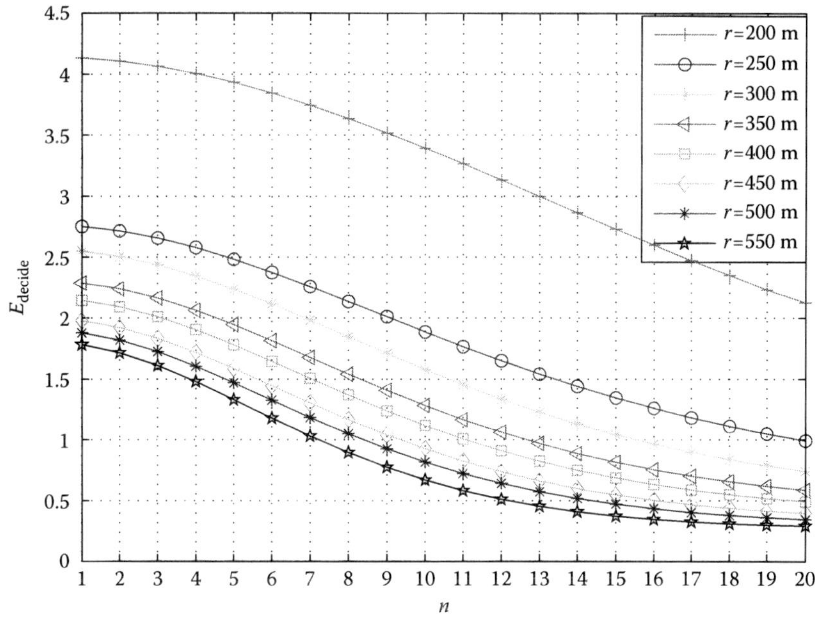

Figure 6.7 Relation between the number of multicast n and E_{decide} under different multicast ranges r.

To discuss the impact of length of CAM packet L_P, which is one of the important network environment parameters, we modify the value of L_P to various values. The result is shown in Figures 6.8 and 6.9.

We can learn from the model of the multicast process we designed here that the rise in the length of CAM packet L_P causes the time increase during data transmission, the accuracy of the CAM packet will decline, and value of packet will decrease. Meanwhile, it will make the packet delivery ratio *PDR* be smaller; as a result, the profit of multicast's value will decrease. Above all, we can conclude that when the multicast time n is in the same value, the larger the L_P is, the smaller the decision value E_{decide} is.

As shown in Figures 6.8 and 6.9, if the threshold value is 1.5, when the value of L_P is taken in the range 80–210 bytes, the time of decision-making process is 6 times, 5 times, 4 times, ..., 2 times. From this, we can show that the larger the length of CAM packet, the lesser the time of decision-making process.

The influence generated by R_d –the system transmission data rate on the result of multicast decision-making can be seen in Figures 6.10 and 6.11. Here we set the value of R_d in different values.

Deriving from the aforementioned definitions, we know that when the transmission data rate increases, the transmission delay will be reduced; thus the cost of multicast C will be reduced and the value of CAM packet R_{value} will decline. As a result, when the time of multicast is unchangeable, the higher the value taken by R_d, the larger the value of E_{decide}. Figures 6.10 and 6.11 show that, when the value of transmission data rates in current network is in 3–8 Mbps, the multicast times is 2 times, 3 times, 4 times, ..., 6 times, 8 times, respectively. Therefore, it can be seen that if the bandwidth in network becomes bigger, the time of a CAM packet being multicasted in the decision-making process will be more.

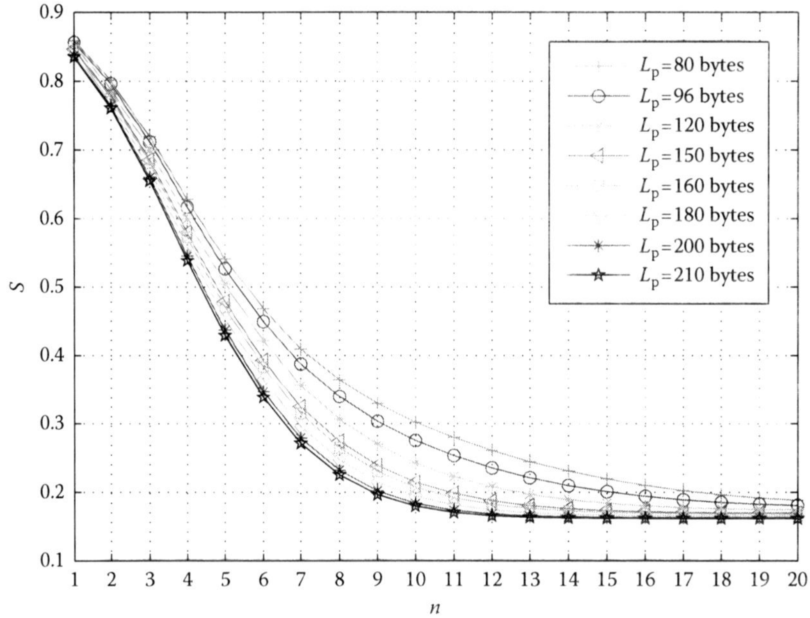

Figure 6.8 Relation between the number of multicast n and S under different packet's lengths L_P.

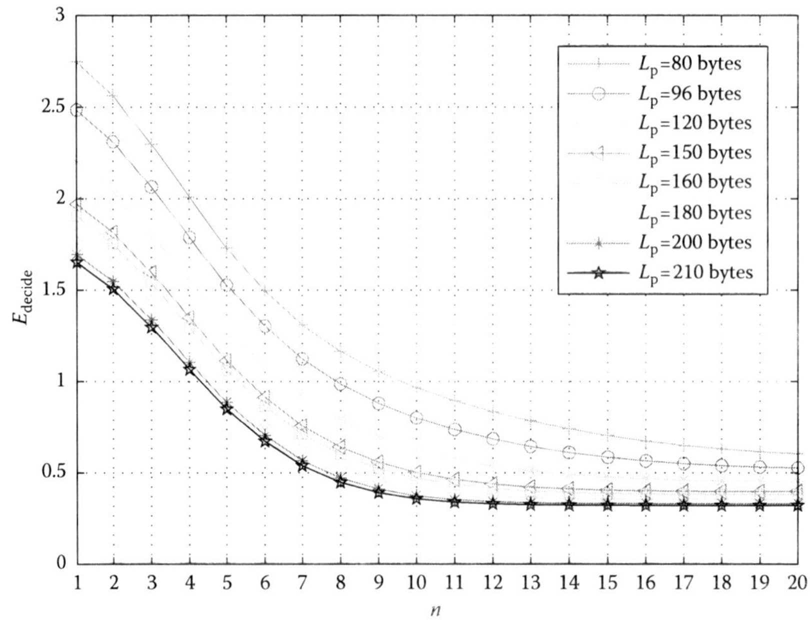

Figure 6.9 Relation between the number of multicast n and E_{decide} under different packet's lengths L_P.

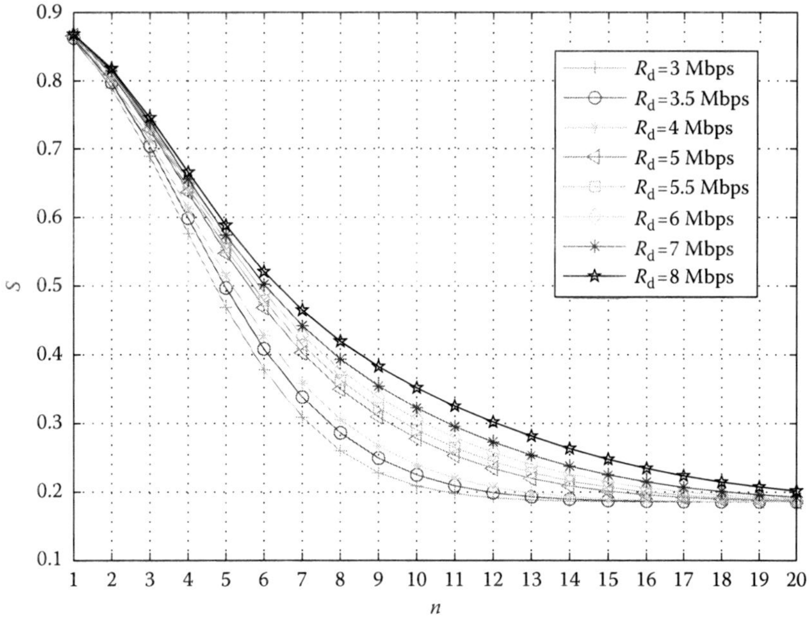

Figure 6.10 Relation between the number of multicast n and S under different transmission data rates R_d.

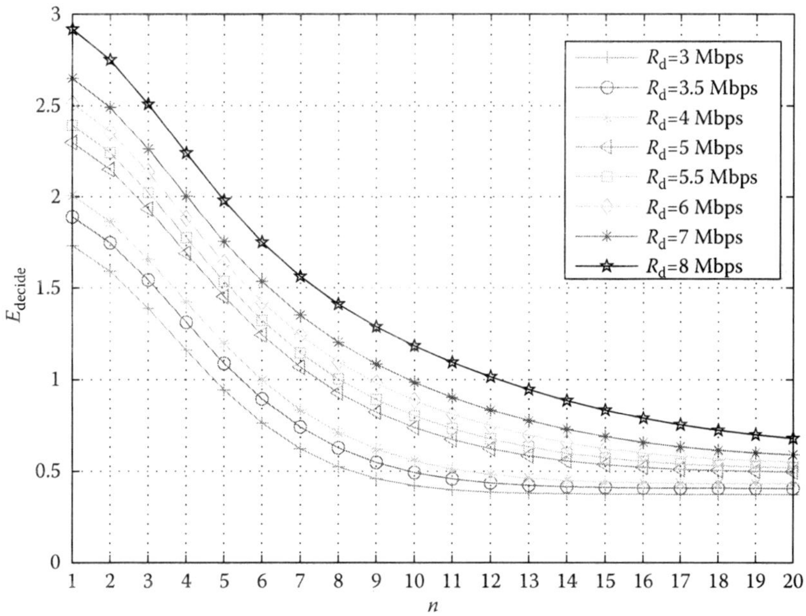

Figure 6.11 Relation between the number of multicast n and E_{decide} under different transmission data rates R_d.

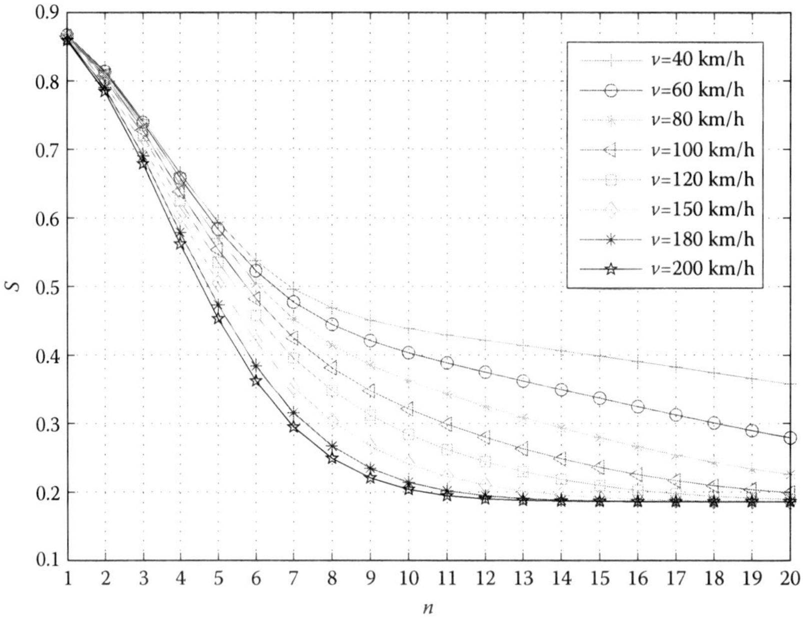

Figure 6.12 Relation between the number of multicast n and S under different vehicle speeds v.

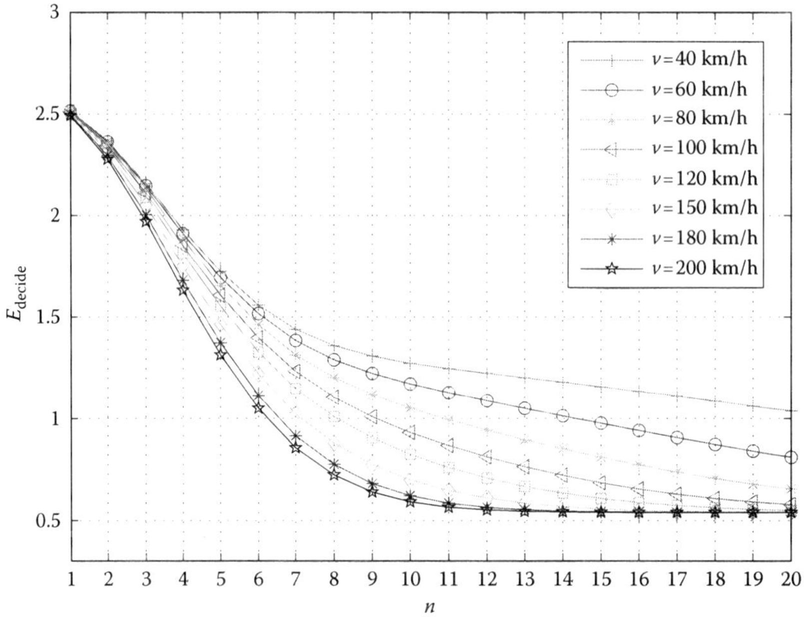

Figure 6.13 Relation between the number of multicast n and E_{decide} under different vehicle speeds v.

Here, through setting the value of vehicle speed in different values, we discuss the speed of vehicle in network's impact to the result of decision-making process in Figures 6.12 and 6.13.

According to the definitions, when there is a rise in the vehicle speed in the current network, during a constant transmission delay duration the larger deviation position will be generated by the initial vehicle, which will cause the degree of accuracy of the packet to decrease; as a result, the decision value E_{decide} will be smaller. Figures 6.12 and 6.13 show that if we set the threshold value $\theta = 2$, when the vehicle speed is in the range of 40–200 km/N, the time of packet be, multicasted is 6 times, 5 times, 5 times, ..., 4 times. From this, we can learn that the faster the vehicle speed, the less the time the packet spends being multicasted.

III

VEHICULAR DATA APPLICATION AND MOBILITY CONTROL

Chapter 7

Road Recognition from Trajectories Collected by Freight Vehicles Based on Least Squares Support Vector Machine Fitting

Daxin Tian, Yue Yang, Xiaolei Ma, and Yunpeng Wang

Beihang University

Zhengguo Sheng

University of Sussex

CONTENTS

7.1	Introduction	116
7.2	Preliminaries	117
	7.2.1 Data cleaning	117
	7.2.2 Subtrajectory partitioning	117
7.3	Generating Maps	121
7.4	Experiments	125
	7.4.1 Experimental data	125
	7.4.2 Centerline fitting	126
7.5	Conclusion	130

7.1 Introduction

It is crucial for navigation systems and all other map-based applications to obtain highly accurate and quickly updated road maps. However, existing methods of generating maps are largely based on geographical surveys [178]. Such methods require a great deal of manpower and time, as it is hard to keep pace with the speed at which new roads are updated, especially in developing countries. A low-accuracy and slowly updated road map often confuses drivers when planning routes and, worse, may result in accidents [84].

In recent years, with the popularization of GPS equipment on mobile objects such as vehicles and mobile phones, the spatiotemporal trajectories of such objects are becoming available. These GPS devices periodically upload information regarding their current positions, speeds, and times to the server via mobile communication networks. Thus, a new approach that automatically generates road maps using GPS-traced data has emerged in the past decade. There are three main categories of these methods: k-means, kernel density estimation (KDE), and trace merging. Refs. [5,55,101,209] represent k-means, which usually work by clustering the samples around roads and generating the centerlines by linking the centers of the clusters. An established practice in KDE, as in Refs. [31,50], is to build a grid of cells denoting the space covered by sensor readings, with road centerlines being generated by the contour lines of the three-dimensional map formed by those cells. Refs. [27,168] are typical of trace merging, and the routine is to incrementally merge the traces to fit the road centerlines.

However, the methods mentioned here always assume trace data sampled at a high-GPS sampling rate (typically at 1 Hz) and with a highly accurate position. The traces for these methods are generally taken from special survey vehicles. Commonly used GPS devices, on the contrary, usually upload data with a much lower accuracy (around tens of meters), and the sampling frequency is relatively low (much lower than 1 Hz) due to bandwidth and storage cost concerns. It is challenging to build high-accuracy maps using these coarse-grained data, because the low-sampling rate and inherent noise of trace data would cause incorrect roads recognition and inaccurate centerlines fitting. For instance, the road segment recognized by two samples may not exist in the real world; besides, those methods might consider each node to be on the roads, while it is highly possible to deviate from a real road. None of the mentioned methods are designed for this kind of trace data. Ref. [140] proposed a method for applying these coarse-grained data to generate a road map: Traces were clustered by road segments, and a shape-aware B-spline fitting was applied to each cluster. While the study focused on recognizing road segments more correctly, the final centerline fitting was not sufficiently accurate. However, the accuracy of the map is crucial for most map-based applications such as navigation systems.

Thus, this chapter aims at proposing an algorithm for recognizing high-accuracy road centerlines by using large-scale but coarse-grained GPS trace data. Emphasis is on improving the accuracy of road centerlines. We process those data by prerunning the original traces to filter out useless data, clustering similar traces to obtain raw lines, and applying the least square support vector machine (LS-SVM) [221] to refine each raw line and obtain the final results. Finally, a typical commercial GPS data set collected from freight vehicles running in JiangSu province was adopted for evaluation of the algorithm, and the result shows that our method significantly outperforms previous work in terms of accuracy.

The rest of the chapter is organized as follows. Section 7.2 introduces the preliminaries of using original data to filter out abnormal data and partition the original traces. Section 7.3 discusses the algorithm for generating maps in detail and presents the effectiveness of applying LS-SVM to fit the centerlines. A comparative study for evaluation is discussed in Section 7.4, and finally, Section 7.5 concludes.

7.2 Preliminaries

This section introduces the process of prerunning the raw data. We first introduce the process of data cleaning to filter out corrupt data. Then a dedicated partitioning process is created to filter out some useless data and to reduce the input to the road-recognition algorithm.

7.2.1 *Data cleaning*

In a set of historical trajectories, there are usually corrupt data caused by errors in sensors or communication equipment, especially in the complex vehicular environment. To obtain the correct analysis results, the raw data needs to be detected to remove inaccurate records from the database. As proposed in Ref. [290], instantaneous data are considered abnormal if they greatly deviate from the center of the distribution. Here, the 3σ rule [182] is adopted to select usable data by using the speed data. Since the 3σ rule is only suitable for those data that follow normal distributions, it is necessary to check whether this condition applies to the data under consideration. If the result returned by the K–S [105] shows that the data set does not follow the normal distribution, it will be transformed into a pseudonormal distribution by the following method [290]:

$$x^{(\alpha)} = \begin{cases} \frac{x^{\alpha}-1}{\alpha}, & \alpha \neq 0, \\ \ln(x), & \alpha = 0. \end{cases} \tag{7.1}$$

To calculate the optimal exponential, α, Ref. [24] proposed a method where by α is the parameter that can return a maximum value for Equation 7.2:

$$l(\alpha) = \max\left(-\frac{M}{2}\ln\left(\frac{1}{M}\sum_{i=1}^{M}\left(x_i^{(\alpha)} - \overline{x}^{(\alpha)}\right)^2\right) + (\alpha-1)\sum_{i=1}^{M}\ln(x_i)\right), \tag{7.2}$$

where M denotes the number of data points. x_i denotes each point's velocity, and $\overline{x}^{(\alpha)}$ is defined as follows:

$$\overline{x}^{(\alpha)} = \frac{1}{M}\sum_{i=1}^{M}\frac{x^{\alpha}-1}{\alpha}. \tag{7.3}$$

The confidence interval of x can be determined using the 3σ rule, $[x^{(\alpha)} - 3\sigma^{(\alpha)}, x^{(\alpha)} + 3\sigma^{(\alpha)}]$, and the data in this confidence interval would be saved. Otherwise, if the data processed by the above method still fail the K–S test, meaning that the pseudonormal distribution of the data does not obey the normal distribution, a threshold (ε) would be found to help select the data. The cumulative probability distribution of the velocity data is adopted. In fact, if the data follow a normal distribution, then $P(\mu - 3\sigma \leq X \leq \mu + 3\sigma) = 99.7\%$, where μ and σ denote parameters of the normal distribution. Thus, the confidence coefficient of these velocity data is defined as $\alpha = 0.15\%$. In this case, the confidence interval of velocity data is defined as $Pr(X \geq \varepsilon_1) = 1 - \alpha$, $Pr(X \leq \varepsilon_2) = 1 - \alpha$. Data in this confidence interval would be reserved, which is actually $[\varepsilon_1, \varepsilon_2]$.

7.2.2 *Subtrajectory partitioning*

After the preliminary process, the effective data are successfully extracted. The whole trajectory of each freight vehicle, in most cases, can be well described by only some of all the trace nodes. Because of this, those particular nodes selected to present the full trajectories would form a series of subtrajectories. This process is called partitioning. It is conceivable that the endpoints of the subtrajectory may denote the road that it passes through.

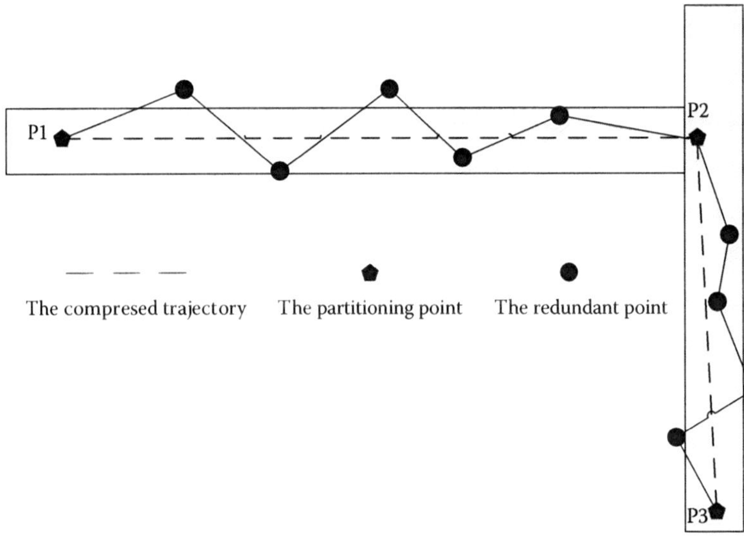

Figure 7.1 Ideal partitioning.

As Figure 7.1 shows, the purpose of partitioning is to describe each road with a minimum number of points. In the next step, recognizing raw road lines, the details of a single road segment are not required. Thus, the clustering of the subtrajectories is enough to recognize the raw road lines. Here, the main purpose of partitioning the original trajectory is to reduce the input to the road-recognition algorithm to save memory storage, time, and cost for large-scale data.

The main work involved in this partitioning process is to find the subtrajectory that represents the whole trajectory. Apparently, since the trajectories from the data set are mainly distributed on the roads, the ideal partitioning (see Figure 7.1) is to have every subtrajectory on the road represented by two trace nodes. Those particular points that contain significant knowledge about the whole trajectory are defined here as partitioning points and correspond to the pentagon points in Figure 7.1. Those points that can be ignored are defined as redundant points, which are the round points in the figure. The solid line represents the original trajectory, and the dashed line denotes the compressed trajectory. The key procedure of partitioning the trajectory should be the selection of partitioning points. If more partitioning points were kept, the subtrajectory would be more similar to the original trajectory, while the number of trajectory partitions would be larger. Thus, it is crucial to establish a new function to take both the similarity and complexity into account. The idea of the minimum description length (MDL) principle is adopted to build this function.

Considering M to be a class of models and D to be the given data sequence, the MDL principle is to find a hypothesis $H \in M$ such that the sum of the description length of H and the length applying H to encode D is minimal [77]. Thus, there are two parts to encoding: $c1$ is in charge of encoding hypothesis H, and $c2$ is used to encode D applying hypothesis H. The MDL principle holds that

$$H_{\text{mdl}} = \text{argmin}\{L_{c_2}(D|H) + L_{c_1}(H)\}, H_{\text{mdl}} \in M, \tag{7.4}$$

where argmin $f(x)$ means choosing a value of x such that $f(x)$ can be minimized; $L_{c_2}(D|H)$ is the encoding length of $c_2(D|H)$, and $L_{c_1}(H)$ is the description length of the hypothesis. The best hypothesis, H, to explain D is the one that minimizes the sum of $c_2(D|H)$ and $L_{c_1}(H)$.

Here, the hypothesis D corresponds to different sets of trajectory partitions. This formulation is actually intended to find the optimal partitioning of a trajectory H from the class D_i. As a result, the problem of finding the optimal partitioning is translated into finding the best hypothesis using the MDL principle. To solve the trajectory-partitioning problem, we define $L_{c_1}(H)$ and $L_{c_2}(D|H)$ as follows:

$$L_{c_1}(H) = \log 2(\sum_{k=1}^{n} \text{len}(N_{i_{k-1}}N_{i_k}), \tag{7.5}$$

where $\{N_0, N_1, \ldots N_{n-1}, N_n\}$ denotes the whole trajectory, and the set of the subtrajectory is:

$$\{\text{Path}_k = N_{i_{k-1}}N_{i_k} | k = 1, 2 \ldots m\}, \tag{7.6}$$

where N_{i_k} represents those partitioning points, and $N_{i_0} = N_1$ and $N_{i_m} = N_n$. $\{N_i | i = 1, ..k...n\}$ is the set of such points. len $(N_{i_{k-1}}N_{i_k})$ denotes the length of a line segment $N_{i_{k-1}}N_{i_k}$, i.e., the Euclidean distance between $N_{i_{k-1}}$ and N_{i_k}, and $L_{c_1}(H)$ denotes the sum of the lengths of the subsequences. In this case, $L_{c_1}(H)$ means the description length of the hypothesis that partitions the trajectory into a set of subtrajectories. Then $L_{c_2}(D|H)$ is defined as follows:

$$L_{c_2}(D|H) = \log 2 \left(\sum_{i=1}^{m} \sum_{j=i_{k-1}}^{i_k} \sqrt{dp(N_jN_{j+1}, N_{i_{k-1}}N_{i_k}) da(N_jN_{j+1}, N_{i_{k-1}}N_{i_k})} \right). \tag{7.7}$$

$L_{c_2}(D|H)$ represents the similarity between a trajectory and a set of its trajectory partitions. To measure the difference, we use the product of the perpendicular and angle distances, which is the area formed by the trajectory partition $N_{i_{k-1}}N_{i_k}$ and a series of original line segments $\{N_lN_{l+1} | l \in (i_{k-1}, i_k)\}$ partitioned by $N_{i_{k-1}}N_{i_k}$.

The measurement of distance is defined as follows. The distance functions mainly contain two components: (1) the perpendicular distance (dp) and (2) the angle distance (da). Components from similarity measures used in the area of pattern recognition [34,71] are adopted. To fit in our requirement, it is necessary to modify these components.

As Figure 7.2 shows, $d1$ is the Euclidean distance between A and E, and $d2$ is the Euclidean distance between B and F. An idea of the arithmetic mean is adopted for the calculation of d_p, such that d_p represents the distance between the two lines more impartially. Then d_p is determined by

$$d_p = \frac{d1 + d2}{2} \tag{7.8}$$

and d_a is determined by

$$d_a = \min(L1, L2) \cos\theta, \tag{7.9}$$

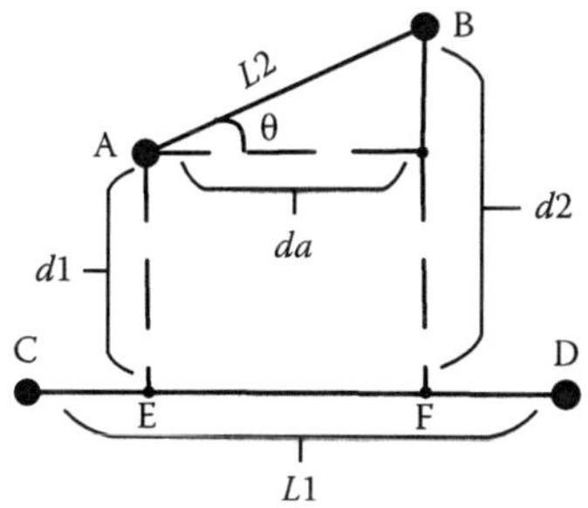

Figure 7.2 Definition of distance.

where θ computes the smallest intersecting angle between lines AB and CD. $L1$ and $L2$ denote the lengths of lines AB and CD, and $\min(L1,\ L2)\cos\theta$ transforms the angular difference into distance, as shown in Figure 7.2.

Based on the above equations, as the number of partitioning points decreases, $L_{c_1}(H)$, the sum of the lengths of the subsequences decreases as well. This means that the complexity of the model decreases. On the other hand, $L_{c_2}(D|H)$, the sum of the area, increases, meaning that the difference between the hypothesis and the original data is growing. As mentioned before, the aim of this algorithm is to find an optimal H_{mdl} to achieve the best trade-off between complexity and precision. In fact, it is difficult to solve this problem, since the global result is hard to obtain. Here, the algorithm is aimed at finding the so-called partitioning points at which the behavior of a trajectory changes rapidly to represent the subtrajectories. Thus, this optimal problem can be translated into finding the partitioning points. The factor $L_{c_2}(D|H)$ plays a more important role than $L_{c_1}(H)$ when the behavior of a trajectory changes rapidly. Thus, we find those points at which $L_{c_2}(D|H)$ changes rapidly. When designing the algorithm to carry out the iteration necessary to find the partitioning points, it is important to avoid errors caused by the offset of the GPS and some other factors. For instance, as Figure 7.3 shows, if P1 is considered to be the initial point of the iteration, the algorithm would find P4 as the point at which $L_{c_2}(D|H)$ changes rapidly. It may regard P3 as the partitioning point, while if this iteration continues, we would find that P5 or some point behind P5 is a better choice as a partitioning point. Thus, we aim to find a method to alleviate the disturbance caused by the offset of GPS and some other similar factors. Here the idea of simulated annealing (SA) [119] is adopted.

SA is a generic probabilistic metaheuristic for the global optimization problem of locating a good approximation to the global optimum of a given function in a large search space. It is often used when the search space is discrete (e.g., all tours that visit a given set of cities). The following pseudocode presents the modified SA for the partitioning algorithm, where $l_{c_2}(D|H)$ denotes the encoding length $l_{c_2}(D|H)$ of a single subtrajectory and P is defined as follows:

$$P = \frac{1}{1 + \mathrm{e}^{-a(k)x+b}}. \tag{7.10}$$

P denotes the influence of annealing temperature, suggesting that as annealing temperature increases, the iteration approaches its optimal value such that the probability decreases. k indicates the steps of iteration, $a(k)$ is defined as $a(k) = a_0 \times r^k$, r determines the speed of annealing, and b is a constant. The difference between the adjacent $l_{c_2}(D|H)$ is defined as variable x.

The key idea of Algorithm 7.2 is to accept a worse solution with a certain probability for a single subtrajectory, and this probability decreases with the growth of the iteration's steps. In the case shown in Figure 7.3, the tendency changes rapidly in P4, while the algorithm can

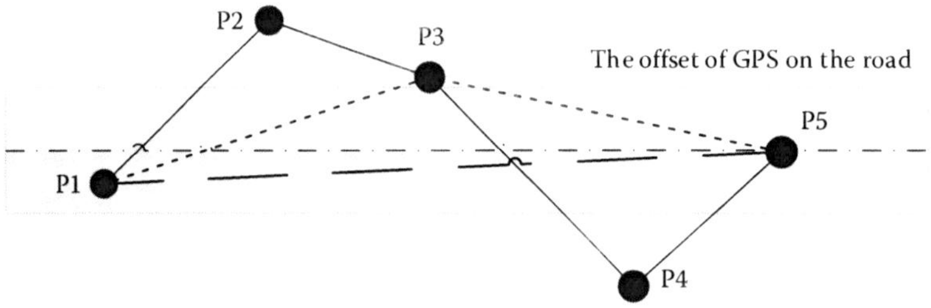

Figure 7.3 Potential problem.

Algorithm 7.2: Modified SA for partitioning

Require: the original whole trajectory S
$i_0 = 0$
for i = 0 through S.length **do**
  select the next point to find the difference $x = l_{c_2}(D|H)new - l_{c_2}(D|H)$, $k = i - i_0$
  if $P(x,k) > random(0,1)$ **then**
    add the previous point to S', i_0 = i
  end if
end for
Ensure: the set of partitioning points S'

still keep the iteration going with a certain probability such that a better partitioning point like P5 could be discovered. So the partitioning algorithm used here can also accept a worse choice with a certain probability that it may miss a true partitioning point. Apparently, this probability is positively correlated with $l_{c_2}(D|H)$: if $l_{c_2}(D|H)$ changes rapidly, then this probability would be large; otherwise, if $l_{c_2}(D|H)$ changes slowly, this probability would be very small. We test this partitioning process and find that the partitioning ratio in an urban area is around 0.6 and that in highways is around 0.4, fitting the requirement of the process.

7.3 Generating Maps

Unlike the method proposed in Ref. [140], this algorithm focuses on the accuracy of road centerlines using coarse-grained data. An acknowledged hypothesis for the road centerlines is that they can best segment the trace nodes distributed on the roads. Thus, the key idea of this algorithm for generating road centerlines for each road segment is to collect the trace nodes distributed on it and then apply a classifier to find the optimal segmentation. Here, we apply LS-SVM as a classifier to segment the data, since LS-SVM can perform this task well both linearly and nonlinearly. By increasing the number of nodes around a single road segment, the accuracy can be increased, since the error of GPS offset would be further eliminated. Thus, the algorithm of recognizing road centerlines can be divided into two main parts: the generation of raw road segments with a relatively low accuracy and the final fitting process to refine them to have higher accuracy.

To generate the raw road map, those subtrajectories are grouped first to form several clusters; then a method called Y-splitting [140] is used to refine those clusters to obtain the raw road map. We next use the raw road map to update the existing digital map and use the Open Street Map (OSM) [83] to test this algorithm. OSM is a collaborative project to create a free, editable map of the world that anyone can edit and update. The data are contributed by volunteers around the world and are totally free. We adopt it as the basic road network, since it has a mature and proven architecture and is open source. The following structure is designed for the date related to each generated raw road segment and the OSM road segments: RoadID as the ID of its road, Sid as its own ID, N_{st} as the first point, and N_{ed} as the last point. Since the raw road segments are not refined, the merging process takes OSM as the basic road network, and each generated raw road segment is matched with OSM; if unmatched, the segment is added to the basic road network.

We select an area to test the algorithm of generating maps. After the aforementioned preliminary step, subtrajectory partitioning and generation of the backbone maps take place. The backbone map is generated, as shown in Figure 7.4a, and then merged with the original OSM map, as Figure 7.4b, to obtain Figure 7.5, which is considered to be a new OSM map.

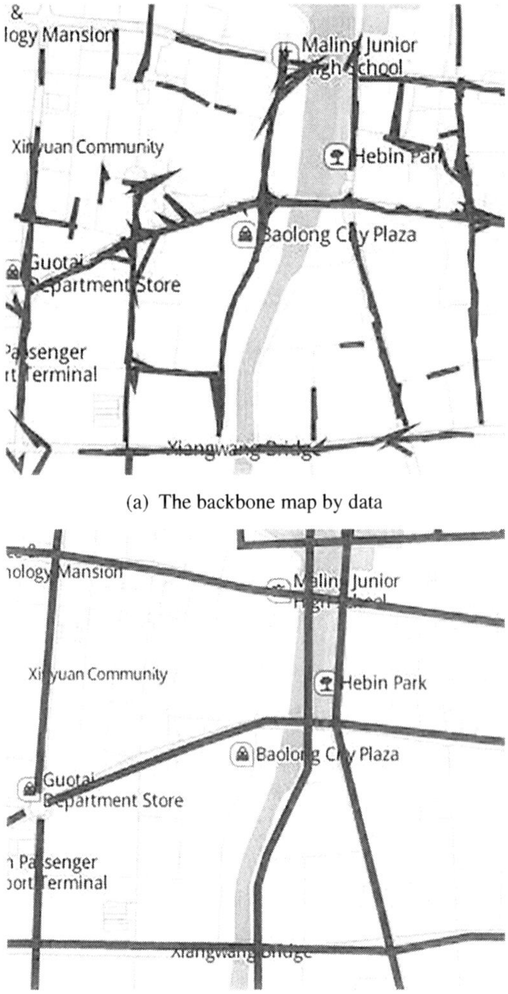

(a) The backbone map by data

(b) The OSM map

Figure 7.4 Map generated by data and OSM map.

After generating the new raw roads by merging the raw road lines, we inferred with OSM, the next step is to collect the trace nodes distributed on each single road segment for the final fitting. For the new road map, the data structure of each road segment is defined as: RoadID as the ID of its road, Sid as its own ID, N_{st} as the first point, N_{ed} as the last point, and N_{arr} as an array containing the middle points. For each original subtrajectory, Algorithm 7.3 traverses the newly generated map and finds if there is a road segment it may belong to. ε and ϑ are the two parameters denoting the threshold of the distance and angle, respectively; if the distance between the two lines is less than ε and the angle between these two lines is less than ϑ, then the match function is used to process it; the two parameters are adopted from Ref. [140], and the distance between the two lines is obtained by the method from Ref. [142]. In this function, the basic method for obtaining the distance between a trace node and its pedal to the line is used to determine if this point should be allocated to the line. The coordinate of the pedal is defined below:

$$x_{iN} = t \times x_{iN_{st}} + (1-t) \times x_{iN_{ed}} \tag{7.11}$$

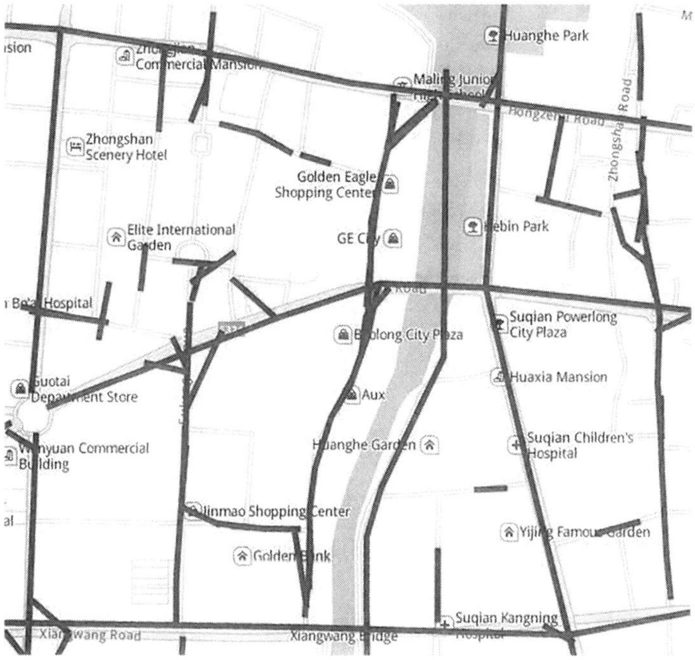

Figure 7.5 Merged map.

Algorithm 7.3: Roads centerline generation by LS-SVM

```
for each line of original traces do
    for each line LN in new merged road collections do
        getdistance(s,LN)
        if getdistance(s,LN) ≤ ε and gettheta(s,LN) ≤ ϑ then
            Match(s,LN)
        end if
    end for
end for
```

Require: Lines, Line LN

```
for each point N_0 in s do
    get the pedal N_d
    if in (7.11) t ∈ [0,1] and the distance between N_0 and N_d ≤ ξ then
        allocate N_0 to N_arr, remove N_0 from s
    end if
end for
```

```
for each line LN in the new merged line collections do
    Using the LS-SVM to refine LN by the nodes list N_arr
end for
```

where x_{iN} denotes the coordinate of the pedal and t is a parameterization representation; if t is between 0 and 1, it means the pedal at that point is on the line segment formed by N_{st} and N_{ed}.

Since the curves of roads in the real world are usually not linear, a linear equation is unsuitable for fitting. Related studies [55,140] adopt the least squares approximation of the B-spline to fit the curves, since the B-spline itself is not flexible and accurate enough. Here, LS-SVM is adopted to fit the curves as it has excellent performance on classification and regression problems. After the aforementioned process, for the reserved massive trace nodes for each new raw road segment, LS-SVMs are applied to fit the curve of the centerlines. LS-SVMs are least squares versions of support vector machines (SVM) [4], which can work well when used for classification and regression analysis. Compared with classical SVMs, the main process of a LS-SVM is to solve a set of linear equations instead of a convex quadratic programming (QP) problem; thus, it can solve regression problems with a larger sample size. The optimal function of LS-SVM is

$$\min(J(w,e)) = \frac{1}{2}w^T w + \gamma \Sigma e_i^2 \tag{7.12}$$

such that

$$y = w^T \varphi(x_i) + b + e_i, i = 1, \ldots, l. \tag{7.13}$$

The final equation of the fitting function of LS-SVM [221] is as follows:

$$f(x) = \sum \alpha K(x, xi) + b. \tag{7.14}$$

In Equation 7.14, α and b denote the solution of the following set of equations:

$$\begin{bmatrix} 0 & y1 & \cdots & yl \\ y1 & y1y1K(x1,x1) + 1/c & \cdots & y1ylK(x1,xl) \\ & & \ddots & \\ yl & yly1K(xl,x1) & \cdots & y1ylK(xl,xl) + 1/c \end{bmatrix} \times \begin{bmatrix} b \\ a1 \\ \\ al \end{bmatrix} = \begin{bmatrix} 0 \\ 1 \\ \\ 1 \end{bmatrix}. \tag{7.15}$$

It is clear that the algorithmic complexity of LS-SVM is O(n^2) through these equations; thus, LS-SVM cannot handle regression problems with too many samples, even though its algorithmic complexity is still better than that of SVM, since it transfers a quadratic optimization problem into a set of linear equations. A radial basis function (RBF) is adopted here as the kernel function:

$$K(x, x_i) = \exp(-\|x - x_i\|^2/\sigma^2). \tag{7.16}$$

To obtain the LS-SVM model, two extra parameters are needed: γ as the regularization parameter denoting the trade-off between training error minimization and the smoothness of the function, and σ^2 as the parameter of RBF. We apply the cross-validation method to obtain these two parameters.

In fact, the mechanism of LS-SVM determines that it can perform better than the typically used B-spline [140]. The B-spline aims to find a best-fitting line to minimize the sum of the deviation of the samples from the fitting line. However, the least squares approximation of the B-spline needs to set the highest power and number of control points before the two parameters that limit the final fitting result, which suggests that this fitting may not be optimal and that the B-spline might generate over-fitting. For LS-SVM, it also aims to find a best segment, while there is an optimal process for the trade-off between the deviation minimum and the curvature of the line. Thus, LS-SVM is more flexible and accurate than the B-spline, as will be proved by evaluation in Section 7.4.

In some cases, if there are enough trace nodes allocated to a line, it is possible to use these data to obtain the road boundary. In this case, a road is considered to be a two-dimensional

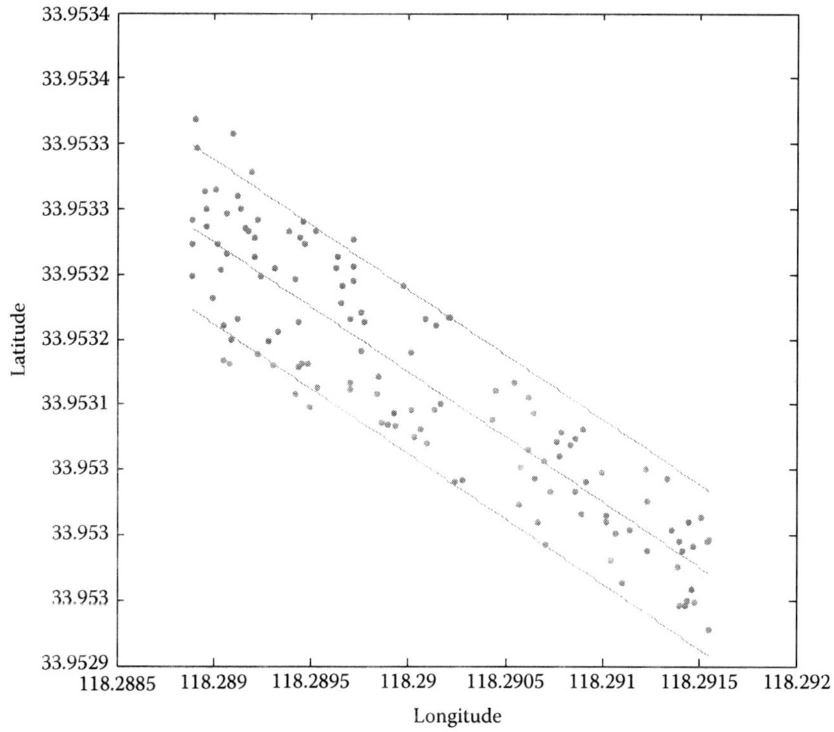

Figure 7.6 Borders and centerline of the roads.

object, and the trace nodes around this road can fill it up. Then LS-SVM is applied to find the envelope curve of this road. It is clear that with the increment of the number of samples, the border curves will become more reliable and accurate. Here some effort is made to find the border of the centerlines of the roads. In most cases, it can be demonstrated that the border of the road is parallel to its centerline; thus, the problem of finding the border is translated into finding the two parallel lines that contain most trace nodes. The fitting function as follows has the same α as the centerlines:

$$f(x) = \sum \alpha K(x, xi) + b'. \tag{7.17}$$

The 3σ principle is then adopted to find the confidence interval of the error of original trace nodes between the fitting centerlines. By neglecting the nodes that would disturb the final result, the minimum b' is obtained to satisfy the requirement of containing most of the trace nodes. This is meaningful in that researchers can use the border to obtain the widths of roads and thus the number of lanes, since the width of each lane is standard. Figure 7.6 shows an example of fitting the centerlines and obtaining the border.

7.4 Experiments

7.4.1 Experimental data

This section introduces the massive amount of experimental data used here. All GPS data are taken from vehicular communication terminals, which can receive data from speed sensors, GPS devices, and other sensors attached to the trucks. These terminals in trucks generally send data to the server every 60 s, and then those historical data are saved in our database. Since the frequency of the transmission of data is around 0.167 Hz, an ideal data set from one truck over a month should contain more than 43,000 records. In fact, trucks seldom send so much data, since they sometimes have their sensors and GPS devices turned off. Furthermore, trucks cannot move constantly; on occasion, they must stop for a variety of reasons, while the effective data are only those whose velocity is above the threshold value. Here we extend the limitations of the data. It selects those trucks that have more than 5000 effective records, which corresponds to about seven work days, and considers the data set of these trucks to be effective, since the trajectories over a week are suitable for road recognition [139]. After the above process, about 500 effective trucks are selected for the algorithm. Here, we mainly select the trace nodes distributed in Jiangsu province for empirical research.

7.4.2 Centerline fitting

In this section, we evaluate the algorithm. As mentioned earlier, the key idea is to propose an algorithm for improving the accuracy of road centerline fitting. It is actually difficult to compute the error of the recognized roads since the map is truly accurate. We try to evaluate the fitting algorithm using an idea adopted from Ref. [248], which proposed a method for comparing the accuracy of different digital maps with large-scale probe vehicle GPS data. Ref. [248] proposed that the median of the absolute offset can be used to measure the accuracy of road maps. Similarly, we calculate the distance of each node and the nearest point on the roads, and we use the average value of the absolute offset between the nodes and their nearest points to measure the accuracies of the centerlines fitted by our algorithm and previous work; the error is defined as follows:

$$\text{Error} = \text{mean}(\text{dis}(P_i(x,y), P_{it}(x_t,y_t))), \tag{7.18}$$

where dis means the Euclidean distance between $P_i(x,y)$ and $P_{it}(x_t,y_t)$. $P_{it}(x_t,y_t)$ is the nearest point on the road for node $P_i(x,y)$, which is shown in Figure 7.7. In the figure, the circle denotes the detection range, and the empirical value of the radius is set as 50 m since most of the nodes are within this range.

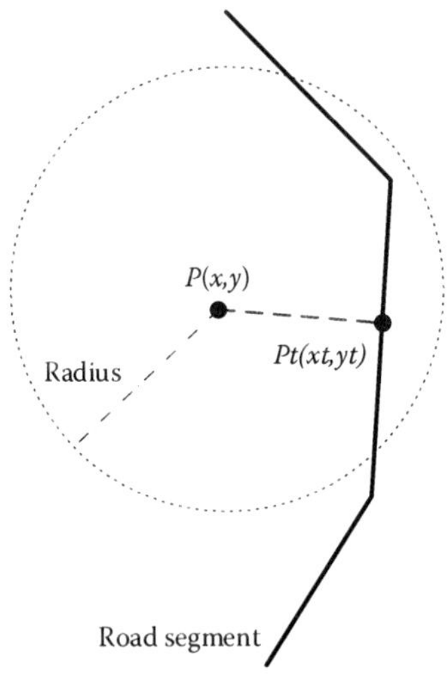

Figure 7.7 Nearest point on the road for each node.

Ref. [248] assumed that the distribution of the probe nodes on roads should be close to Gaussian distribution so as to make it obvious that more accurate fittings of centerlines would be in the middle of all trace nodes, and the absolute offsets of the nodes from the lines should be close to zero.

In this section, we compare our method of using LS-SVM with the least squares approximation of B-splines used in Refs. [55,140]. For a typical example of centerline fitting as shown in Figure 7.8, where the average deviation is 3.056×10^{-5} and the average deviation of the latter is 8.4726×10^{-5}, it is clear that LS-SVM performs much better than B-spline in this complex case. On the whole, it is clear that our method of using LS-SVM better describes the shape of the roads. Then we analyze the details of this instance. Figures 7.9 through 7.11, respectively, denote the amplified figure for the three ellipses in Figure 7.8 from left to right. Figure 7.9 denotes the part of fitting nonlinear segment; at this part, the line fitted by LS-SVM is generally distributed on the middle of the nodes, while B-spline has improper curvature, and the worse result, at this part, is that the average deviation of the former is 3.4327×10^{-5} and that of the latter one is 1.1379×10^{-4}. Figure 7.10 denotes the part of fitting linear segment; at this part, B-spline totally deviates from the nodes distributed on the road segment, while LS-SVM is flexible and it consistently segments the nodes. The error of the former in this part is 2.3934×10^{-5} and that of the latter one is 2.197×10^{-4}. Figure 7.11 shows that the two methods have the maximum difference between one other; it is clear that B-spline is very far away from the trajectories. The error of the former is 2.5344×10^{-5} and that of the latter is 7.4328×10^{-4}.

Since Refs. [55,140] use a cubic B-spline for all roads, it is possible to render the linear roads as "curving," since the choice of the control points would significantly influence the final fitting result. In most cases, the control points obtained by the least squares algorithm would not imply a linear fitting. As Ref. [140] shows, more than one-third of the roads in Shanghai

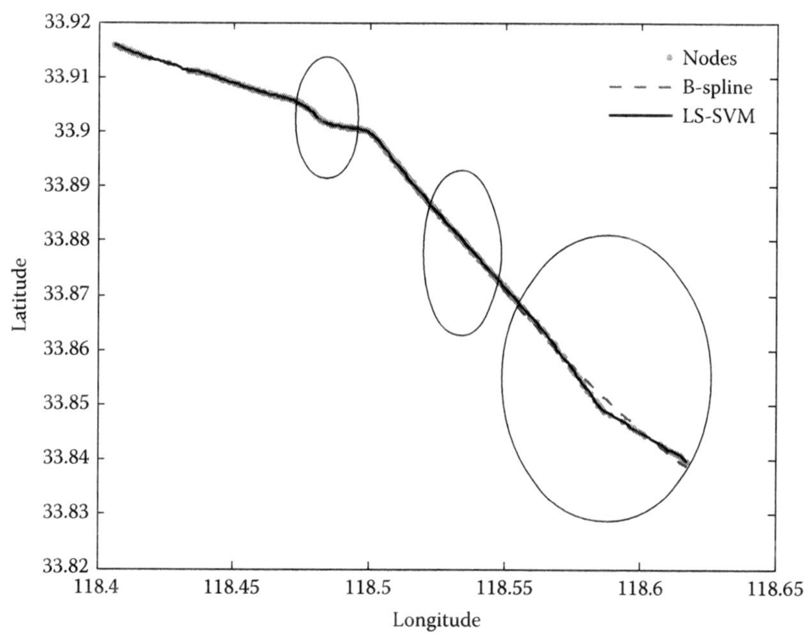

Figure 7.8 Fitting result of LS-SVM and B-spline.

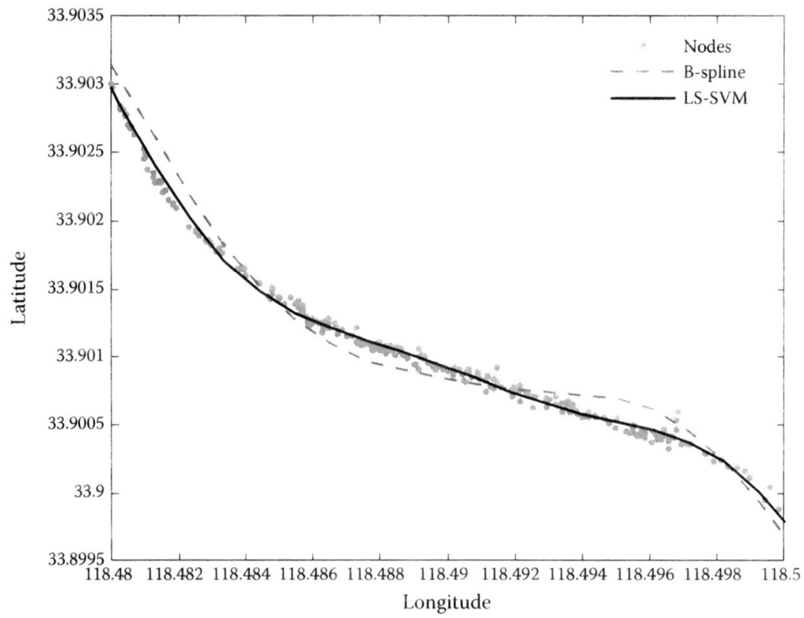

Figure 7.9 Leftmost marked part.

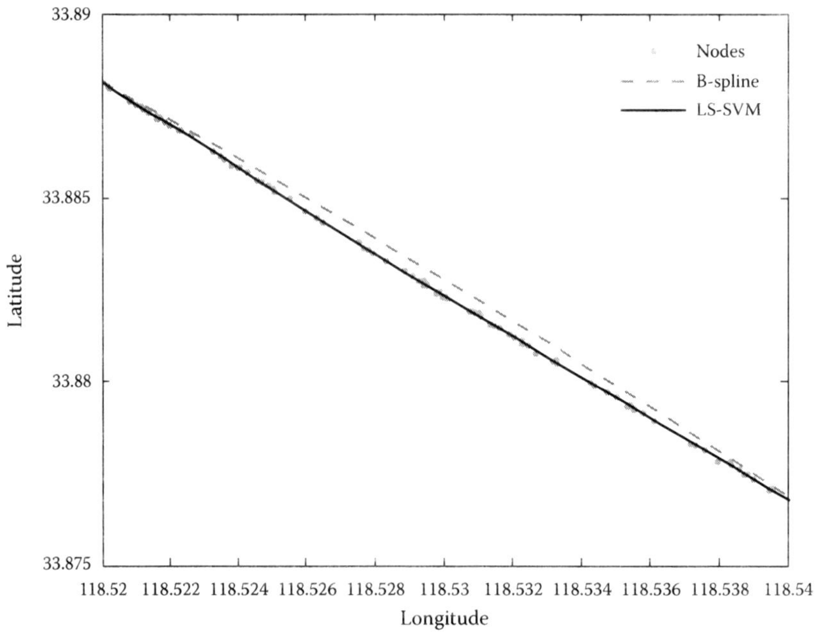

Figure 7.10 Middle marked part.

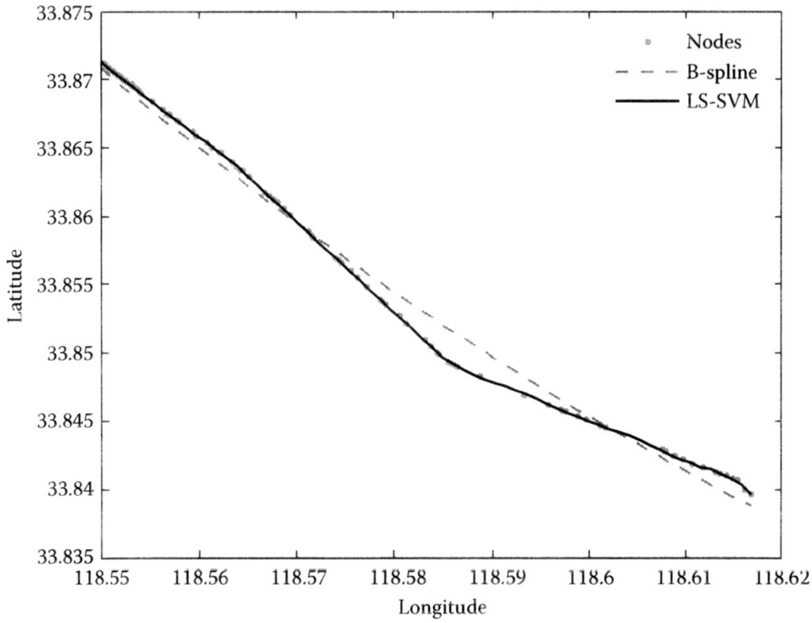

Figure 7.11 Rightmost marked part.

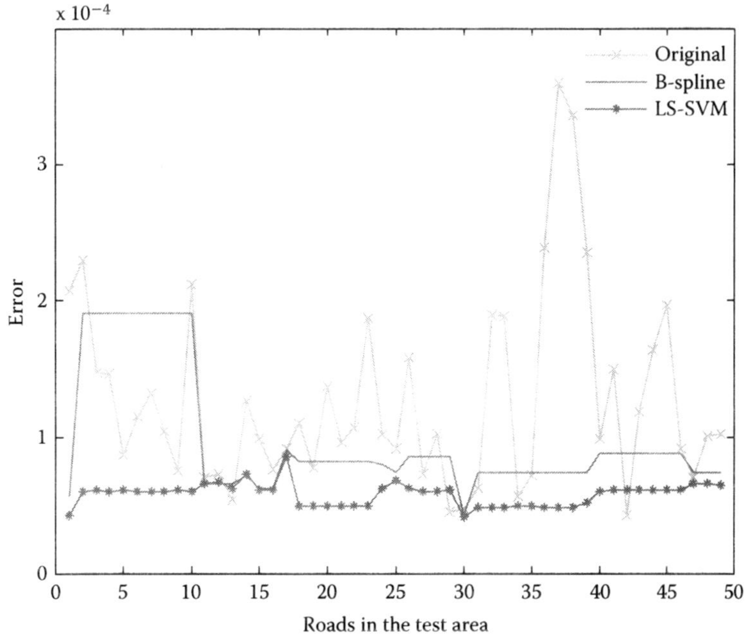

Figure 7.12 Deviation of lines fitted by the two methods and the original lines.

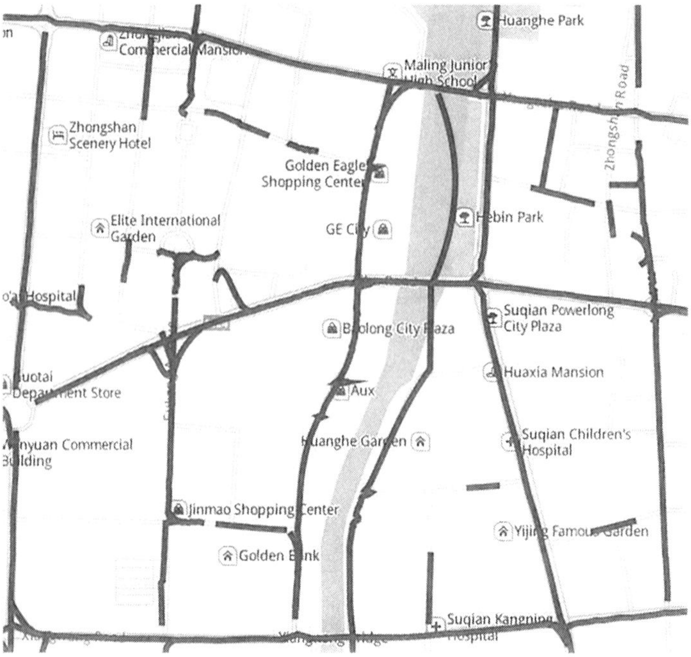

Figure 7.13 Final fitting centerlines of the test area.

are totally linear; thus, it is a crucial problem for those linear roads that curved fitting lines are not suitable. Our algorithm using LS-SVM can solve this problem and can be used not only in linear cases but also in nonlinear ones. The algorithm is more adaptive to different road shapes.

Using the road segments with enough trace nodes allocated to them in the test area, we test the accuracy of the fitting result by LS-SVM, B-spline, and the original backbone lines as Figure 7.5 shows; their deviation is given in Figure 7.12. Compared with the original backbone lines, the method of fitting lines using LS-SVM and B-spline with coarse-grained data can both revise the road map and improve the fitting accuracy, while our method clearly performs better than the B-spline. The average deviations are, respectively, 1.4×10^{-4}, 0.9×10^{-4}, and 0.5×10^{-4}. The result of our method is nearly half that of previous work. This proves that our method of using LS-SVM shows better performance when dealing with road centerline fitting problems and thus improves the final accuracy of recognized roads, which is crucial for map-based applications. The final results of the refinement process for the backbone road map in the test area using our method are shown in Figure 7.13.

7.5 Conclusion

In this chapter, we proposed an approach for automatically recognizing a more accurate digital road map using GPS data collected from freight vehicles. Those data are typical commodity GPS data that are large-scaled, have low accuracy and low frequency. The 3σ principle was applied for the filtering of original data. Then the subtrajectory partitioning process was applied to obtain subtrajectories for the next clustering process. We proposed that the generated map should be merged with the existing maps to update it. Finally, LS-SVM was adopted to fit the curves for a more accurate mapping of the roads. The experimental results prove that our approach achieves higher accuracy for coarse-grained GPS data in the evaluation section.

Chapter 8

DSRC for Vehicular Positioning

Yunpeng Wang, Xuting Duan, Daxin Tian, Xuejun Zhang, and Yingrong Lu
Beihang University

CONTENTS

8.1 Introduction 132
8.2 Related Work 134
8.2.1 Vehicular positioning 134
8.2.2 Hybrid TOA/DOA estimation 134
8.2.3 Bayesian compressive sensing 135
8.3 Network Performance of DSRC and Its Effect on Traffic Capacity 136
8.3.1 Throughput and delay limits of 802.11p 136
8.3.2 The highway capacity limits of the condition using V2V communication 138
8.4 DSRC Implementation and Field Testing 143
8.4.1 V2X communication system 143
8.4.2 Performance evaluation test bed 143
8.4.2.1 Test bed 144
8.4.2.2 Performance evaluation system 144
8.5 DSRC for Vehicular Positioning 146
8.5.1 Hybrid TOA/DOA data model 146
8.5.2 Multitask Bayesian compressive sensing–based DOA estimation 147
8.5.3 Experiments 148
8.5.3.1 Hybrid TOA/DOA location procedure 148
8.5.3.2 Energy threshold analysis 149
8.5.3.3 Performance analyzing 151
8.6 Conclusion 158

8.1 Introduction

Wireless communication technology is the foundation of a vehicle network and can achieve information interaction between vehicle-to-vehicle (V2V) and vehicle-to-infrastructure (V2I) technologies. Different communication technologies are applied to suitable application scenarios, such as a low-delay vehicle safety application, a wide range of vehicle group management, and convenient entertainment services. But for vehicle safety application in particular, broadcasting an emergency message requires a low-latency and high-transmission-rate communication network, which is under a high-speed vehicle environment and must adapt to frequent changes in the network topology [96,146].

Modern intelligent transportation systems (ITS) aim at applying information and communication technologies for the development of future transportation systems, which will be safer, more effective, more conserving of energy, and environmentally friendly [64]. The IEEE 802.11p standard is exclusively considered for dedicated short-range communications (DSRC), a technology employed by vehicular ad hoc networks (VANETs) to achieve vehicular communications. Its modulation format is based on orthogonal frequency-division multiplexing (OFDM). Forward error correction (FEC), the structure of preamble sequences, and pilot-symbol schemes are exactly the same as IEEE 802.11a. Moreover, all 802.11 class standards share the same medium access scheme, well known as carrier sense multiple access with collision avoidance (CSMA/CA) [271]. The overhead in the media access control (MAC) determines a throughput upper limit (TUL) and a delay lower limit (DLL) when the data rate increases [277]. The U.S. Federal Communication Commission (FCC) has allocated the 75 MHz spectrum for DSRC at 5.9 GHz to be exclusively used for V2V and V2I communications. It can provide transmission rates ranging from 3 to 27 Mbps with 10 MHz bandwidth. The supported communication distance ranges up to 1000 m under a variety of circumstances, such as urban, suburban, rural, and freeway, where the vehicular mobility can be low or high (even at 30 m/s) [80].

V2V communication can improve the safety of Vehicle Infrastructure Integration (VII) by allowing vehicles to exchange information. In emergence, they exchanged speed and braking messages, in order to coordinate their acceleration (deceleration) rates. A research result shows that all vehicles that used both V2V communication and sensors could increase highway capacity by about 273% compared with all the manual vehicles [175]. After nearly 8 years of research, Europe and the United States have launched a series of projects ranging from constructed fundamental facility and developed vehicle network standards to decreased energy consumption, reduced gas emission, advanced vehicle safety, and improved road capacity. But now the systems are encountering many problems in practice; the most popular methods used to estimate benefits involve simulation [269]. Large consortium projects such as Connected Vehicles, CVIS, SAFESPOT, Strategic Platform for Intelligent Traffic Systems (SPITS), and Car-2-Car have demonstrated the feasibility of DSRC systems [174]. A Safety Pilot Model Deployment is being conducted in Ann Arbor, MI, which involves installing devices in about 2800 vehicles, including cars, trucks, and buses [262].

Traffic safety and efficiency benefits can be achieved through the development and deployment of VANETs. The global navigation satellite systems (GNSSs), such as global positioning system (GPS), are a primary means to provide position information for vehicular applications in VANETs or telematics sensor systems. However, there are many challenges for vehicular positioning in most urban canyons of dense traffic flows, such as wireless environmental complexity, multiple sensing signal interference, and the possibility of GPS signal interruption. Additionally, the existing effect of multiple propagation paths on wireless sensing makes it difficult to locate an object with high accuracy. Therefore, a single positioning technology based on GPS is not enough for some positioning applications in complex environments. For instance, vehicle safety applications in most real-life traffic scenarios always require a relatively high level of accuracy in vehicular position information, so as to achieve the objectives of collision avoidance, lane-changing assistance, danger warning, etc. Even differential GPS cannot provide the specific accuracy required by these vehicle safety applications due to serious multipath effects existing in actual signal sensing environments [111]. For these reasons some innovative approaches, combined with mobile communication technologies in vehicular networks, have been developed to enhance positioning accuracy [162,163].

Apart from the traditional GPS-based approaches, a lot of emerging positioning solutions are implemented based on spatial radio frequency (RF) [170,195], which include those methods based on the measurement of the time of arrival (TOA) [263], time difference of arrival (TDOA) [183], direction of arrival (DOA) [145], and received signal strength (RSS) [107,141] of radio signals transmitted between the mobile terminal (MT) and base stations (BSs), etc. In the past decades, many approaches for spectral estimation have been used in DOA estimation [2], such as the multiple signal classification (MUSIC) [203], estimation of signal parameters via rotational invariance techniques (ESPRIT) [188], and the maximum-likelihood estimation (MLE) [218]. But a main drawback of most of these methods is high computation cost and poor precise estimation when there is not a prior knowledge of the number of signals, or they are applied in the specific situations with low signal-to-noise ratio (SNR) conditions [217]. Accordingly, these methods are not so practical to be embedded in some small onboard devices for vehicular positioning. Several computational efficiency approaches without an estimation of signal magnitude covariance matrix and its eigenvalues decomposition are presented in Refs. [145,217]. Alternatively, for achieving a better performance, hybrid methods are proposed to use a combination of available range, range-difference, and angle measurements for positioning [43,90,144,284].

A wide-area, Wi-Fi-based positioning system for metropolitan areas has been presented in Ref. [267]. It adopts an received signal strength indication (RSSI)-fingerprinting approach, which is based on a beacon map of radio sources in a specific environment. As the work in Ref. [267] shows, a wide-area Wi-Fi-based positioning system device can localize itself with a rough median error of 13 m. Similarly, a Global System for Mobile communications (GSM)–based approach in Ref. [191] is able to expand the sensing range, but it performs worse than that in Ref. [267]. Effective methods for measuring data, such as TOA and DOA in wireless channels of non-line-of-sight (NLOS) propagation, have been presented in Refs. [194,270]. The authors of Refs. [194,270] have conducted some experiments to illustrate that their solution achieves an improvement of accuracy in positioning estimation. But these aforementioned methods are well suited to implementation in low-mobility and stable environments rather than in general vehicular situations of high mobility. They have not been tested for instantaneous vehicular positioning, so they are not appropriate or potentially may fail to support the vehicular safety applications relevant to VANETs. On the other hand, a system has been proposed to estimate the relative location between vehicles and the roadside units (RSUs), which uses the carrier frequency offset (CFO), a synthesis feature based on Doppler frequency shift (DFS) [162]. The method can give a good performance for positioning in high-speed scenarios. However, as argued in Ref. [162], the standard deviation (STD) of positioning error provided by the system increases when the speed of the vehicle decreases. This is because DFS is too difficult to be captured with a high degree of accuracy when the speed of the vehicle is too low. In addition, Ref. [48] proposes a DSRC-based positioning solution in which a Markov localization algorithm is also presented to process low-accuracy GPS data. The authors of Ref. [48] have validated their approach through both simulations and actual experiments. However, although the approaches in Ref. [48] can do a good job at vehicular positioning under certain conditions, they require that all mobile nodes be equipped with GPS and depend heavily on the performance of the relevant GPS hardwires. But it is not realistic to fully satisfy this requirement in real-life situations.

8.2 Related Work

8.2.1 Vehicular positioning

Vehicular positioning is one of the most important applications for traffic systems, and accurate positioning information on a vehicle is essential for support of vehicular safety applications. Ideally, a vehicle would get knowledge of its own position from GPS. However, those GPS devices with good accuracy but high cost are not widely deployed, so the related applications dependent on high-precision, GPS-based positioning technologies are limited in actual urban canyons. On the other hand, RSU-assisted positioning technologies are some good alternatives in terms of achieving better trade-offs between the cost of onboard devices and the demand for accuracy in positioning. As many existing articles [162,211] have argued, RSU-assisted positioning technologies or hybrid positioning technologies are more likely to be implemented in some complex traffic environments, which is an inevitable trend in the development of wide-area vehicular sensor networks and mobile ad hoc networks in cities. RSUs deployed along the roads are usually called anchors, reference nodes, or base stations. They can provide a priori information on their own position with high precision and use self-position information to determine the relative position of a large number of vehicles (also named as unlocalized mobile nodes) moving on the roads. Measurements between multiple unlocalized nodes and anchors could be any physical reading, such as TOA, DOA, RSS, or the connectivity between those nodes (it is worth pointing out here that the connectivity represents the connection between any two nodes via wireless communication), which can be used to deduce the absolute or relative positions of those mobile nodes. It should be noted that these aforementioned measurements could be attained via different modalities (e.g., Bluetooth [268], ultra wideband (UWB) [3,81], Cellular [148,194], or radio frequency identification (RFID) [195,242]).

8.2.2 Hybrid TOA/DOA estimation

TOA is a measurement of the propagation time of the signal movement from a source to a receiver, which requires that the system clocks of all involved BSs and MTs are precisely synchronized. Here, we treat a BS as a signal source and an MT as a receiver. Once the TOA between any one source and one receiver is obtained, the distance between these two nodes can be consequently determined by multiplying the TOA by the known propagation speed. In most practical applications, GPS is widely used to provide TOA information, which is named as the GPS-TOA positioning technology. It is usually assumed that the positions of all the relevant BSs are known and the TOA information is noise free. In addition, at least three BSs are needed in a GPS system, and they are required to operate in a collaborative manner so as to achieve the triangulation in the two-dimensional (2D) case [284].

The DOA method originates in classical radio direction-finding techniques [147]. Generally, an antenna array is needed for the estimation of DOA as well as a certain signal-processing algorithm that should be employed at the relevant BSs. Aiming at positioning an MT, two or more BSs are required to send the detection signals and receive the corresponding reflected signals via their antenna arrays so as to provide the DOA information related to this MT. The

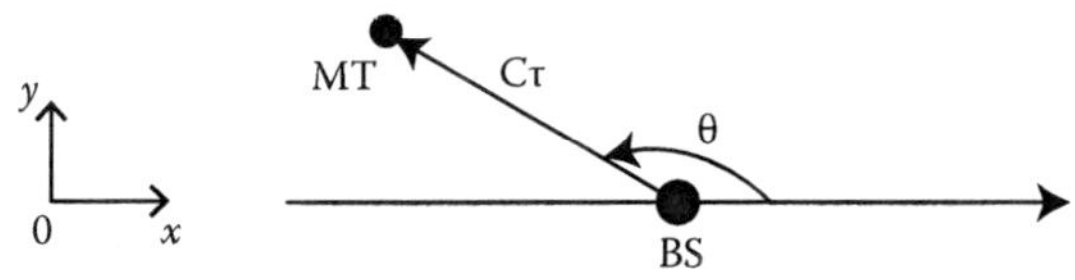

Figure 8.1 The hybrid TOA/DOA method.

position of MTs in the 2D plane is then estimated by determining the intersection of the signal paths. However, the DOA estimation is not easy to accomplish, because neither the number of signals simultaneously impinging on an antenna array nor their corresponding directions are known in advance. Furthermore, these received signals are always disturbed by noise. Therefore, sufficient antenna arrays are needed to improve the estimation accuracy. Usually, the number of antenna arrays in use should be larger than the number of received signals.

Both TOA and DOA estimation techniques require that the receivers should be able to accurately estimate the bearings or arrival times of the signals impinging on antenna arrays. Their accuracy highly depends on the propagation conditions of the wireless channels. If the line-of-sight (LOS) path between BS and MT exists, high positioning accuracy can be achieved. However, this estimation will be heavily distorted by both additive environmental noises and by multipath signals resulting from reflection or diffraction between the BSs and MT [147]. Therefore, a joint estimation method based on both TOA and DOA has been developed to deal with the aforementioned problem. This joint estimation method is known as hybrid TOA/DOA estimation, which can enhance positioning precision [295]. Hybrid TOA/DOA estimation only needs a single BS and can offer many advantages over multiple-BS estimations. Specifically, it does not require time synchronization between MTs and BSs and does not require any additional information about the environment, so it can significantly reduce the additional network overhead [147,164,284]. The basic concept of the hybrid TOA/DOA estimation is given in Figure 8.1. By assuming that the MT is in LOS with the BS, the hybrid TOA/DOA estimation determines the MT position via a geometric relationship between the MT and the BS as illustrated in Figure 8.1 (In Figure 8.1, the parameter c denotes the signal speed, which is identical to the light speed; the parameter τ denotes the time lag when the wireless signal travels from the BS to the MT; and the parameter θ is used to represent the signal direction in the given X–O–Y coordination system.).

8.2.3 Bayesian compressive sensing

The basic Bayesian compressive sensing (BCS) technique has been proposed based on the machine learning theory [47,68,193]. It provides an efficient compressive sensing framework that employs a Bayesian inference formalism for estimating underlying signals. On the basis of BCS, one need not directly measure all the dimensions of the signals of interest but rather a smaller number of dimensions. Since BCS takes into account a certain quantity of additive noise in compressive-sensing measurements and incorporates machine learning ability, it can provide a reliable estimation of the noise variance. Some BCS-based solutions for estimating the multidimensional data of interest have many appealing features, such as high computational efficiency, accuracy, and robustness to environmental noises. Additionally, BCS has been widely adopted in many other research fields, such as medical image processing, microwave imaging, and signal array synthesis.

8.3 Network Performance of DSRC and Its Effect on Traffic Capacity

8.3.1 Throughput and delay limits of 802.11p

The 802.11p parameters are defined in Table 8.1 [238]. For 802.11p, all the nodes have one data rate and control rate homogenous pair among (3, 3),(4.5, 3),(6, 6),(9, 6),(12, 12),(18, 12),(24, 12), and (27, 12).

IEEE 802.11 MAC layer employs CSMA/CA protocols called DCF and PCF. DCF defines a basic access mechanism and an optional RTS/CTS (Request-To-Send/Clear-To-Send) mechanism. This chapter only focuses on the basic access mechanism; the RTS/CTS mechanism is similar. To derive the TUL and DLL, we first need to derive two performance metrics: the achievable MT (maximum throughput) and the achievable MD (minimum delay). Because the channel exists in noisy interference, throughput is lower than MT and delay is higher than MD in the practical environment. This chapter analyzes the TUL, DLL, MD, and MT of 802.11p in ideal conditions. The assumed conditions are as follows:

1. The channel is an ideal channel without errors
2. At any transmission cycle, there are no hidden node and exposure node problems.

In the basic access mechanism, a transmission cycle includes a DIFS deferral, backoff, data transmission, short interframe space (SIFS) deferral, and acknowledgment (ACK) transmission.

The symbol notation is defined in Table 8.1; the average backoff time $\overline{CW}$ is formulated by

$$\overline{CW} = \frac{CW_{\min} T_{\text{slot}}}{2}. \tag{8.1}$$

For IEEE 802.11p, the data transmission delay $T_{\text{D\_DATA}}$ and the ACK transmission delay $T_{\text{D\_ACK}}$ are expressed as follows:

Table 8.1 Parameters of IEEE 802.11p

Parameters	*IEEE 802.11p*	*Notations*
T_{slot}	13 μs	A slot time
τ	2 μs	Propagation delay
T_{P}	32 μs	Transmission time of the physical preamble
T_{DIFS}	58 μs	DIFS time
T_{SIFS}	32 μs	SIFS time
$CW_{\min}$	15	Minimum backoff window size
T_{PHY}	64 μs	Transmission time of the PHY header
T_{SYM}	8 μs	Transmission time for a symbol
$L_{\text{H\_DATA}}$	28 bytes	MAC overhead in bytes
L_{ACK}	14 bytes	ACK size in bytes
$T_{\text{H\_DATA}}$		Transmission time of MAC overhead
T_{ACK}		ACK transmission time
L_{DATA}		Payload size in bytes
T_{DATA}		Transmission time for the payload
Mbps	3,4.5,6,9,12,18,24,27	Million bits per second
N_{DBPS}	24,36,48,72,96,144,192,216	Data bits per OFDM symbol

The function Ceiling accords to the upper integer of absolute value. For example, Ceiling $(0.5) = 1$, Ceiling $(-0.5) = -1$.

$$T_{\text{D\_DATA}} = T_{\text{P}} + T_{\text{PHY}} + T_{\text{SYM}} \times \text{Ceiling}\left(\frac{16+6+8L_{\text{H\_DATA}}+8L_{\text{DATA}}}{N_{\text{DBPS\_DATA}}}\right) \tag{8.2}$$

$$T_{\text{D\_ACK}} = T_{\text{P}} + T_{\text{PHY}} + T_{\text{SYM}} \times \text{Ceiling}\left(\frac{16+6+8L_{\text{ACK}}}{N_{\text{DBPS\_CONTROL}}}\right) \tag{8.3}$$

Packet delay is defined as the time elapsed between the transmission of a packet and its successful reception. The MT and the MD are expressed as follows:

$$\text{MT} = \frac{8L_{\text{DATA}}}{T_{\text{D\_DATA}} + T_{\text{D\_ACK}} + 2\tau + T_{\text{DIFS}} + T_{\text{SIFS}} + \overline{CW}} \tag{8.4}$$

$$\text{MT} = T_{\text{D\_DATA}} + \tau + T_{\text{DIFS}} + \overline{CW}. \tag{8.5}$$

It is easy to see that the throughput (delay) is an increasing (decreasing) function of the data rate. From Equations 8.1 through 8.5, letting the data rate go to infinite, we get the TUL and DLL based on Limit Theorem [277].

$$\text{TUL} = \frac{8L_{\text{DATA}}}{2T_{\text{P}} + 2T_{\text{PHY}} + T_{\text{DIFS}} + T_{\text{SIFS}} + \overline{CW}} \tag{8.6}$$

$$\text{DLL} = T_{\text{P}} + T_{\text{PHY}}\tau + T_{\text{DIFS}} + \overline{CW}. \tag{8.7}$$

Figure 8.2 shows that the TUL upper bounds of all the MTs for IEEE 802.11p. When the payload size is 1000 bytes, the MT for 27 Mbps is 11.2 Mbps and the TUL is 20.9 Mbps. The MT for 27,000 Mbps with the same set of overhead parameters almost reaches the TUL. Figure 8.3 shows that the DLL lower bounds of all the MDs for IEEE 802.11p. The DLL is the same for all payload sizes, i.e., 253.5 s. When the payload size is 1000 bytes, the MD for 27 Mbps is 565.5 s. The MD for 27,000 Mbps with the same set of overhead parameters almost reaches the DLL.

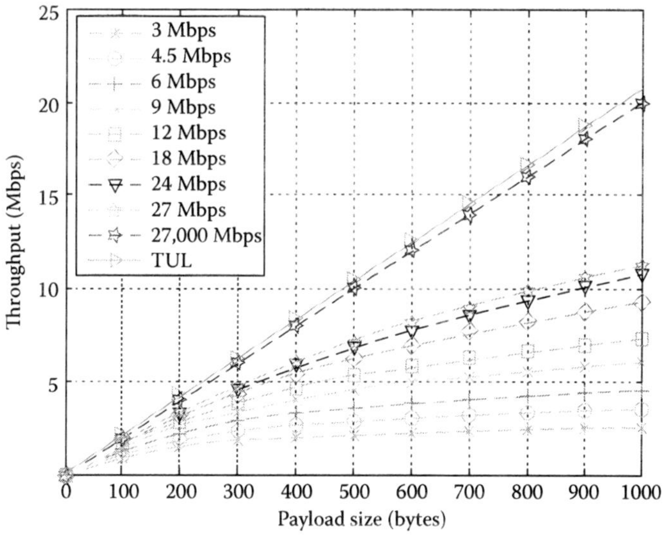

Figure 8.2 Maximum throughputs and TUL of 802.11p.

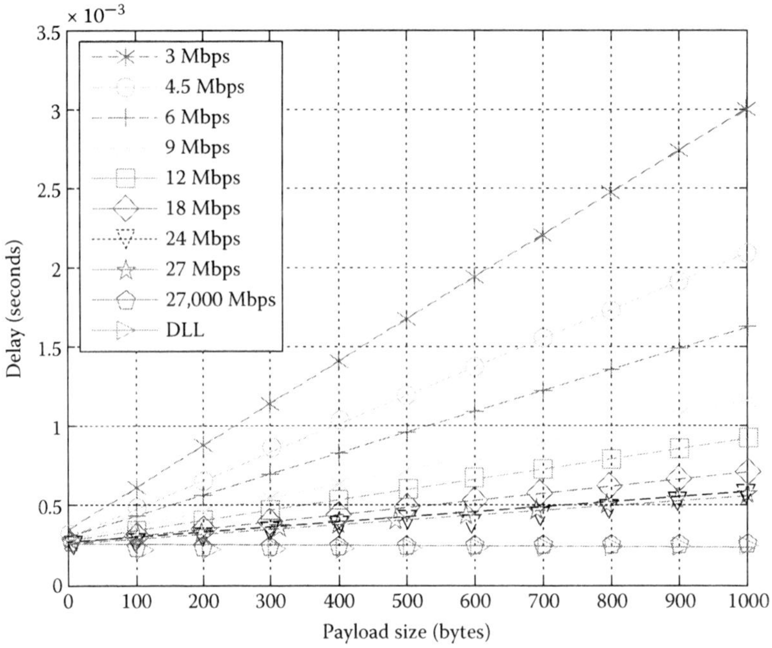

Figure 8.3 Minimum delays and DLL of 802.11p.

8.3.2 The highway capacity limits of the condition using V2V communication

We defined D_{f} as the safe following distance in meters that the vehicle maintains between it and the next the preceding vehicle.

$$D_{\mathrm{f}} = (T_{\mathrm{d}} \times V/3.6) + \left[V^2/25.92\,|a_0|\right] - \left[V^2/25.92\,|a_{\max}|\right], \tag{8.8}$$

where T_{d} is the delays from vehicles detect an emergence to the brake operation is carried out automatically. This delay consists of detection time, V2V communication time, and mechanical response time by an automobile braking system. The constant 3.6 is used to convert vehicle speed V from km/h to m/s. The terms $25.92\,|a|$ are from $2\,|a|$ times 12.96, which is the constant to convert V^2 from $(\mathrm{km/h})^2$ to $(\mathrm{m/s})^2$. The symbol a represents deceleration rates, which are in the range of $[a_{\min}, a_{\max}]$. The symbol $a0$ is the deceleration rates of a target vehicle. In this chapter, we use $a_{\min} = -5\ \mathrm{m/s^2}$ $a_{\max} = -8.5\ \mathrm{m/s^2}$.

We calculated the theoretical highway capacity under ideal conditions. We assumed the conditions as follows:

1. All vehicles on the highway are equipped with the 802.11p communication unit.
2. Vehicles will brake automatically upon receiving emergency messages.
3. All vehicles become aware of an emergency and then brake with the negotiated deceleration rates Cx.

In Figure 8.4, after vehicles receive the emergency messages, they brake at the negotiated deceleration rates Cx, which is the minimum value in absolute. We define the minimum value as ac,

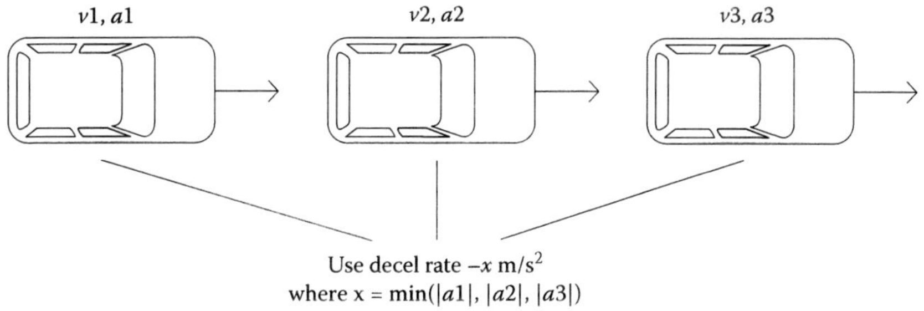

Figure 8.4 Negotiated deceleration rates used by communication vehicles.

the V2V communication time as T_c, and the detection and reaction time T_r as 0.1 s, as in [150]. When there exist some interference sources in the V2V communication environment, the occurrence of losing package in the wireless channel increases the delays of the V2V network. We define the total number of losing packages as a percentage of transmission packages as PLR and T_d is given by:

$$T_d = \frac{T_c}{(1-\text{PLR})+T_r}, \tag{8.9}$$

where PLR = [0%, 5%, 10%, and 15%]. The vehicle safety following distance for V2V communication is given by:

$$D_c = \frac{T_d \times V}{3.6}. \tag{8.10}$$

Highway capacity definition is the capacity of a facility as the maximum hourly rate at which persons or vehicles reasonably can be expected to traverse a point or a uniform section of a lane or roadway during a given time period under prevailing roadway, traffic, and control conditions [252]. From this definition, highway capacity (C) in vehicles/hour/lane can be estimated as:

$$C = 3600 \times V/\left[3.6 \times (L+\overline{D})\right] = 1000 \times V(L+\overline{D}), \tag{8.11}$$

where vehicle length (L) average is 4.3 m as in [97] and $\overline{D}$ is the average vehicle safe following distance. In this chapter, we discuss the situation where all the vehicles use V2V communication, and $\overline{D}$ was calculated by formulation 8.10 as D_c.

Figure 8.5 shows the highway capacity change at speed 120 km/h in different PLR for V2V communication vehicles. In Figure 8.5, the delay is set to DLL of 802.11p and the payload size is set to 1000 bytes. The capacity improves significantly when the vehicle speeds up. However, PLR has little impact. When the speed of all the vehicles is 120 km/h, PLR is 0%, the capacity is increased to 15703 vehicles/h/lane. Figure 8.6 shows the partial enlarged details in Figure 8.5 near speed 118 km/h. When the speed of all the vehicles is 118 km/h, PLR is 0%, the capacity is 15,555 vehicles/hour/lane and when PLR is 15%, the capacity is 15,552 vehicles/h/lane.

Figure 8.7 shows the highway capacity change at speed 120 km/h in different bitrates for V2V communication vehicles. In Figure 8.7, the payload size is set to 1000 bytes and PLR is set to 0%. The capacity improves significantly when the vehicle speeds up. However, bitrates have little impact. When the speed of all the vehicles is 120 km/h, delay is MD of 27 Mbps, the capacity is increased to 15,682 vehicles/h/lane. Figure 8.8 shows the partial enlarged details in Figure 8.7 near speed 118 km/h. When the speed of all the vehicles is 118 km/h, delay is MD

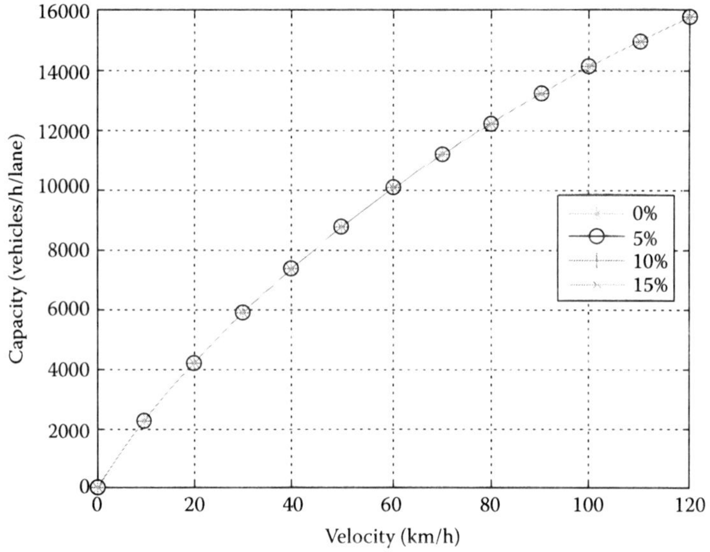

Figure 8.5 Highway capacity with varied PLRs at different speeds (delay = DLL, payload size = 1000 bytes).

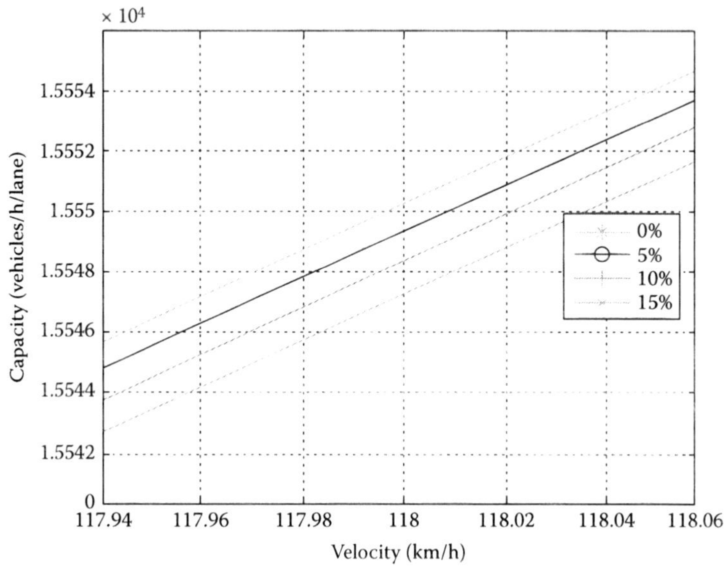

Figure 8.6 Details of highway capacity with varied PLRs at different speeds (delay = DLL, payload size = 1000 bytes).

of 27 Mbps, the capacity is 15,534 vehicles/h/lane, and delay is MD of 3 Mbps, the capacity is 15,372 vehicles/h/lane.

Figure 8.9 shows the highway capacity change at speed 120 km/h in different PLR for V2V communication vehicles. In Figure 8.5, the delay is set to 50 ms according to DSRC technology and the payload size is set to 1000 bytes [160]. The capacity improves significantly when the

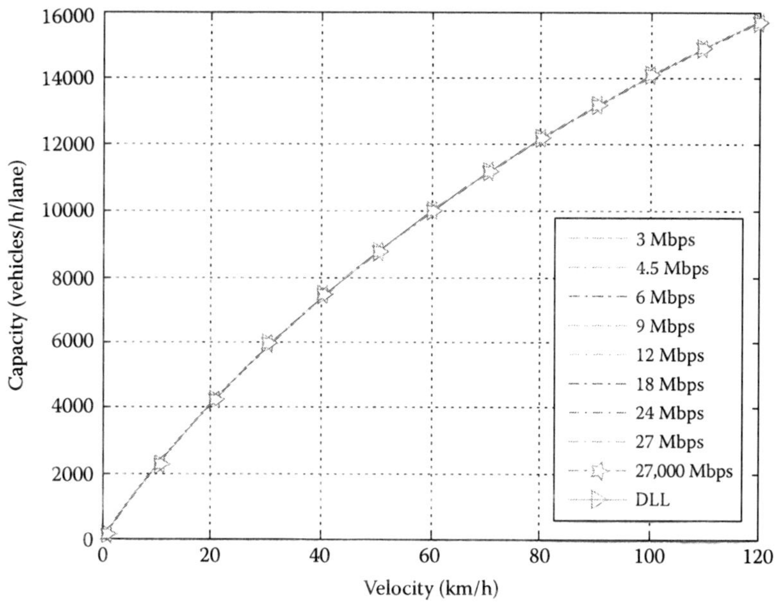

Figure 8.7 Highway capacity with varied bitrates at different speeds (payload size = 1000 bytes, PLR = 0).

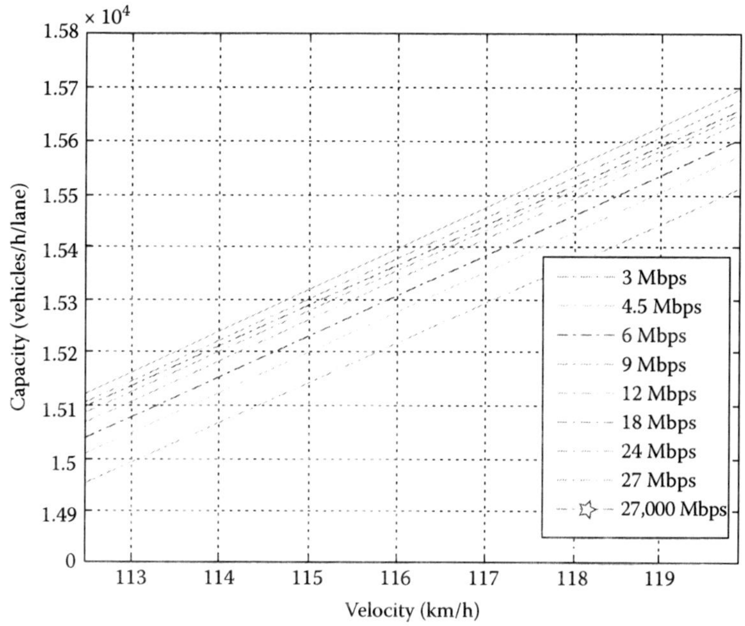

Figure 8.8 Details of highway capacity with varied bitrates at different speeds (payload size = 1000 bytes, PLR = 0).

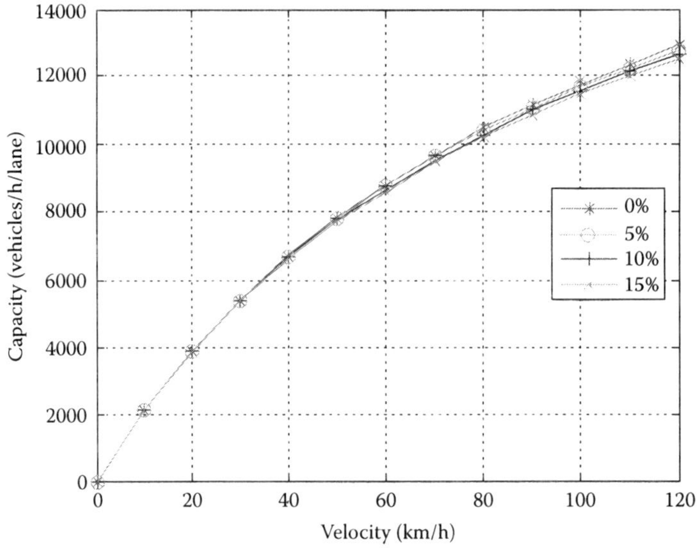

Figure 8.9 Highway capacity with varied PLRs at different speeds (delay = 50 ms, payload size = 1000 bytes).

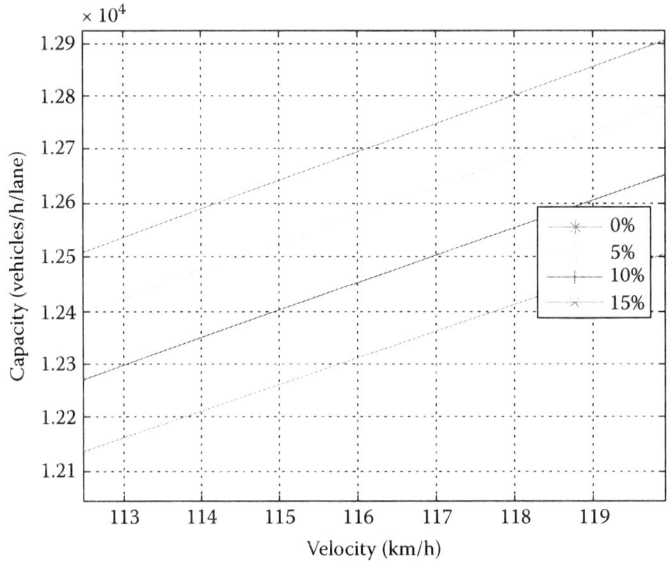

Figure 8.10 Details of highway capacity with varied PLRs at different speeds (delay = 50 ms, payload size = 1000 bytes).

vehicle speeds up. However, in a certain scope, PLR have a more obvious influence. When the speed of all the vehicles is 120 km/h, PLR is 0%, the capacity is 12,903 vehicles/h/lane. Figure 8.10 shows the partial enlarged details in Figure 8.9 near speed 118 km/h. When the speed of all the vehicles is 118 km/h, PLR is 0%, the capacity is 12,803 vehicles/h/lane, and PLR is 15%, the capacity is 12,413 vehicles/h/lane. The capacity decreases because the delay of DSRC is higher than the theoretical delay of 802.11p.

When all the vehicles are at a speed of 100 km/h, the highway capacity for all manual vehicles is 2868.98 vehicles/h/lane [175]. With the same speed and 0% PLR, the achievable highway capacity upper limit of 802.11p for V2V communication is 14,097 vehicles/h/lane, which is 4.9 times the capacity in the case of all manual vehicles.

8.4 DSRC Implementation and Field Testing

8.4.1 V2X communication system

We have developed a prototype of DSRC units, including onboard units (OBUs) and RSUs (see Figure 8.11), each OBU with one IEEE 802.11p interface (an RSU with three interfaces), one Ethernet connector, and one serial console (RSUs are same as OBUs). The operation of the computer inside the vehicle group can send User Datagram Protocol (UDP) packets via standard socket Application Programming Interfaces (APIs), such as sendto() and recvfrom() in C language format, and DSCR unit can convert UDP packets to non-Internet Protocol (non-IP)–based DSRC short messages, which can be broadcasted to anyone within the communication coverage (see Figure 8.11).

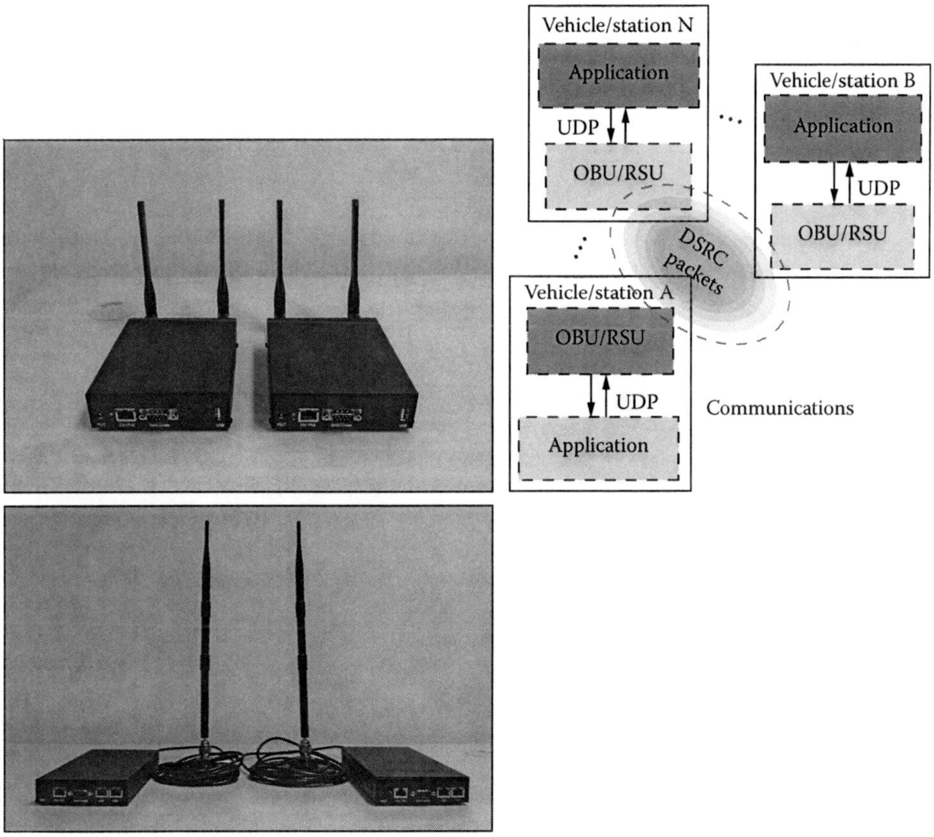

Figure 8.11 The mobility of multiple vehicles in different traffic scenarios.

8.4.2 Performance evaluation test bed

8.4.2.1 Test bed

To evaluate our developed DSRC system, a testing case was set up involving two vehicles equipped with a single DSRC OBU and laptops for data logging. The vehicles were driven along different roads in Beihang University circularly (dotted circular line in Figure 8.12) with different speeds for about 120 min. In addition, two RSUs with laptops were placed at the locations of the points A and B (as shown in Figure 8.12), respectively, which was about 1 m off the road (the lane width is about 3.5 m), and the height difference between the vehicle and the RSUs was about 0.5 m.

Both the vehicles and RSUs were broadcasting and receiving DSRC packets. The default packet size is set to 64 bytes. The QoS measurement function was deployed in both RSUs, and the testing metrics included communication range, throughput, packet loss rate, delay, and jitter (see Figure 8.12). In this experiment, the vehicles first moved in the same direction and then against each other, at different speeds, relative speeds, and distances. The maximum relative speed was 40 km/h, and the maximum distance between the two vehicles was about 256 m, as well as the maximum distance between vehicles and RSUs was about 850 m. More than 100,000 DSRC packets were transmitted or received, and the available data used for metrics measurements were more than 50,000 series in each estimation case, which provided enough data to validate the efficiency and effectiveness of our developed DSRC system.

The performance of our units is shown in Figure 8.13. In Figure 8.13, the averaged delay is centered at about 48 ms. The direction of movement and the vehicle speed does not have obvious influences on the delay. In Figure 8.13, the averaged throughput (the packet size is set to 400 bytes) is about 300 kbps. And similarly, the direction of movement and the vehicle speed do not have obvious influences on the throughput.

8.4.2.2 Performance evaluation system

As aforementioned, we have validated our developed V2X communication units with the developed performance evaluation system in the test bed. Most importantly, the developed

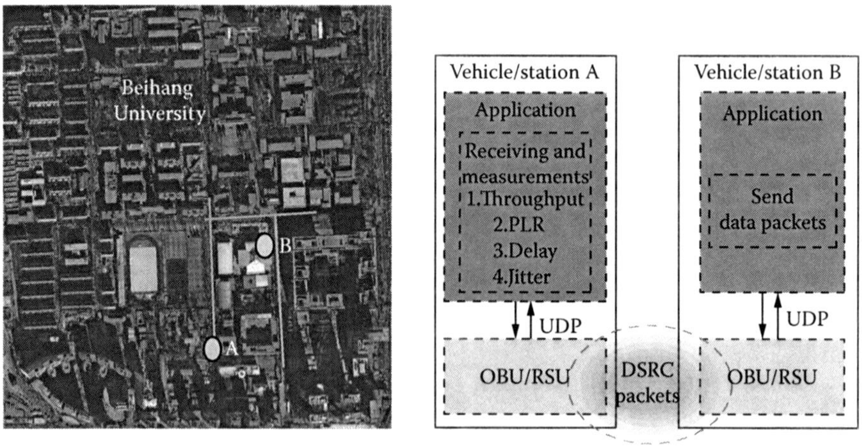

Figure 8.12 DSRC test bed and its performance evaluation system.

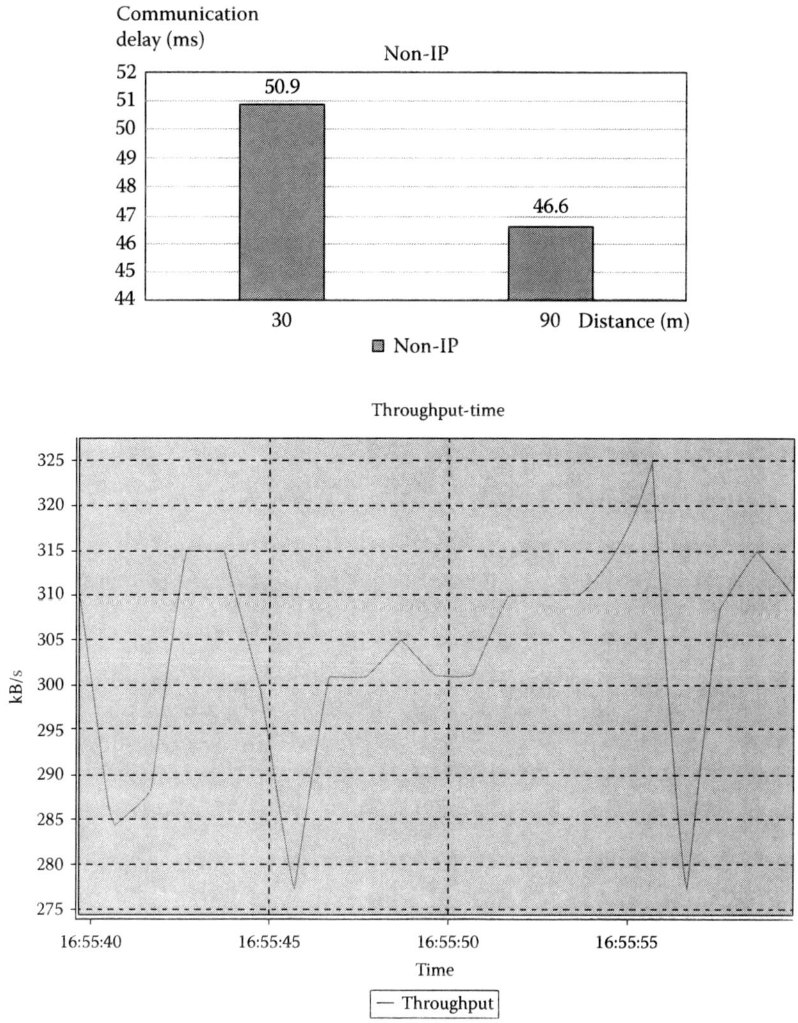

Figure 8.13 The evaluation results at different speeds and distances.

performance evaluation system can validate efficiency and effectiveness of other general DSRC-based V2X communication units.

8.4.2.2.1 Maximum range

The maximum communication range of the system, defined as the upper limit of the wireless communication distance between the DSRC packet transmitter and the receiver, can be estimated by calculating the relative positions of these two vehicles at the time instant when they do not exactly receive each others' packets.

8.4.2.2.2 Delay

The delay is measured as the time lapse from the time instant when the packet is generated by the transmitter to the time moment when it is received by the receiver. It is worth pointing out that since the packet generator and the receiver are deployed at the laptops of the upper layer

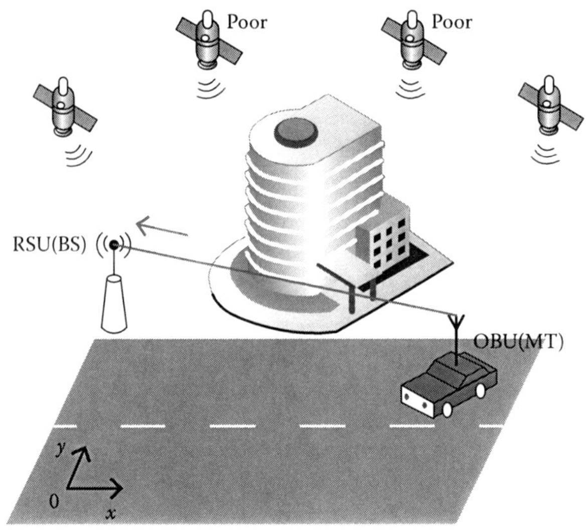

Figure 8.14 Problem definition and the initial idea for the solution.

but not at the DSRC units, the actual value of a packet delay represents the time during which a packet is generated and transmitted to the DSRC unit at the transmitting side via wired link, passed through the MAC layer and the physical layer, propagated in free space and received by the DSRC unit at the receiving side, and finally arrives at the other laptop.

8.5 DSRC for Vehicular Positioning

The problem to be solved is estimating the position of a vehicle when it passes a certain section of the road in dense urban areas where GPS signals are poor or not available. Figure 8.14 shows the situation our work mainly focuses on: one RSU is assumed to be equipped with an antenna array and deployed at the roadside. It can receive those signals that are periodically broadcasted from vehicular OBUs. Under this condition, we would like to estimate the vehicle position. Additionally, we assume that the position of the RSU is precisely known in advance, and we also set up a specific global coordination system whose X- and Y-axes are, respectively, parallel and orthogonal to the road. We consider that the RSU and the OBU are on the same 2D plane, so that the propagation path of DSRC packets between RSU and OBU is always LOS in our concern.

8.5.1 Hybrid TOA/DOA data model

We assume that a set of K DSRC signals transmitted from the vehicle to the RSU is in our LOS arriving from unknown bearings $\theta_1, \theta_2, \ldots, \theta_K$ on a linear array. The uniform linear array, mounted on the RSU for receiving the DSRC signals, is composed of M antennas arranged along the X-axis with uniform inter-element spacing d (see Figure 8.11); each element of the antenna array is isotropic in pattern. K signals are supposed to be a narrow band and characterized by the same frequency content. Accordingly, the vehicle position can be determined by computing a simple geometrical relationship (see Figure 8.15). The geometrical relationship

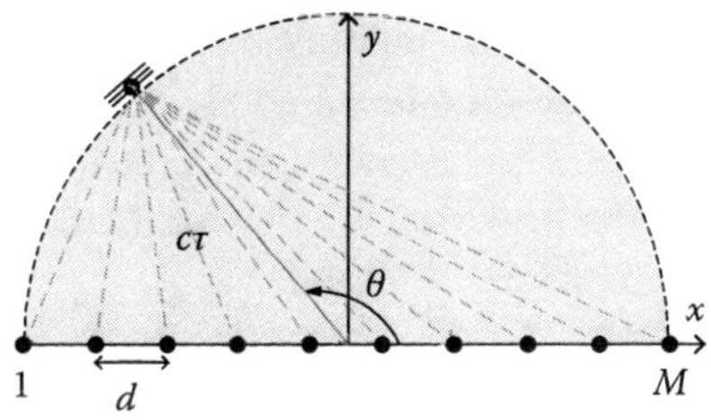

Figure 8.15 Impinging signals on uniform linear antenna array and the hybrid TOA/DOA positioning method.

can be estimated by using the TOA and DOA of the incoming DSRC signals. The model of impinged signals on an M-element antenna array be expressed as [65]

$$h(t) = A(\theta)s(t-\tau) + g(t), t = 1,2,\ldots,N, \tag{8.12}$$

where $h(t) = [h_1(t), h_2(t), \ldots, h_M(t)]^T \in C^{M1}$ is the noisy data vector received at array and T denotes the transpose operator, $A(\theta) = [a(\theta_1), a(\theta_2), \ldots, a(\theta_K)] \in C^{MK}$ is the matrix of transfer vectors, $a(\theta_k) = \left[1, e^{-j2\pi d \cos\theta_{k/\lambda}}, \ldots, e^{-j2\pi(M-1)d\cos\theta_{k/\lambda}}\right]^T \in C^{M1}$ is a so-called transfer vector between the kth signal and the received signals vector $h(t)$, d is the distance between each adjacent element of the antenna array, λ is the wavelength of DSRC signals, $s(t-\tau) \in C^{M1}$ is the vector of signals amplitudes, τ is the time delay (TOA) of the received signals transmitted from the vehicle to the RSU, $g(t) \in C^{M1}$ is an additive noise vector with zero mean and σ^2 variance, and N is the number of snapshots.

8.5.2 Multitask Bayesian compressive sensing–based DOA estimation

The signals impinging on the antenna array are intrinsically sparse in a three-dimensional space. To apply the BCS approach, the visible angular is discretized into $W \gg K$ samples (see Figure 8.14), such that $\tilde{A}\left(\tilde{\theta}\right) \in C^{M\times W}$ in (8.12). If W is large enough, we consider it is a true proposition that the DOA of the incoming signals belongs to the set of the W bearings $\tilde{\theta}_w, w = 1,2,\ldots,W$. So, the estimation problem turns out to be that of recovering the sparse signal vector $\tilde{s} \in C^{W*1}$ coincided with the W-sample of the angle scope, $\left[\tilde{\theta}_1, \tilde{\theta}_2, \ldots, \tilde{\theta}_w\right]$ (if $\hat{\theta}_w = \theta_k$ the corresponding element in $\tilde{s}$ is nonzero).

To deal with complex data using the Multitask Bayesian Compressive Sensing (MT-BCS) method, Equation 8.12 is rewritten as:

$$\begin{bmatrix} \Re\{h(t)\} \\ \Im\{h(t)\} \end{bmatrix} = \begin{bmatrix} \Re\{\tilde{A}(\tilde{\theta})\} & -\Im\{\tilde{A}(\tilde{\theta})\} \\ \Im\{\tilde{A}(\tilde{\theta})\} & \Re\{\tilde{A}(\tilde{\theta})\} \end{bmatrix} \begin{bmatrix} \Re\{\tilde{s}(t-\tau)\} \\ \Im\{\tilde{s}(t-\tau)\} \end{bmatrix} + \begin{bmatrix} \Re\{g(t)\} \\ \Im\{g(t)\} \end{bmatrix}, \tag{8.13}$$

where $\Re\{.\}$ and $\Im\{.\}$ are the real and imaginary parts, respectively. The sparse signal vector $\tilde{s}$ are given by [145]:

$$\hat{s}_{\text{MT\_BCS}} = \frac{1}{N}\sum_{t=1}^{N} \arg\left\{\left[\max_{\hat{s}(t-\tau)} \mathcal{P}\gamma([\hat{s}(t-\tau), \Gamma]\,|h(t))\right]\right\}, \tag{8.14}$$

where $s^{t-\tau}$, with $t = 1,2,\ldots,N$, are of some level of similarity, because $\tilde{s}$ at the different snapshots have a high statistical correlation [145,192]. As mentioned before, $h(t)$ is the noisy data vector received at array and N is the number of snapshots. $\mathcal{P}\gamma\{.\}$ is a probability function. $\tilde{s}_{\mathrm{MT-BCS}}$ represents the amplitude of received signals estimated by MT-BCS. Based on Ref. [68], we constructed the following equation to determine the optimal value denoted as Γ_{MT-BCS} that makes the $\ell(\Gamma)$ in the Equation 8.15 attain its maximum.

$$\Gamma_{\mathrm{MT\_BCS}} = \arg\max_{\Gamma} \left\{ \ell^{\mathrm{MT\_BCS}}(\Gamma) \right\}, \tag{8.15}$$

where $\ell^{\mathrm{MT-BCS}}(\Gamma)$ is the logarithm of the optimal value Γ, and it is computed through RVM given by Ref. [55]:

$$\ell(\Gamma) = -\frac{1}{2}\sum_{t=1}^{N} \left\{ \log(|Z|) + (W+2\phi_1)\log\left[h^T(t)(Z)h(t)+2\phi_2\right] \right\}, \tag{8.16}$$

where $Z \triangleq I + \tilde{A}(\tilde{\theta})\Lambda(\Gamma)^{-1}\tilde{A}(\theta)^T$ and ϕ_1, ϕ_2 are user-defined coefficients [192], Λ is a diagonal matrix. Subsequently, $\tilde{s}_{\mathrm{MTBCS}}$ in Equation 8.14 has been improved as follows [145] :

$$\hat{s}_{\mathrm{MT\_BCS}} = \frac{\sum_{t=1}^{N} \left\{ \left[\tilde{A}(\tilde{\theta})^T \tilde{A}(\tilde{\theta}) + \Lambda(\Gamma_{MT\_BCS})\right]^{-1} \tilde{A}(\tilde{\theta})^T h(t) \right\}}{N}, \tag{8.17}$$

where $\tilde{A}(\tilde{\theta})$ is referred to as a project matrix in BCS. It can be expressed as follows:

$$\tilde{A}(\tilde{\theta}) = \begin{bmatrix} \Re\{\tilde{A}(\tilde{\theta})\} & -\Im\{\tilde{A}(\tilde{\theta})\} \\ \Im\{\tilde{A}(\tilde{\theta})\} & \Re\{\tilde{A}(\tilde{\theta})\} \end{bmatrix}. \tag{8.18}$$

8.5.3 Experiments

The objectives of this section are twofold: On the one hand, it provides guidelines for applying the MT-BCS method to location estimation based on hybrid TOA/DOA. On the other hand, it accesses the method's effectiveness in both improving location accuracy and robustness. As for the MT-BCS, the user-defined parameters ϕ_1, ϕ_2 are assigned as in Ref. [68].

8.5.3.1 Hybrid TOA/DOA location procedure

In a time-synchronized network, the TOA estimator $\tilde{\tau}$ is determined by the time delays from targets to RSUs. In this chapter, we consider the time estimation deviation $\Delta\tau = \tau - \tilde{\tau}$ according to norm distribution, and its standard deviation σ_{TOA} satisfies the condition $c\sigma_{TOA} = 0.01\,\mathrm{km}$ [33] and the DOA estimator $\tilde{\theta}$ is estimated through MT-BCS.

Indeed, in the DOA estimation procedure, thanks to the sparse feature, the estimated number $\tilde{K}$ of incoming signals in three-dimensional space can be confirmed simply by counting nonzero elements of the reconstruction vector of signal amplitudes $\tilde{s}(t-\tau)$. Nevertheless, in $\tilde{s}(t-\tau)$, many elements are close but not equal to zero that is generated by noise, which does not correspond to any true signals. In particular, the true signals correspond to a few of the largest elements with high probability. So, an energy threshold has been identified to improve the credibility of the DOAs estimation. Based on energy content, the elements of $\tilde{s}(t-\tau)$ are listed in descending order as α_i, $i = 1, W$, where $\alpha_1 = arg[\max_{\alpha_i} \tilde{s}(t-\tau)], \alpha_W = arg[\min_{\alpha_i} s(t-\tau)]$. We set the energy threshold η meeting the comparison condition as follows

$$\sum_{i=1}^{\tilde{K}} \alpha_i^2 < \eta \cdot \sum_{i=1}^{W} \alpha_i^2. \tag{8.19}$$

The process for estimating DOAs is shown in Procedure 1 in Table 8.2.

Table 8.2 Procedure 1: The DOAs, $\hat{\theta}^{\mathbf{i}}$, Estimation

Step 1	$\tilde{K} = 1$
Step 2	$If \sum_{i=1}^{\tilde{K}} \alpha_i^2 < \eta \cdot \sum_{i=1}^{W} \alpha_i^2, \hat{\theta}_i = \hat{\theta}_w$
	$i = i + 1$
	else stop
Step 3	$\tilde{K} = \tilde{K} + 1$, *go to Step* 2.

Next, we define the actual target's position as $\Phi_t := (x_t, y_t)$, the RSU position as $\Phi_0 := (x_0, y_0)$, and the estimated target's position as $\tilde{\Phi}_t := (\hat{x}_t, \hat{y}_t)$, which are yielded in the estimated TOA and DOA values $\hat{\tau}$ and $\hat{\theta}$ as

$$(\hat{x}_t, \hat{y}_t) : \begin{cases} \hat{x}_t = x_0 + c\hat{\tau}\cos\hat{\theta} \\ \hat{y}_t = y_0 + c\hat{\tau}\cos\hat{\theta} \end{cases} \tag{8.20}$$

8.5.3.2 Energy threshold analysis

To determine the optimal energy threshold $\hat{\eta}^{\text{opt}}$, the following benchmark test case has been considered in Table 8.3. The minimum angular distance between the DOAs of two adjacent incoming signals has been identified as $\Delta\theta_{\min} = 1°$, while the angular scope has been uniformly discretized into $W = 181$ samples (Figure 8.16). The SNR of measured data is defined (Equation 8.21):

$$\text{SNR} = 10\log\left(\frac{\| h^{\text{Noiseless}}(t) \|_2^2}{M\sigma^2}\right), \tag{8.21}$$

where $h^{\text{Noiseless}}(t)$ is the noise-free signals received at RSU, $\|\cdot\|_2$ is a ℓ_2 norm operator. Since the DOAs are randomly chosen, different scenarios Q = 250 have been considered for the condition that $M = 10$, $N = 10$, $K = 4$, and SNR $\in [0 : 5 : 20]$ to give a consistency statistics

Table 8.3 Parameters Setting

Incoming Signals		*RSU's Antenna Array*	
Center frequency	5.890 GHz	Frequency range	5.850–5.925 GHz
Modulation modes	BPSK	Radiation pattern	Isotropic
Angles scope ($\hat{\theta}_w$)	$[0°, 180°]$	Element spacing (d)	$\lambda/2$
Bandwidth (W_B)	10 MHz		

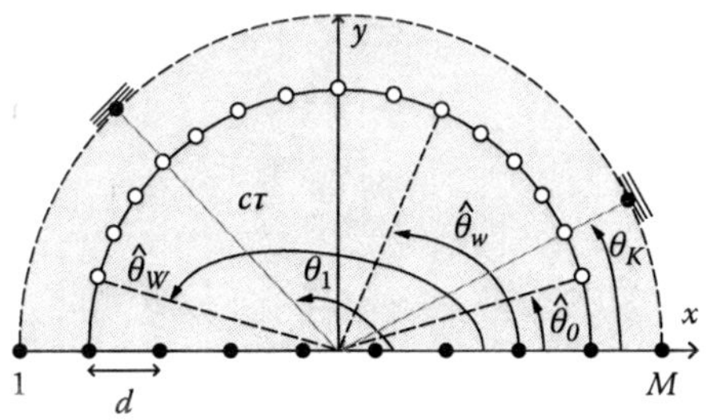

Figure 8.16 Angular scope discretization.

validation. The value scopes of η are defined as [0:0.0001:1]. The reason why the snapshot N is set at 10 times is that the processing time of those values is about 100 ms, which may achieve a higher positioning frequency than 10 Hz for real-time vehicular positioning [265].

In MT-BCS–based estimation, the optimal threshold has been identified by the values $\hat{\eta}^{\text{opt}}$ (Algorithm 8.4) that minimize the modified mean relative error (MRE), which is defined in Equation 8.22. The values of the MRE averaged over Q simulations for each scenario at hand have been assumed to be a reliable indicator for finding the optimal value of $\hat{\eta}^{\text{opt}}$, which occurs at the minimum values of MRE. The optimal value of $\hat{\eta}^{\text{opt}}$ different SNRs conditions are presented in Table 8.4. More specifically, the MRE has been computed without prior knowledge of the number of incident signals, K. The data in Table 8.4 show that the first K signals tend to have higher strength when the SNRs increase, because the selected optimal thresholds $\hat{\eta}^{\text{opt}}$ in high SNR conditions are higher than those in low SNR conditions.

$$\text{MRE}|(ts,M,N,\text{SNR},K) = \frac{1}{Q}\sum\nolimits_{Q}\left|\frac{\left[\tilde{K}(q)|(ts,M,N,\text{SNR},K)\right] - K}{K}\right| \tag{8.22}$$

Algorithm 8.4 find $\eta^{opt} \in$ *different variables* M,N,SNR

Ensure: $\tilde{A}(\tilde{\theta}), SNR = [2,5,10,15,20], K = 4$
Require: $\hat{s}_{MTBCS}, \hat{\eta}^{opt}$

Get initial $h(t)$
for $q = q_{begin}$ to Q **do**
  $q_{begin} = 1$
  $\hat{s}_{MT-BCS}(q) = f_{MT-BCS}(\tilde{A}(\tilde{\theta}), h(t))$
  Listed the elements of $\hat{s}_{MT-BCS}(q)$ in descending order as α_i
  for $ts = ts_{begin}$ to TS **do**
    $\tilde{K}$=1
    $ts_{begin} = 0.0001$
    while $\sum_{i=1}^{\tilde{K}} \alpha_i^2 < ts * \sum_{i=1}^{W} \alpha_i^2$ **do**
      $\tilde{K}++$
    end while
    break
    return $\tilde{K}(q_{begin})|(ts_{begin}, M, N, SNR, K)$
    $ts_{begin++}$
  end for
  return $\tilde{K}(q_{begin})|(ts_{begin}, M, N, SNR, K), ts = 0.0001, ..., TS$
  $q_{begin++}$
end for
return $\tilde{K}(q)|(ts_{begin}, M, N, SNR, K), ts = 0.0001, ..., TS, q = 1, ..., Q$
$\hat{\eta}^{opt}|(M,N,SNR,K) = arg\{\min_{ts}[MRE|(ts,M,N,SNR,K)]\}, q = 1, ..., Q,$
$ts = 0.0001, ..., TS$

Table 8.4 $\hat{\eta}^{\text{opt}}$ in Different SNRs (M = 10, N = 10, K = 4)

SNR(dB)	2	5	10	15	20
$\hat{\eta}^{\text{opt}}$	0.8963	0.9115	0.9541	0.9814	0.9941

To draw more general conclusions on the behavior of the proposed method, further experiments have been carried out varying the element number of the array, M, the snapshot number, N, and the SNR. The values of P_K averaged over $Q = 250$ simulations for each scenario. Similar to MRE behavior, the P_K improves when the element number of the array, M, the snapshot number, N, and the value of SNRs increases.

$$P_K = \frac{1}{Q}\sum_Q P_k^{(0)}\left(\hat{\eta}^{opt}|(M,N,\text{SNR},K)\right) \tag{8.23}$$

$$P_K = \begin{cases} 1, & \text{if } \tilde{K} = \text{K} \\ 0, & \text{otherwise} \end{cases} q = 1,\ldots,Q. \tag{8.24}$$

With reference to the representative test case with $K = 2$, $M[5:5:25]$, $N[10:10:100]$, and SNR$[0:5:20]$ (Figure 8.17), Figure 8.17 shows that, as expected, increasing the element number of the array, M, and the snapshot number, N, drastically improves the estimation performance of the number of incident signals. Similar conclusions hold true for the analyses whose results are shown in Figure 8.17. Such simulation results effectively state that a key feature of the MT-BCS method, that is, its high reliability even when no a priori information on the scenario is available.

8.5.3.3 Performance analyzing

8.5.3.3.1 Hybrid TOA/DOA MT-BCS–based location method

The hybrid TOA/DOA, MT-BCS–based location method has been used in ten consecutive time instant (snapshots). Thanks to the high statistic correlation, intermultiple snapshots [145,192], the bulk of added noise could be eliminated through multimeasurement on incoming signals. Figure 8.18 shows the estimate of DOAs based on MT-BCS, where there is no prior knowledge within each scenario of signal numbers K. The value of $\hat{\eta}^{\text{opt}}$ is given in Table 8.4, while the corresponding experiments' results showing the MSE values are given in Table 8.5. For illustrative purposes, the "+" symbols and the gray dots indicate the actual DOAs and those estimated after optimal energy filtering, respectively. As expected, the estimation accuracy is inversely to K/M and is lowered on the condition that the adjacent bearings in space are too similar or too close to X-axis. On the other hand, the estimation is accurate in the directions of signals far from X-axis. As the ratio K/M is near one-to-ione, the MSE is too large to retrieve OR identify vehicular positions.

As shown in Table 8.6, dealing with vehicles number K = 4, the optimal energy threshold is more robust for either a highly similar or highly discriminated DOAs collection (see Figure 8.19).

Table 8.5 DOA Estimation Based on MT-BCS and MSE Values

$SNR = 10\,dB$, $\hat{\eta}^{\text{opt}} = 0.9541$			
K	$\theta(°)$	$\hat{\theta}(°)$	MSE
2	37, 115	37, 115	0.00
4	51, 75, 147, 165	51, 75, 149, 164	0.75
6	12, 73, 97, 108, 122, 155	7, 73, 96, 97, 107, 153	5.67
8	21, 25, 58, 81, 119, 120, 151, 167	16, 24, 58, 59, 81, 119, 120, 151	14.25

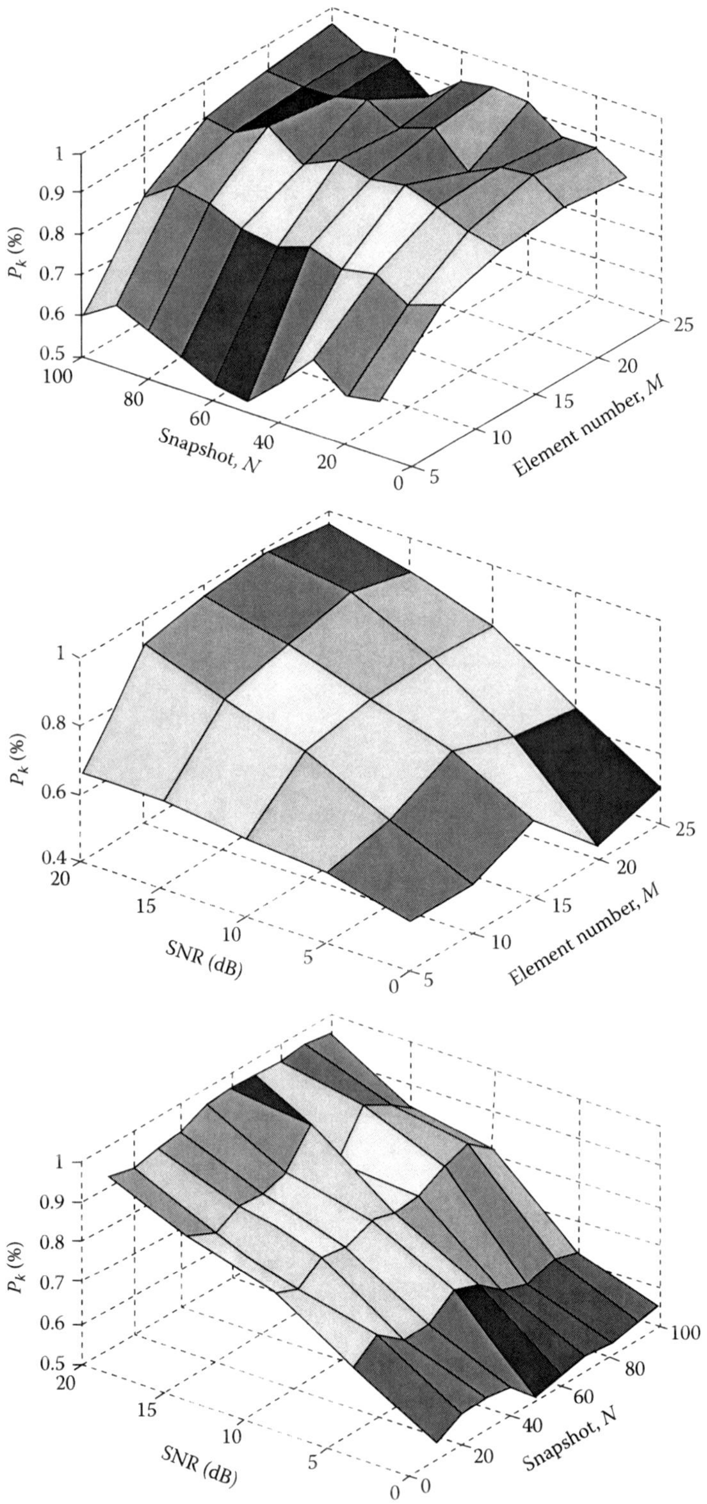

Figure 8.17 Plots of the number of received signals ($d = /2$; $K = 2$): (a) $M \in [5:5:25]$, $N \in [10:10:100]$; (b) $M \in [5:5:25]$, SNR $\in [0:5:20]$ dB; (c) $N \in [10:10:100]$, SNR $\in [0:5:20]$ dB.

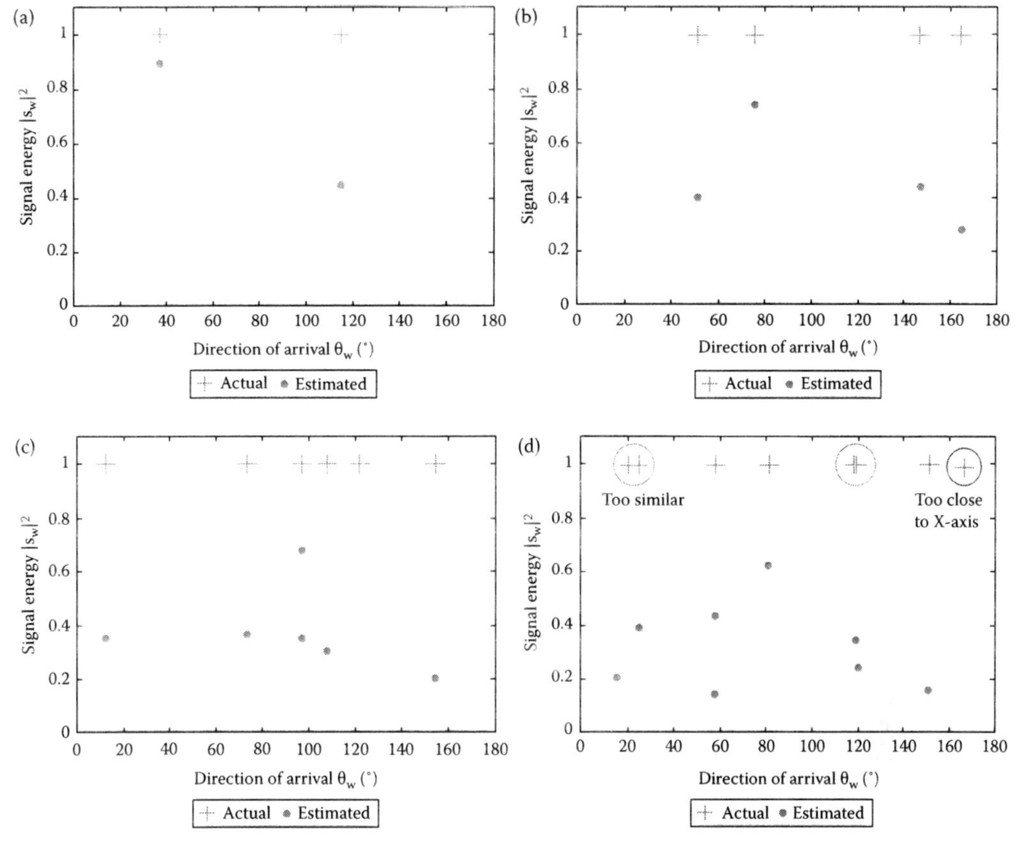

Figure 8.18 MT-BCS–based DOAs.

Table 8.6 DOA Estimation Based on MT-BCS ($K = 4$) and MSE Values

$M = 10$, $N = 10$, $SNR = 10dB$, $\eta^{opt} = 0.9541$, $K = 4$		
$\theta(°)$	$\hat{\theta}(°)$	MSE
45, 60, 62, 63	346, 60, 61, 62	0.00
92, 95, 126, 130	93, 94, 124, 130	1.00
51, 73, 118, 15060, 75, 97, 118	51, 73, 118, 150	0.00
60, 75, 97, 118	60, 75, 97, 118	0.00

The root mean square error in each scenario, $\text{RMSE}^{(q)}$, is an indicator of the reliability of the method at predicting the q-th scenario. The latter considers both the errors in estimating the number of incident signals, $\tilde{K}^{(q)}$ and the responding DOAs $\widehat{\theta^{(q)}}_k, k = 1, \ldots, \tilde{K}^{(q)}$.

$$\bar{\theta}_j^{(q)} = \arg\left\{ \min_{\theta_K, k\in[1,K]} \left| \theta_K - \hat{\theta}^{(q)} \right| \right\} \tag{8.25}$$

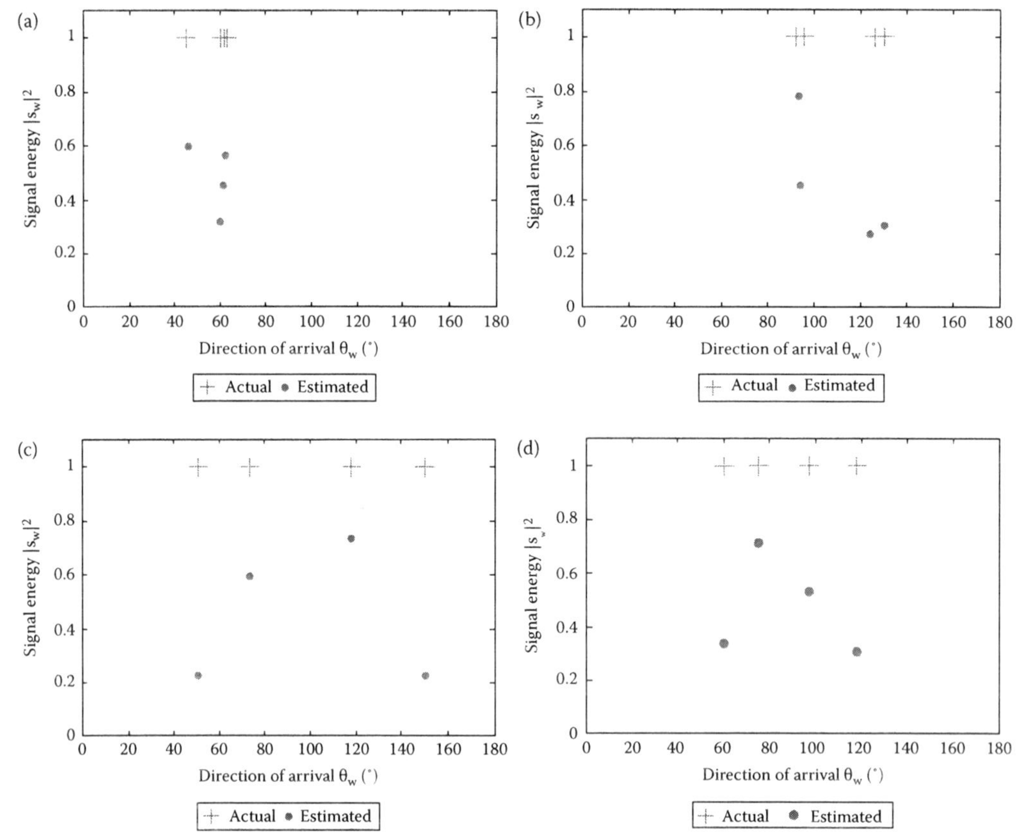

Figure 8.19 Actual and estimated DOAs.

$$\mathrm{RMSE}^{(q)} = \begin{cases} \sqrt{\dfrac{\left\{\sum_{k=1}^{\tilde{K}^{(q)}} |\theta_k - \hat{\theta}_k^{(q)}|^2 + |K - \tilde{K}^{(q)}|(\Delta\theta_{max})^2\right\}}{K}}, & \text{if} \quad \tilde{K}^{(q)} \leq K \\ \sqrt{\dfrac{\left\{\sum_{k=1}^{\tilde{K}^{(q)}} |\theta_k - \hat{\theta}_k^{(q)}|^2 + \sum_{j=K+1}^{\tilde{K}^{(q)}} |\tilde{\theta}^{(q)} - \tilde{\theta}_j^{(q)}|^2\right\}}{K}}, & \text{if} \quad \tilde{K}^{(q)} > K \end{cases} \tag{8.26}$$

$$\overline{\mathrm{RMES}} = \frac{1}{Q}\sum\nolimits_Q \mathrm{RMSE}^{(q)} \tag{8.27}$$

The number of signals is defined in Equation 8.26, where θ_k and $\theta^{\widehat{(q)}}{}_k$ are the kth actual and closest estimated DOAs (the values of θ_k and $\theta^{\widehat{(q)}}{}_k$ minimize the values of $\left|\theta_k - \theta^{\widehat{(q)}}{}_k\right|$ among the $\tilde{K}^{(q)}$ estimates), while $\Delta\theta_{\max}$ is a penalty term equal to the maximum admissible localization error ($\Delta\theta_{\max} = 180$) and used only in the situation where the estimated number of incident signals is smaller than the actual signals. $\overline{\theta_j}^{(q)}$ defined in Equation 8.25 is applied when the estimated number of incident signals is bigger than the actual one.

The numerical analysis is concerned with a comparative assessment of MT-BCS and state-of-the-art approaches such as Root-MUSIC [184] and ESPRIT [188]. Figure 8.20 plots the $\overline{RMSE}$ 8.27 averaged over $Q = 250$ simulations for each scenario without any prior knowledge of K and yielded by the MT-BCS, and the two reference methods as a function of SNR, varying from 0 to 20 (step in this case is 5). Because there is a penalty term $\Delta\theta_{\max}$ in 8.26, the simulation

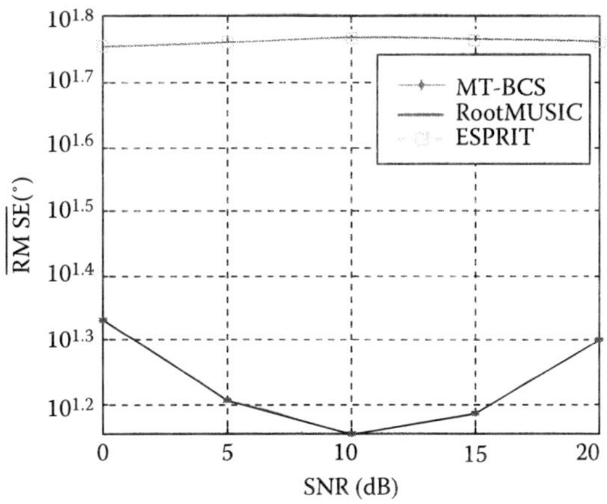

Figure 8.20 Multiple snapshots DOA estimation (M = 10; N =10; d = /2; K = 2) Plots of $\overline{RMSE}$ when SNR[0:5:20].

results do not achieve a better performance when the SNR increases. Moreover, the MT-BCS performs better than Root-MUSIC and ESPRIT under the same SNR condition.

The hybrid vehicular location method has been tested at an urban intersection where the street width is 40 m and the RSU is located at the center [0, 0] of the [−100,100][−100,100] positioning coverage region. Considering the wireless propagation path from vehicles to the RSU in a LOS environment, the antenna array mounted on the RSU should be at the same horizontal level as the antenna mounted on vehicles.

8.5.3.3.2 Cramer-Rao bound for hybrid TOA/DOA-based location estimation

By denoting with the location estimator $\Phi_t := (\hat{x}_t, \hat{y}_t)$ of targets (vehicles), the result location RMSE is

$$\text{RMES} = \sqrt{E\left\{\left\|\hat{\Phi}_t - \Phi_t\right\|_2^2\right\}} = \sqrt{E\left\{(\hat{x}_t - x_t)^2 + (\hat{y}_t - y_t)^2\right\}} \tag{8.28}$$

Thus, its lower bound $\text{RMSE}_{\Phi_{t,\min}}$ is given by Ref. [65]

$$\text{RMES}_{\Phi_t,\min} = \sqrt{c^2\tau^2\text{CRB}_\theta + c^2\text{CRB}_\tau}, \tag{8.29}$$

where CRB_τ and CRB_θ is

$$\begin{cases} \text{CRB}_\tau = \frac{1}{2M\upsilon W_B^2} \\ \text{CRE}_\theta = \frac{3\lambda^2}{4\pi^2 d^2 \sin^2\theta (M-1)M(2M-1)} \end{cases} \tag{8.30}$$

υ denotes SNR 8.21. Apparently, $\text{RMSE}_{\Phi_{t,\min}}$ is dominantly dependent on the signal bandwidth WB and array configuration parameters. Figure 8.21 depicts the minimum RMSE of location estimates by the hybrid TOD/DOA-based approaches. The white dot at $\Phi_0 := (0,0)$ in Figure 8.21 denotes the position where RSU is located.

Obviously, the position estimation error is closely related to the vehicular position. The estimated RMSE worsens when the DOAs at RSU are close to 0 or π. The DOA estimation performance is better when the vehicles are within the region close to $\pi/2$ (relative to the direction of the positive X-axis) or the vehicles are near the RSU (the values of $c\tau$ are small).

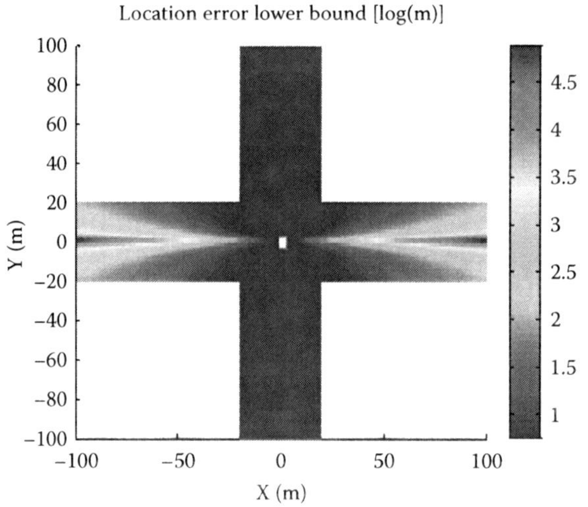

Figure 8.21 RMSE of hybrid TOA/DOA-based location estimation.

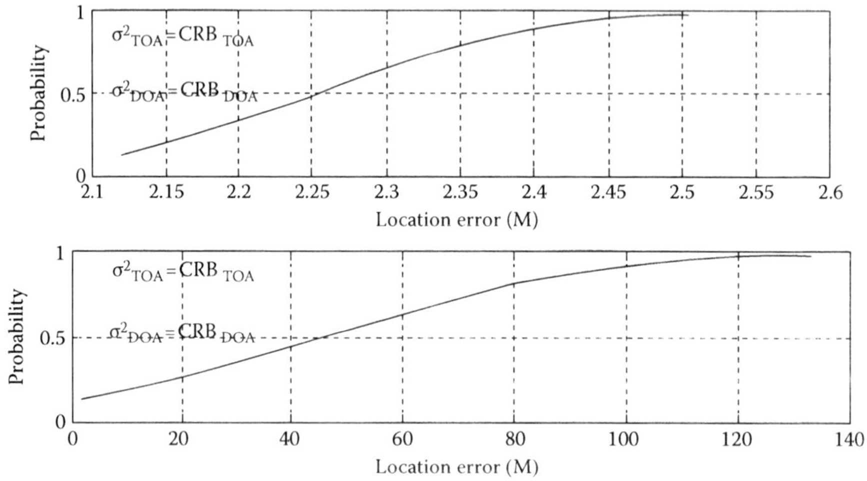

Figure 8.22 CDF of the location error (CRB for TOA/DOA of one single RSU).

The cumulative distribution functions (CDFs) of the location error related to the TOA/DOA in orthogonal directions are plotted and compared in Figure 8.22. As expected, when the vehicles move along the Y-axis $(\pi/2)$, the proposed estimation method based on hybrid TOA/DOA achieves better performance in terms of location accuracy.

8.5.3.3.3 Simulation for real-time vehicular positioning

We additionally simulate two cars moving along the vertical road shown in Figure 8.20 and use those three methods to estimate the real-time position of cars. The speeds of both cars are fixed at 60 km/h and their travel distance is set to be 160 m. Then we calculate the absolute position error (it is worth pointing out here that the absolute position error can be computed by using Equation 8.28 between the actual position data and the estimated outcomes obtained by our proposed method (marked as MT-BCS) and by the other two methods (marked RootMUSIC and ESPRIT). These results are illustrated in Figure 8.23. According to these results, it can be see that the absolute position errors related to both cars stochastically fluctuate. This is because there exists a certain simulated noise in the estimation. In the simulation, we assume that the distances between the cars and the RSU are contaminated with Gaussian white noise during the process of different estimations. The standard deviation of the simulated noise is set to be

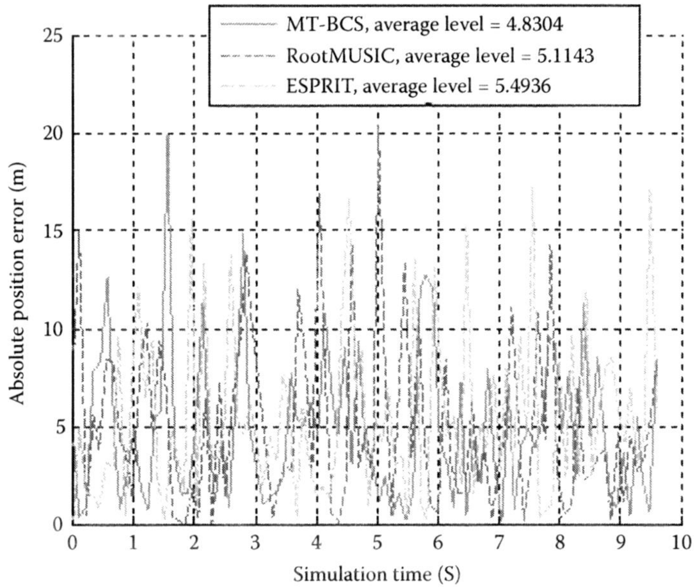

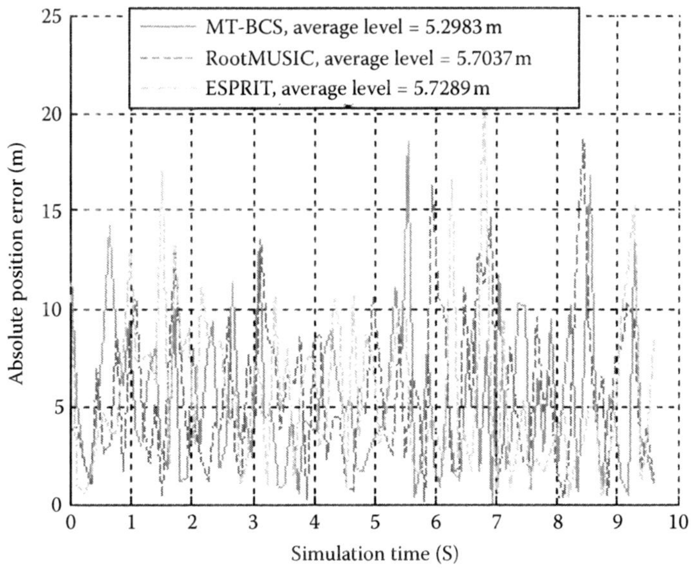

Figure 8.23 The absolute position error related to the two simulated cars, obtained by comparison methods.

10 m. On the other hand, our proposed method can achieve a lower absolute position error for both cars on average when compared to the other two methods. These results strengthen the advantage of our proposed method.

8.6 Conclusion

In this chapter, we analyzed the 802.11p limit and achievable value of throughput and delay. According to delay, we calculated the limit and achievable value of highway capacity. The results show that V2V communication can significantly improve highway capacity. In a certain scope, PLR has little impact on high capacity. Vehicle speed is the primary factor to increase highway capacity. Also, we propose a new vehicular location method based on hybrid TOA/DOA estimation. Beginning with a hybrid data model and an MT-BCS estimation method, this work analyzes the amplitudes of received signals in 3D spaces at an antenna array based on the BCS algorithm, so as to estimate TOA and DOA information. By computing a simple geometrical relationship between vehicles and the RSU, vehicular locations can be determined via combining both TOA and DOA estimators provided by the RSU. Cramer-Rao bound of vehicular location measurement has been analyzed, and the performance of the aforementioned position estimation method is illustrated through extensive experiments. The simulation results show several good features of the proposed method, which include: (1) an accurate and reliable target detection rate and DOA estimation without any prior knowledge on the number of incoming signals in 3D space; (2) a robust and accurate estimation of multisnapshot observation in different SNR conditions; and (3) low time complexity without computing the correlation matrix and its eigenvalue decomposition. In addition, the comparative simulation also demonstrates that our proposed method outperforms the two traditional approaches: Root-MUSIC and ESPRIT.

Chapter 9

Vehicular Positioning Enhancement

Yunpeng Wang, Xuting Duan, Daxin Tian, and Xuejun Zhang
Beihang University

Min Chen
Huazhong University of Science and Technology

CONTENTS

9.1 Introduction 160
9.2 A DSRC-Based Vehicular Positioning Enhancement Model Using a Mulitple-Model KF 161
9.2.1 Vehicular positioning enhancement model 161
9.2.2 Connected vehicles–enhanced IMM-KF for vehicular positioning 163
9.2.2.1 CV-IMM-KF for estimating the TVs state 164
9.2.3 Numerical study 166
9.2.3.1 Simulation scenario 166
9.2.3.2 Simulation results 167
9.3 A DSRC-Based Vehicular Positioning Enhancement Model Using a Distributed Multiple-Model KF 168
9.3.1 System model and localization enhancement 168
9.3.1.1 System model 169
9.3.1.2 The CV-enhanced DIMM-KF for mobile vehicle localization 172
9.3.2 General performance analysis 174
9.3.2.1 CRLB 175
9.3.2.2 SPEB 175
9.3.2.3 EFIM 176
9.3.2.4 The mSPEB 176
9.3.2.5 Insights into factors affecting the CP performance 179
9.3.3 Numerical results 180

9.3.3.1 Simulation setup ... 180
9.3.3.2 Single-trial analysis ... 181
9.3.3.3 Monte Carlo results ... 181
9.4 Conclusion ... 185

9.1 Introduction

The availability of high-accuracy location awareness is essential for a diverse set of vehicular applications including intelligent transportation systems, location-based services (LBS), navigation, as well as a couple of emerging cooperative vehicle infrastructure systems (CVIS) [6]. Typically, as an important technique, the real-time vehicle positioning system has drawn great attention in the fields of transportation and mobile communications [186]. However, it still faces a big challenge in areas with inconsistent availability of satellite networks, especially in dense urban areas where the stand-alone global navigation satellite systems (GNSSs) (e.g., global positioning system [GPS]) cannot work well. Even though a set of high-precision location equipment (e.g., differential global positioning system (DGPS)) is deployed, the positioning performance is adversely impacted in non-line-of-sight (NLOS) (e.g., buildings, walls, trees, vehicles, and more obstructions) scenarios, or by the severe multipath effect in urban canyon environments [284].

In vehicle ad hoc networks (VANETs), it is expected that all vehicles with wireless communication capability will be able to accurately sense each other and contribute to vehicular collision avoidance, lane departure warning, and intersection safety enhancements [204,212,294]. Apart from GPS, a lot of emerging location systems relying on the spatial radio frequency, such as wireless communication signals (e.g., Wi-Fi, cellular, radio frequency identification) or inertial navigation system (INS), are implemented [7,40,154,245]. In Refs. [10,41,147,186,276,279], the fundamental techniques in positioning systems have been presented based on the real-time measurements of time of arrival (TOA), time difference of arrival (TDOA), direction of arrival (DOA), received signal strength (RSS), Doppler frequency shift (DFS), fingerprinting, and wireless channel state information (CSI) techniques. Especially, the cloud-based wireless network proposed in Ref. [36] is expected to provide flexible virtualized network functions for vehicular positioning. Recent researches indicate that these measurements are challenged by some drawbacks varying from complexities of time synchronization, occupations of high bandwidth, to huge costs on the implementations [284]. Although there already exist some location systems, such as those presented in Refs. [10,12], which can achieve lane-level location performance, these systems require the accurate detection on unique driving events through smart phones or the deployment of lane anchors. So they dramatically depend on the accuracy in real-time event data provided by smart phones, social network, and the roadside anchors [35,287,288].

To resolve these drawbacks, a new class of vehicular cooperative positioning (CP) methods has been presented in recent years [147,219]. On the basis of vehicle-to-vehicle (V2V) and vehicle-to-infrastructure (V2I) communications and data fusion technologies [7,200,227,257], CP is able to further enhance the accuracy and the precision performance of the vehicle localization systems. Dedicated short-range communication (DSRC), with a bandwidth of 75 MHz at the 5.9-GHz band, is designed for wireless access in vehicular environments (WAVEs) to ensure a maximum communication range up to 1000 m under line-of-sight (LOS) conditions, or up to 300 m in high-mobility environments, and to provide the capacity of 50 millisecond-delays on the end-to-end communication and a data rate from 3 to 27 Mb/s [249]. Due to the aforementioned properties, DSRC has become an attractive technology for the connected vehicle (CV) applications that aim to establish an interconnected system among intelligent vehicles, and make incremental improvements in traffic safety, transport efficiency, and environmental contaminants. To set up the fundamental framework on the cooperative localization systems, insightful explorations have been presented in Refs. [257,281] from the fundamental theories to the real-world applications, including the theoretical limits, the optimized algorithms, and the advanced technologies. Specifically, the field-testing researches indicate that some DSRC-based CP techniques achieving lane-level accuracy can profoundly benefit many applications related to traffic safety [10,227,293].

In this chapter, we present a framework of DSRC-based enhancement for mobile vehicle localization using the DSRC physical layer data and the coarse position and velocity data provided by the commodity GPS. The enhancement is achieved by sharing and combining multilateral information of local vehicles through DSRC. The main contributions of this chapter are summarized as follows:

- A motion state of each vehicle is represented by its real-time position and velocity. Using the first-order Taylor series approximation, we have developed a linearized system model to formulate the relationship between the real-time vehicular motion state and the physical layer measurements, including the DFS and the RSSI and obtained a transition matrix that reveals the benefit of information interaction among local vehicles into cooperative localization enhancements.
- With the linearized system model aforementioned, we have further proposed an interactive multiple-model (IMM)–based Kalman filter (KF) that is used to track the covariances of the DFS measurements under different situations. In addition, we have proposed a distributed interactive multiple-model (DIMM) KF as well, which can be applied to track variations of acceleration of vehicles and the covariances of the DFS and the RSSI measurements under different situations. The KF is used to achieve local information fusion among vehicles in an online distributed manner, such that it can enhance the position performance of vehicle localization systems.
- We integrate the GPS measurements from both a target vehicle and its neighboring vehicles into a vehicular positioning filter and set a relatively conservative number of neighboring vehicles for the basic set, which can reduce the dispensable computation complexity. Simulation results show that the information fusion provides a great enhancement compared with GPS-only localization.
- We derive a novel theoretical lower bound limiting the positioning estimation performance, named the mSPEB. The closed form of the mSPEB needs to process a lower dimensional equivalent Fisher information matrix (EFIM) and calculate the bound for the minimum eigenvalue of a high-dimensional Fisher information matrix (FIM), such that it is with lower complexity, when compared to the SPEB used in current literature [8,10,91] that has to calculate the inverse of the high-dimensional FIM directly.

The remainder of the chapter is organized as follows. The DSRC-based vehicular positioning enhancement model using the IMM-KF is presented in Section 9.2. The DSRC-based vehicular positioning enhancement model using the DIMM-KF is presented in Section 9.3. Finally, the conclusions are discussed in Section 9.4.

9.2 A DSRC-Based Vehicular Positioning Enhancement Model Using a Mulitple-Model KF

9.2.1 Vehicular positioning enhancement model

It is noted that DFS is too difficult to be extracted from noise when the speed of the vehicle is quite low, and thus for DFS vehicular positioning methods, the standard deviation (STD) of positioning error increases as the relative speed between the target vehicle and the other vehicles decreases. Here, we investigate the method to overcome this problem. We focus on the scenario that the neighbor vehicles travel in the opposite direction of the target vehicle (TV), for this case can show an obviously detectable Doppler Effect. The problem to be solved is to estimate the position of a TV moving on a 2D road, where there are many other neighboring vehicles around the TV. Drivers of all vehicles are able to know their own state information (position, velocity, etc.) provided by a coarse GPS receiver, and they can know the neighboring vehicles' state information via vehicular communications as well. Consequently, this case can be treated as a simple but practical CV scenario. A TV is considered as a research object for positioning enhancement based on CV, and a neighboring vehicle is considered as a vehicle that is within a certain communication range to the TV and is moving in the opposite direction of the TV. Each vehicle is assumed to be with an onboard unit (OBU) providing both the DSRC and the DFS measurements.

Considering the ith moving vehicle at time instant k with a state vector $\theta_k^i = [p_{x,k}^i, p_{y,k}^i, v_{x,k}^i, v_{y,k}^i]^T$, $i = 1, \dots, n_p$, where $(p_{x,k}^i, p_{y,k}^i)$ and $(v_{x,k}^i, v_{y,k}^i)$ denote the ith vehicles position and velocity, respectively, and n_p is the total number of the vehicles driving on the road, and T is a transpose operator. The dynamic state can be modeled by the following system:

$$\theta_k^i = F_{k-1}^i \theta_{k-1}^i + G_{k-1}^i (\mu_{k-1}^i + \omega_{k-1}^i), \tag{9.1}$$

where F_{k-1}^i is the state transition matrix, and G_{k-1}^i is the noise distribution matrix. $\mu_{k-1}^i$is the control vector, and ω_{k-1}^i is zero-mean white Gaussian noise with covariance matrix Q_{k-1}^i.

For the dynamic model presented by (9.1), the following observation model of the TV can be defined:

$$\psi_k = h(\theta_k) + \gamma_k(r_e), \tag{9.2}$$

where $\theta_k^i = [p_{x,k}^i, p_{y,k}^i, v_{x,k}^i, v_{y,k}^i, \omega_k^1, \dots, \omega_k^j]^T$ is a nonlinear observation vector in terms of θ_k. ω_k^j is the DFS of the received signal from the jth neighboring vehicles, $j = 1, \dots, n_k$, and n_k is the total number of the neighboring vehicles on the road. $\gamma_k(r_e)$ is the observation noise that can be used to describe the M types of observation errors by assuming a set of another covariance matrixes. The transition among M types of the errors is generally modeled as a first-order M-state homogeneous Markov chain r_e, $e = 1, 2, \dots, M$.

Specifically, assuming that the DFS measurements from the OBU can be modeled in a derivative form of the DSRC carrier frequency, f, as follows:

$$\omega_k^j = -\frac{f}{c} \nabla_t (d_k^j) + \theta_k^j(r_e) \tag{9.3}$$

and

$$d_k^j = \sqrt{(p_{x,k} - p_{x,k}^j)^2 + (p_{y,k} - p_{y,k}^j)^2}, \tag{9.4}$$

where c is the speed of light, d_k^j is the relative distance between the TV and its neighboring vehicle j, and θ_k^j is the DFS observation noise of neighboring vehicle j. Substituting (9.4) into (9.3) yields

$$\omega_k^j = -\frac{f}{c} \left[\frac{(p_{x,k} - p_{x,k}^j)(v_{x,k} - v_{x,k}^j) + (p_{y,k} - p_{y,k}^j)(v_{y,k} - v_{y,k}^j)}{\sqrt{(p_{x,k} - p_{x,k}^j)^2 + (p_{y,k} - p_{y,k}^j)^2}} \right] + \theta_k^j(r_e), \tag{9.5}$$

where $(p_{x,k}^i, p_{y,k}^i)$ and $(v_{x,k}^i, v_{y,k}^i)$ are the position and velocity vector of the neighboring vehicle j. To solve this nonlinear observation function, as an extended Kalman filter (EKF) method presented in Ref. [9], with the first-order Taylor expansion of Equation 9.5 around an arbitrary state vector, h can be transformed to a fixed form of matrix, in which all components are supposed to obtain from both the GPS and OBU. As a result, the observation model of the TV can be reformulated as a linear one:

$$Z_k = H_k \theta_k + \gamma_k(r_e) \tag{9.6}$$

and with the observation transition matrix:

$$H_k = \begin{bmatrix} 1 & 0 & 0 & 0 \\ 0 & 1 & 0 & 0 \\ 0 & 0 & 1 & 0 \\ 0 & 0 & 0 & 1 \\ h_k^{11} & h_k^{12} & h_k^{13} & h_k^{14} \\ \vdots & \vdots & \vdots & \vdots \\ h_k^{j1} & h_k^{j2} & h_k^{j3} & h_k^{j4} \end{bmatrix}, \tag{9.7}$$

where

$$h_k^{j1} = \nabla_{p_{x,k}}(\omega_k^j) = -\frac{f}{c}\frac{(p_{y,k}-p_{y,k}^j)\left[(p_{y,k}-p_{y,k}^j)(v_{x,k}-v_{x,k}^j)-(p_{x,k}-p_{x,k}^j)(v_{y,k}-v_{y,k}^j)\right]}{(d_k^j)^3}, \tag{9.8}$$

$$h_k^{j2} = \nabla_{p_{y,k}}(\omega_k^j) = -\frac{f}{c}\frac{(p_{x,k}-p_{x,k}^j)\left[(p_{x,k}-p_{x,k}^j)(v_{y,k}-v_{y,k}^j)-(p_{y,k}-p_{y,k}^j)(v_{x,k}-v_{x,k}^j)\right]}{(d_k^j)^3}, \tag{9.9}$$

$$h_k^{j3} = \nabla_{v_{x,k}}(\omega_k^j) = -\frac{f}{c}\frac{(p_{x,k}-p_{x,k}^j)}{d_k^j}, \tag{9.10}$$

$$h_k^{j4} = \nabla_{v_{y,k}}(\omega_k^j) = -\frac{f}{c}\frac{(p_{y,k}-p_{y,k}^j)}{d_k^j}. \tag{9.11}$$

On the basis of the aforementioned models from Equations 9.1 and 9.6, it is reasonable to assume that F_{k-1}^i and G_{k-1}^i in the system model are invariable at each time instant and vehicle. Therefore, the position estimation of the TV can be formulated as the problem of linear filtering for M-state jump Markov systems, and the model can be simplified as:

$$\begin{cases} \theta_k^i = F\theta_{k-1}^i + G(\mu_{k-1}^i + \omega_{k-1}^i) \\ Z_k = H_k\theta_k + \gamma_k(r_e) \end{cases}. \tag{9.12}$$

Because H_k can be estimated by data fusion from both the GPS and OBU at each time instant k, a CV-enhanced IMM-KF can be deployed.

9.2.2 Connected vehicles–enhanced IMM-KF for vehicular positioning

In this section, we adopt the IMM approach to propose a vehicular positioning enhancement algorithm based on CV. The structure of the vehicular positioning system is illustrated in Figure 9.1.

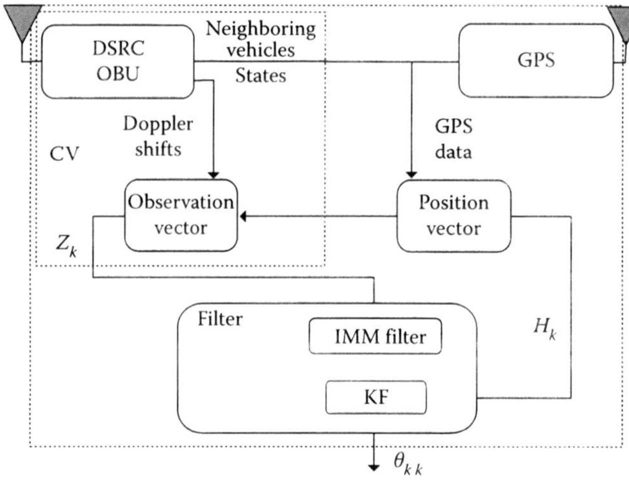

Figure 9.1 Vehicular positioning system with information fusion of the DFS and GPS measurements from both itself and the other neighboring vehicles.

9.2.2.1 CV-IMM-KF for estimating the TVs state

Step 1. Mixing probabilities and state estimates

$$\mu_{k+1,x|t} = \pi_s t \mu_{k.s}/c_t, \tag{9.13}$$

where $\mu_{k+1,x|t}$ is known as the mixing probability in the IMM estimator, $\mu_{k.s}$ is the probability of the event that the sth motion model is in effect at the time step k, $s,t = 1,2,\ldots,M$, correspond to the s,tth mode of the Markov chain r_e, and

$$c_t = \sum_{s=1}^{M} \pi_{st}\mu_{k.s}, \tag{9.14}$$

where c_t is a normalization constant and

$$\theta^0_{k|k,t} = \sum_{st}^{M} \mu_{k+1,s|t}\theta^{0,0}_{k|k,s,t}, \tag{9.15}$$

$$P^0_{k|k,t} = \sum_{st}^{M} \mu_{k+1,s|t} \times \{P^{0,0}_{k|k,s,t} + [\theta^{0,0}_{k|k,s,t} - \theta^0_{k|k,s,t}][\theta^{0,0}_{k|k,s,t} - \theta^0_{k|k,s,t}]^T\} \tag{9.16}$$

Step 2. Mode update and prediction Steps

Calculate $h_k^{j1}, h_k^{j2}, h_k^{j3}, h_k^{j4}$ according to Equations 9.8 through 9.11 and then update the observation transition matrix H_k defined in Equation 9.7.

The CV-IMM-KF gain is given by

$$K_k = P_{k|k-1,t}H_k^T(\varphi_k(N)) \times \left\{H_k(\varphi_k(N))P_{k|k-1,t}H_k^T(\varphi_k(N)) + R_k(r_e(N))\right\}^{-1}, \tag{9.17}$$

where $\varphi_k(N)$ and $r_e(N)$ are functions of N and can change the dimension of the observation transition matrix H_k and the covariance matrix R_k, respectively.

The CV-IMM-KF update steps are given by

$$\theta_{k|k,t} = \theta_{k|k-1,t} + K_k\{Z_k - H_k(\varphi_k(N))\theta_{k|k-1,t}\}, \tag{9.18}$$

$$P_{k|k,t} = P_{k|k-1,t} - K_k\{H_k(\varphi_k(N))P_{k|k-1,t}H_k^T(\varphi_k(N)) + R_k(r_e(N))\}K_k^T. \tag{9.19}$$

The CV-IMM-KF prediction steps are given by

$$\theta_{k+1|k,t} = F\theta^0_{k|k,t} + G\mu_{k,t}, \tag{9.20}$$

$$P_{k+1|k,t} = FP^0_{k|k,t}F^T + GQG^T. \tag{9.21}$$

The likelihood function $\Lambda_{k,t}$ and predicted mode probability $\mu_{k,t}$ are given by

$$\Lambda_{k,t} = \aleph(Z_k - H_k(\varphi_k(N)))\theta_{k|k-1,t}; 0, H_k(\varphi_k(N))P_{k|k-1,t}H_k^T(\varphi_k(N)) + R_k(r_e(N)), \tag{9.22}$$

$$\mu_{k,t} = \Lambda_{k,t}c_t/c, \tag{9.23}$$

where c is a normalizing constant defined as follows:

$$c = \sum_{t=1}^{M} \Lambda_{k,t}c_t. \tag{9.24}$$

Step 3. Estimates combination

$$\theta_{k|k} = \sum_{t=1}^{M} \mu_{k,t}\theta_{k|k,t}, \tag{9.25}$$

$$P_{k|k} = \sum_{t=1}^{M} \mu_{k,t} \times \{P_{k|k,t} + [\theta_{k|k,t} - \theta_{k|k}][\theta_{k|k,t} - \theta_{k|k}]^T\}. \tag{9.26}$$

The overall CV-IMM-KF algorithm for vehicular positioning is described in Algorithm 9.5.

Algorithm 9.5 One Trial of CV-IMM-KF Algorithm

Require: Given $\theta^{0,0}_{k|k,s,t}, P0, 0_{k|k,s,t}, \theta_{k|k-1,t}, P_{k|k-1,t}, \mu_{k,s}, \pi_{st}$, the GPS and DFS measurements $Z_k(p_{x,k}, p_{y,k}, v_{x,k}, v_{y,k})$, $Z_k(p^j_{x,k}, p^j_{y,k}, v^j_{x,k}, v^j_{y,k})$, $Z_k\omega^j_k$

1: **if** $n_k > N$ **then**

2: Extract the largest N values from $\{|\omega^1_k|, \ldots, |\omega^j_k|\}$ to form the basic set $\Omega(k) = \underset{\omega^j_k}{argN\,max}\{|\omega^1_k|, \ldots, |\omega^j_k|\}$, $j = 1, \ldots, n_k, N \leq n_k$.

3: **else**

4: Extract all the values from $\{|\omega^1_k|, \ldots, |\omega^j_k|\}$ to form the basic set

5: **end if**

6: Calculate $h^{j1}_k, h^{j2}_k, h^{j3}_k, h^{j4}_k$ via (9.8)-(9.11)

7: Set $H_k(\varphi_k(N)), R_k(r_e(N))$

8: **for** $t = 1, \ldots, M$ **do**

9: Update $\theta_{k|k,t}$ and $P_{k|k,t}$ via (9.18) (9.19)

10: Predict $\theta_{k+1|k,t}$ and $P_{k+1|k,t}$ via (9.20) (9.21)

11: **end for**

12: Combine the estimated states $\theta_{k|k,t}$ and $P_{k|k,t}$ via (9.22)-(9.26).

9.2.3 Numerical study

9.2.3.1 Simulation scenario

A basic set with N neighboring vehicles for the TV can be formed through Algorithm 9.5. Considering a section of urban roads, which is with a width of four lanes (each one is 3.5 m wide) and a length of 1 km. It is assumed that the traffic density of the road section is 20 vehicles/km and the average speed of traffic is generated stochastically in duration from 50 to 60 km/h following a uniform distribution. The initial positions of the neighboring vehicles are generated stochastically on the road following a uniform distribution as well. The vehicle dynamics described in Equation 9.12 is with

$$F = \begin{bmatrix} I & \triangle t I \\ o & I \end{bmatrix}, \quad G = \begin{bmatrix} \frac{1}{2} \triangle t^2 I \\ \triangle t I \end{bmatrix}, \tag{9.27}$$

where I is a 2×2 identity matrix, o is a 2×2 zero matrix, and $\triangle t$ is the sampling period. The control vector in 9.12 is $u^i = [0, 0.01]^T$. The noise vector $w^i_{k-1} = [\sigma_{ay,k-1}, \ \sigma_{ax,k-1}]^T$ $\aleph(0, Q)$, with covariance matrix $Q = \text{diag}\ [\sigma^2_{ax}, \sigma^2_{ay}]$, where the elements $\sigma_{ax,k-1} = \sqrt{0.99/2}$ and $\sigma_{ay,k-1} = \sqrt{0.01/2}$ are the acceleration noises along the X- and Y-axes, respectively, with STD in m/s^2. The covariance matrix $R(r_e)$ of observation noise $\gamma_k(r_e)\ \aleph(0, R(r_e))$ is described as a first-order Markov chain switching between two models $R(r_1) = \text{diag}[\sigma^2_{px}, \ \sigma^2_{py}, \ \sigma^2_{vx}, \ \sigma^2_{vy}, \ \sigma^2_{\omega_1}(r_1), \ldots, \sigma^2_{\omega N}(r_1)]$ and $R(r_2) = \text{diag}[\sigma^2_{px}, \ \sigma^2_{py}, \ \sigma^2_{vx}, \ \sigma^2_{vy}, \ \sigma^2_{\omega_1}(r_2), \ldots, \ \sigma^2_{\omega N}(r_2)]$, of which the elements are with STDs in units of m,m/s, and Hz. The transition probability for this Markov chain is $F = \begin{bmatrix} 0.9 & 0.1 \\ 0.1 & 0.9 \end{bmatrix}$ and their initial probability is $\mu_0 = [0.5, 0.5]$. According to the achievable performance discussed in Refs. [6,9], as the number of the neighboring vehicles is increasing, the performance enhancement can be less obvious and lead to more additionally computational burden. Therefore, we set $N = 4$, which is a relatively conservative number of the neighboring vehicles for the basic set, which is mentioned in Algorithm 9.5.

As shown in Figure 9.2, the vehicles that start at the first two lanes from the bottom move toward the positive direction of the X-axis. On the contrary, the vehicles that start at the first two lanes from the top move toward the negative direction of the X-axis. The black dot and gray dots denote the initial positions of the TV and the neighboring vehicles, respectively. The TV starts at position (300,5), with the initial velocity vector (60,0) in m/s. However, because of the system noise and control vector set in the dynamics model, the actual velocity is changing slightly over the sampling period.

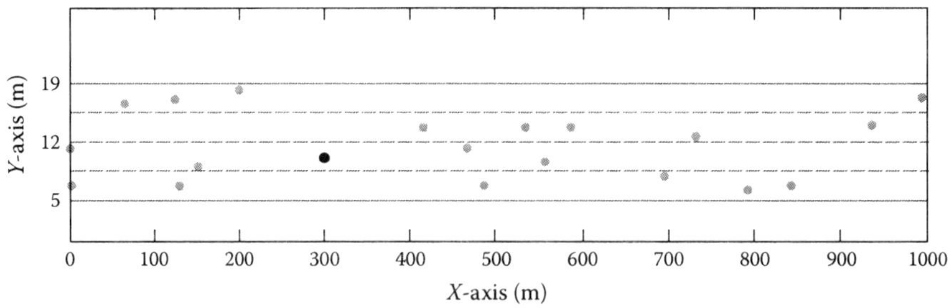

Figure 9.2 Initial scenario.

In the simulations, the sampling period and length are taken to be 0.2 s and 100, respectively, and the communication range of the DSRC is 300 m. As the DFS measurements presented in Refs. [9,202], the Probability Density Function (PDF) of the DFS is approximately zero-mean asymmetric Gaussian with the left and right STDs of 100 and 120 Hz, when the vehicles travel at the speed of 60 km/h, broadcasting the DSRC packets with a frequency of 5.89 GHz and a rate of 100 packets/s. It is worth noting that the PDF of the DFS remains a fairly consistent estimation from LOS to NLOS. Consider the noise of the DFS measurements as zero-mean Gaussian with two states of STDs: $\sigma_{wN}r_1 = 100\,\text{Hz}$ and $\sigma_{wN}r_2 = 120\,\text{Hz}$. Specifically, the state of the observation noise remains unchanged in r_1 between 0 to 6 s, and changes in the following 10 s to r_2. Finally, the state changes back to r_1 for another 4 s. The position and velocity measured by GPS are assumed to be added noise with the variance ($\sigma_{px} = \sqrt{200/2}m, \sigma_{py} = \sqrt{200/2}m$) and ($\sigma_{vx} = \sqrt{15/2}m, \sigma_{vy} = \sqrt{15/2}m$), respectively.

9.2.3.2 *Simulation results*

To quantify the performance of the proposed approach, the root mean square error (RMSE) of vehicular positioning is calculated to assess the closeness of the estimated trajectory $(p_{x,k}, p_{y,k})$ to the true trajectory $(p_{x,k}, p_{y,k})$ at each time instant over $N_m = 500$ Monte Carlo simulations. In Equation 9.28, $(p_{x,k}(m), p_{y,k}(m))$ denotes the estimated position vector in the mth Monte Carlo run at the kth step.

$$\text{RMSE} = \sqrt{\frac{1}{N_m}\sum_{m=1}^{N_m}[(p_{x,k}(m) - p_{x,k})^2 + (p_{y,k}(m) - p_{y,k})^2]}. \tag{9.28}$$

The performance comparison between the proposed CV-IMM-KF and the GPS-only approach is shown in Figure 9.3 with respect to the RMSE in distance. It is obvious that the proposed CV-IMM-KF method outperforms the GPS-alone localization. To indicate the enhancement of vehicular positioning of the proposed approach, the enhancement indicator μ is calculated as follows:

$$\mu = \left(1 - \frac{A\_\text{RMSE}}{B\_\text{RMSE}}\right) \times 100\%. \tag{9.29}$$

The enhancement of vehicular positioning is shown in Table 9.1. Compared to the GPS-based localization, the proposed CV-IMM-KF approach achieves the enhancement of $\mu = 37.48\%$.

Remark 1 If $A_$RMSE is better than $B_$RMSE, μ will be greater than zero. And the increase of μ is linked to the good performance of $A_$RMSE.

Remark 2 By describing the transition of the measurement noise as a first-order M-state jump Markov chain, the proposed CV-IMM-KF approach has been proved to achieve better performance in a scenario that is similar to a practical one.

Table 9.1 CV-IMM-KF and GPS Error Comparison

Method	*RMSE*	*Enhancement*
GPS	14.1426	N/A
CV-IMM-KF	8.8420	37.48%

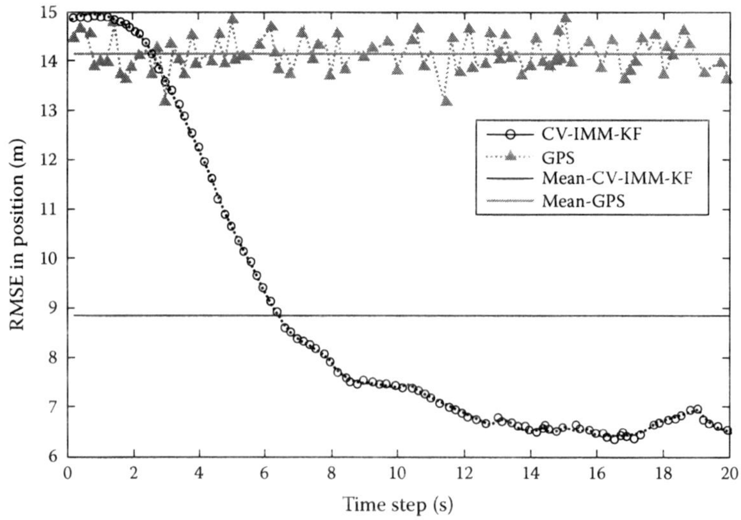

Figure 9.3 CV-IMM-KF and GPS performance in positioning error.

9.3 A DSRC-Based Vehicular Positioning Enhancement Model Using a Distributed Multiple-Model KF

9.3.1 System model and localization enhancement

The problem to be solved is to estimate the position of a TV moving on a road section where there are many other moving neighboring vehicles around the TV. Assume that a part of participated vehicles in the CV scenarios are able to know their own state information, including position, velocity provided by the GPS receiver. Meanwhile, it should be noted that the neighboring vehicles' state information is easy to obtain from the DSRC links from which the DFS and the RSSI measurements can be extracted as well. We define CV penetration to represent the percentage of vehicles who hold the CV abilities on the simulated road section. In this scenario, the TV is considered as a research objective for positioning enhancements, and the neighboring vehicles are considered as the vehicles who are within the coverage of the DSRC networks of the TV (Figure 9.4).

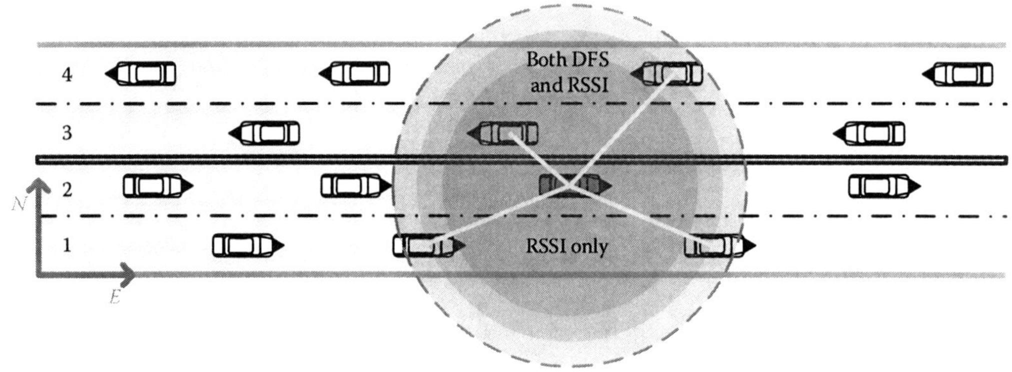

Figure 9.4 Positioning enhanced by the DFS and the RSSI measurements.

9.3.1.1 System model

Consider a CV scenario consisting of the moving vehicles, where each vehicle is equipped with a GPS receiver providing coarse data to set up the state vector $\theta_k = [m_{x,k}, m_{y,k}, \dot{m}_{x,k}, \dot{m}_{y,k}]^T$. The position and the velocity components of the vehicle are denoted by $(m_{x,k}, m_{y,k})$ and $(\dot{m}_{x,k}, \dot{m}_{y,k})$, respectively. The x and y subscripts denote the orientation along the East (E) and the North (N) axes, respectively. The subscript k denotes the time step, and T is the transpose operator. The dynamic procedure of the moving vehicles can be considered as the following motion model [132,254]:

$$\theta_{k+1} = \mathbf{F}\theta_k + \mathbf{G}(\varphi_\mathbf{k} + \zeta_\mathbf{k}), \tag{9.30}$$

with

$$\begin{aligned} \mathbf{F} &= \begin{bmatrix} \mathbf{I}_2 & \Delta t\mathbf{I}_2 \\ \mathbf{O}_2 & \mathbf{I}_2 \end{bmatrix}, \varphi_k = \begin{bmatrix} a_{x,k} \\ a_{y,k} \end{bmatrix}, \\ \mathbf{G} &= \begin{bmatrix} \frac{1}{2}\Delta t^2\mathbf{I}_2 \\ \Delta t\mathbf{I}_2 \end{bmatrix}, \quad \zeta_k = \begin{bmatrix} \zeta_{x,k} \\ \zeta_{y,k} \end{bmatrix}, \end{aligned} \tag{9.31}$$

where φ_k is the discrete-time command process and ζ_k is the system noise modeled as zero-mean Gaussian noise with a covariance matrix $\mathbf{Q_k}$. $\mathbf{F}$ is the system transition matrix describing the movement of the TV between two consecutive time steps. $\mathbf{G}$ is the transition matrix that models the acceleration-related state and the system noise changes. $\mathbf{I}_\mathcal{M}$ denotes a $\mathcal{M} \times \mathcal{M}$ identity matrix, and $\mathbf{O}_\mathcal{M}$ denotes a $\mathcal{M} \times \mathcal{M}$ zero matrix. Correspondingly, $\zeta_{x,k}$ and $\zeta_{y,k}$ are the acceleration noise along the E and the N axes, respectively, and Δt is the sampling period.

The command process φ_k is a time-homogeneous Markov chain with a finite state space that takes a set of acceleration values $\varphi = \{a_1, ..., a_L\}$. The transition probability matrix for the different acceleration states in φ is defined as $\Pi^\varphi = [\pi^\varphi_{pq}]$ with the transition probability $\pi^\varphi_{pq} = \mathbb{P}\{\varphi_k = a_p | \varphi_{k-1} = a_q\}$ where $0 \leq \pi^\varphi_{pq} \leq 1$, $\sum_{q=1}^{L} \varphi_{pq} = 1$, $p, q = 1, \ldots, L$. It should be noted that the system model 9.30 with Markovian switching systems has been widely used to characterize the state variations of the dynamic object [28,91,254]. In the considerable scenario, it is reasonable to utilize the Markov chain in model 9.30 to represent the process that vehicles suffer from sudden changes caused by various traffic incidents, such as stop signs, or traffic lights switching. Moreover, the set φ can be considered as a sectional-continuous function during each instant time interval (the sampling period). Such a model with the acceleration switching among different nonzero means is more effective to characterize the vehicle movements in the real scenario than the motion models only with a zero-mean white Gaussian noise in general [10,28,30,254]. In terms of the system model described as Equation 9.30, the measurement model can be defined as follows:

$$\mathbf{z}_k = \mathbf{h}(\theta_k) + \vartheta_\mathbf{k}(\phi_k), \tag{9.32}$$

where $\mathbf{h} = [m_{x,k}, m_{y,k}, \dot{m}_{x,k}, \dot{m}_{y,k}, \rho_k^1, ..., \rho_k^i, r_k^1, ..., r_k^j]^T$ is a nonlinear measurement vector associated with $\theta_\mathbf{k}$, and $\vartheta_\mathbf{k}$ is the measurement noise modeled as zero-mean white Gaussian noise with varying covariance matrix $\mathbf{R}_k$ determined by ϕ_k. ϕ_k is a time-homogeneous Markov chain with two states to represent the switching modes $\phi = \{s_1, s_2\}$, where s_1 is assigned to the event "LOS", and s_2 is assigned to the event "NLOS". Correspondingly, the transition probability matrix is defined as $\Pi^\phi = [\pi^\phi_{uv}]$ with transition probability $\pi^\phi_{uv} = \mathbb{P}\{\phi_k = s_u | \phi_{k-1} = s_v\}$, where $0 \leq \pi^\phi_{uv} \leq 1$, $\sum_{v=1}^{2} \phi_{uv} = 1$, $u, v = 1, 2$.

Assume that there are j neighboring vehicles within the DSRC coverage of the TV, to whom i of j neighboring vehicles are traveling in the opposite direction $(0 \leq i \leq j)$. Signals transmitted from these i neighboring vehicles can be modeled by the deployment of the DFS measurements. For brief descriptions, we let $N^{DFS} = \{1, 2, \ldots, i\}$ denote the set of the neighboring vehicles who provide the DFS measurements, and ρ_k^{α} denotes the measurements obtained from the neighboring vehicle α at the time instant k, $\alpha \in N^{DFS}$, which can be formulated as follows [10]:

$$\rho_k^{\alpha} = -\frac{f}{c}\frac{d(d_k^{\alpha})}{dt} + \vartheta_k^{\alpha}, \tag{9.33}$$

$$d_k^{\alpha} = \sqrt{(m_{x,k} - m_{x,k}^{\alpha})^2 + (m_{y,k} - m_{y,k}^{\alpha})^2}, \tag{9.34}$$

where f is the transmission frequency of DSRC, and c is the speed of light. d_k^{α} is the relative distance between the TV and its neighboring vehicle α, and ϑ_k^{α} is the DFS-related observation noise. Correspondingly, $(m_{x,k}^{\alpha}, m_{y,k}^{\alpha})$ denotes the position of the neighboring vehicle α. Substituting Equation 9.34 into 9.33, Equation 9.33 can be reformulated as in Equation 9.35, where $(\dot{m}_{x,k}^{\alpha}, \dot{m}_{y,k}^{\alpha})$ is the velocity vector of the neighboring vehicle α.

$$\rho_k^{\alpha} = -\frac{f}{c}\left[\frac{(m_{x,k} - m_{x,k}^{\alpha})(\dot{m}_{x,k} - \dot{m}_{x,k}^{\alpha}) + (m_{y,k} - m_{y,k}^{\alpha})(\dot{m}_{y,k} - \dot{m}_{y,k}^{\alpha})}{\sqrt{(m_{x,k} - m_{x,k}^{\alpha})^2 + (m_{y,k} - m_{y,k}^{\alpha})^2}}\right] + \vartheta_k^{\alpha}, \tag{9.35}$$

Correspondingly, we let $N^{RSSI} = \{1, 2, \ldots, j\}$ denote the set of the neighboring vehicles who provide the RSSI measurements. The received power r_k^{β} corresponding to that measurements from the neighboring vehicle β at the time instant k, $\beta \in N^{RSSI}$, is an important metrics obtained from the DSRC physical layer. According to the log-distance path loss model defined in Refs. [185,237], the laws to model the path-loss behavior of DSRC propagation between vehicles can be formulated as follows:

$$r_k^{\beta} = C - 10\gamma \lg(d_k^{\beta}) + \vartheta_k^{\beta}, \tag{9.36}$$

$$d_k^{\beta} = \sqrt{(m_{x,k} - m_{x,k}^{\beta})^2 + (m_{y,k} - m_{y,k}^{\beta})^2}, \tag{9.37}$$

where C is a constant with regard to the transmission power and $\gamma \in [2, 5]$ is the path-loss exponent. d_k^{β} is the relative distance between the TV and its neighboring vehicle β, and ϑ_k^{β} is the RSSI-related measurement noise. Correspondingly, $(m_{x,k}^{\beta}, m_{y,k}^{\beta})$ denotes the position of the neighboring vehicle β. The transition process between the LOS and the NLOS conditions could be sharply modeled as a first-order Markov chain with two states $\{s_1, s_2\}$ [30,253–255]. As a result, a zero-mean white Gaussian noise is considered with a variance matrix $\mathbf{R}_k^{LOS}$ in the LOS condition, whereas a variance matrix $\mathbf{R}_k^{NLOS}$ is employed in the NLOS condition. Specifically, channel modeling in V2V communication environments is a significant issue without concluding a common sense, so the fundamental log-distance path loss model has been used to depict the V2V channel for simplicity.

Note that the set φ and the set ϕ are two independent Markov chains specifying the behavior of the sudden changes of the acceleration and the transition between the LOS and the NLOS conditions, respectively. It should be mentioned that the state metrics in the measurement model depends on the quality of the DSRC links between the TV and its neighboring vehicles. Moreover, it is with great probability that the measurement vector consists of the metrics measured from both the LOS and the NLOS conditions. Particularly, the neighboring vehicle $\alpha \in N^{DFS}$ could contribute to both the DFS and the RSSI measurements, while the neighboring vehicle $\beta \in N^{RSSI}$ could be functionally divided into two portions. One of them following the set $\beta \in N^{RSSI}/N^{DFS}$ could just benefit the RSSI measurements and the other portion could be with the same function as the neighboring vehicle $\alpha \in N^{DFS}$.

To solve the nonlinear observation problem presented in model 9.32, an EKF method has been used in Ref. [10]. Applying the first-order Taylor expansion to Equation 9.32 around an arbitrary state vector, **h** can be transformed to a stereotyped matrix in which all components are supposed to obtain from the GPS and the DSRC OBU. Subsequently, model 9.32 can be reformulated as follows:

$$\mathbf{z}_k \cong \mathbf{H}_k\theta_k + \vartheta_k(\phi_k), \tag{9.38}$$

where

$$\mathbf{H}_k = \begin{bmatrix} 1 & 0 & 0 & 0 \\ 0 & 1 & 0 & 0 \\ 0 & 0 & 1 & 0 \\ 0 & 0 & 0 & 1 \\ \mathcal{H}_k^{11} & \mathcal{H}_k^{12} & \mathcal{H}_k^{13} & \mathcal{H}_k^{14} \\ \vdots & \vdots & \vdots & \vdots \\ \mathcal{H}_k^{i1} & \mathcal{H}_k^{i2} & \mathcal{H}_k^{i3} & \mathcal{H}_k^{i4} \\ \mathcal{G}_k^{11} & \mathcal{G}_k^{12} & 0 & 0 \\ \vdots & \vdots & \vdots & \vdots \\ \mathcal{G}_k^{j1} & \mathcal{G}_k^{j2} & 0 & 0 \end{bmatrix}, \tag{9.39}$$

with the transition components formulated as Equations 9.40 through 9.45.

$$\begin{aligned} \mathcal{H}_k^{\alpha 1} &= \frac{\partial \rho_k^\alpha}{\partial m_{x,k}} = -\frac{f}{c}\frac{(m_{y,k} - m_{y,k}^\alpha)}{(d_k^\alpha)^3} \times \\ &\quad [(m_{y,k} - m_{y,k}^\alpha)(\dot{m}_{x,k} - \dot{m}_{x,k}^\alpha) - (m_{x,k} - m_{x,k}^\alpha)(\dot{m}_{y,k} - \dot{m}_{y,k}^\alpha)], \end{aligned} \tag{9.40}$$

$$\begin{aligned} \mathcal{H}_k^{\alpha 2} &= \frac{\partial \rho_k^\alpha}{\partial m_{y,k}} = -\frac{f}{c}\frac{(m_{x,k} - m_{x,k}^\alpha)}{(d_k^\alpha)^3} \\ &\quad \times [(m_{x,k} - m_{x,k}^\alpha)(\dot{m}_{y,k} - \dot{m}_{y,k}^\alpha) - (m_{y,k} - m_{y,k}^\alpha)(\dot{m}_{x,k} - \dot{m}_{x,k}^\alpha)], \end{aligned} \tag{9.41}$$

$$\mathcal{H}_k^{\alpha 3} = \frac{\partial \rho_k^\alpha}{\partial \dot{m}_{x,k}} = -\frac{f}{c}\frac{(m_{x,k} - m_{x,k}^\alpha)}{d_k^\alpha}, \tag{9.42}$$

$$\mathcal{H}_k^{\alpha 4} = \frac{\partial \rho_k^\alpha}{\partial \dot{m}_{y,k}} = -\frac{f}{c}\frac{(m_{y,k} - m_{y,k}^\alpha)}{d_k^\alpha}, \tag{9.43}$$

$$\mathcal{G}_k^{\beta 1} = \frac{\partial r_k^\beta(0,0)}{\partial m_{x,k}} = \frac{10}{\ln 10}\gamma\frac{m_{x,k}^\beta}{(d_k^\beta)^2}, \tag{9.44}$$

$$\mathcal{G}_k^{\beta 2} = \frac{\partial r_k^\beta(0,0)}{\partial m_{y,k}} = \frac{10}{\ln 10}\gamma\frac{m_{y,k}^\beta}{(d_k^\beta)^2}. \tag{9.45}$$

Let r^β be an infinite differentiable function in some open neighborhood around $(m_{x0}, m_{y0}) = (0,0)$, then according to Multivariate Taylor Expansion theorem, the linear approximation from the Taylor series of $r^\beta(m_x, m_y)$ can be formulated as

$$\begin{aligned} r^\beta(m_x, m_y) &\cong r^\beta(m_{x0}, m_{y0}) + \frac{\partial r^\beta(m_{x0}, m_{y0})}{\partial m_x}(m_x - m_{x0}) \\ &\quad + \frac{\partial r^\beta(m_{x0}, m_{y0})}{\partial m_y}(m_y - m_{y0}). \end{aligned} \tag{9.46}$$

After putting into the corresponding point, (9.46) can be simplified as

$$r^{\beta}(m_x, m_y) - r^{\beta}(0,0) = \frac{\partial r^{\beta}(0,0)}{\partial m_x} m_x + \frac{\partial r^{\beta}(0,0)}{\partial m_y} m_y. \tag{9.47}$$

Hence, the RSSI measurements of the measurement model can be linearized into a block matrix, as shown in Equation 9.39. The similar proof for the DFS measurements is omitted due to space constraint. It should be noted that the RSSI-related measurements in Equation 9.38 are not the true values measured at the receiver, but are the values calculated by the left hand of Equation 9.47, which is the result of the true RSSI measurements minus the value of $r^{\beta}(m_x, m_y)$ at $(m_x = 0, m_y = 0)$.

9.3.1.2 The CV-enhanced DIMM-KF for mobile vehicle localization

The schematic of the CV-enhanced vehicle localization method is shown in Figure 9.5. The proposed CV-enhanced DIMM-KF algorithm handling two switching parameters in Equations 9.30 and 9.38 works as follows:

Step 1. Mixing probabilities calculation

$$\mu_{k+1,l|s} = \mu^{\varphi}_{k+1,p|q} \mu^{\phi}_{k+1,u|v}, \tag{9.48}$$

$$\begin{cases} \mu^{\varphi}_{k+1,p|q} = \pi^{\varphi}_{pq} \mu^{\varphi}_{k,p} / c^{\varphi}_{q} \\ \mu^{\phi}_{k+1,u|v} = \pi^{\phi}_{uv} \mu^{\phi}_{k,u} / c^{\phi}_{v} \end{cases}, \tag{9.49}$$

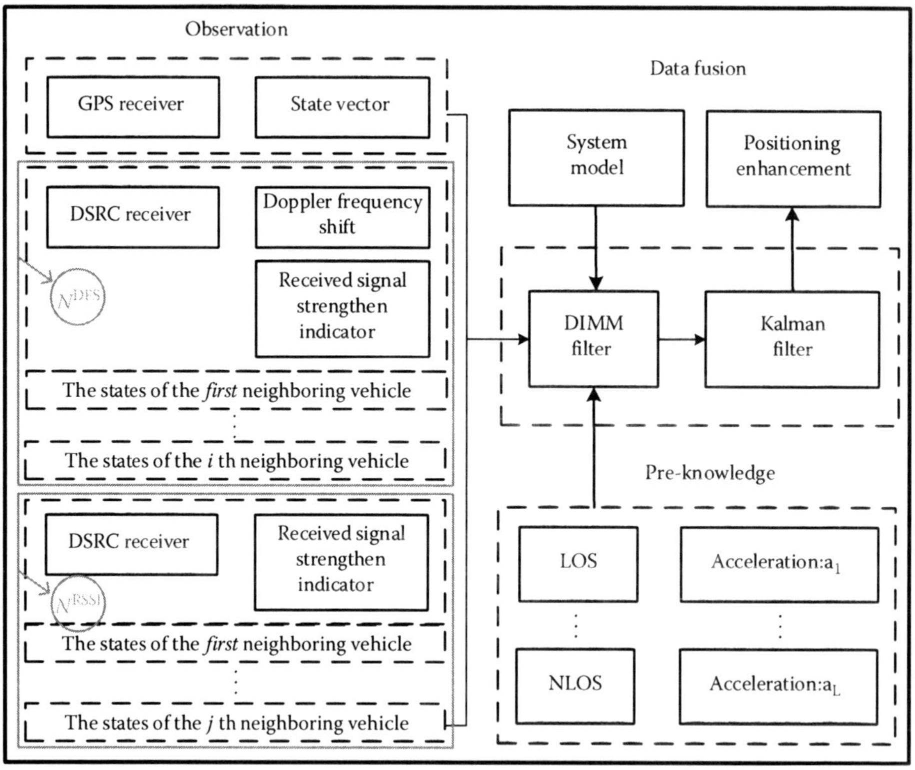

Figure 9.5 The schematic of the proposed vehicular cooperative localization method.

where $\mu^{\varphi}_{k+1,p|q}$ and $\mu^{\phi}_{k+1,u|v}$ are defined as the mixing probabilities that are common in the conventional IMM estimator. Both of them can be obtained from Equation 9.49, where $l,s = 1,2,\ldots,2L$. In (9.49), π^{φ}_{pq} and π^{ϕ}_{uv} represent the transition probabilities of the two aforementioned independent Markov chains, and $\mu^{\varphi}_{k,p}$ and $\mu^{\phi}_{k,u}$ are the probabilities of the event that the pth motion model and the uth channel mode are in effect at the time step k, respectively, where $p,q = 1,2,\ldots,L$, corresponding to the p,qth mode of the Markov chain φ, and $u,v = 1,2$, corresponding to the u,vth mode of the Markov chain ϕ. Consequently, the normalized constant can be formulated as

$$c_s = c^{\varphi}_q c^{\phi}_v, \tag{9.50}$$

where $=$ c^{φ}_q and c^{ϕ}_v are the distributed normalized constants for different Markov chains with the formulation as

$$\begin{cases} c^{\varphi}_q = \sum_{p=1}^{L}\{\pi^{\varphi}_{pq}\mu^{\varphi}_{k,p}\} \\ c^{\phi}_v = \sum_{u=1}^{2}\{\pi^{\phi}_{uv}\mu^{\phi}_{k,u}\} \end{cases}. \tag{9.51}$$

Step 2. Interaction

Mixing the state estimations and the covariance matrices according to the following Equations 9.52 and 9.53, respectively,

$$\theta^{0}_{k|k,s} = \sum_{p=1}^{L}\left\{\mu^{\varphi}_{k+1,p|q}\sum_{u=1}^{2}\left\{\mu^{\phi}_{k+1,u|v}\theta^{0,0}_{k|k,p,u}\right\}\right\}, \tag{9.52}$$

$$\mathbf{P}^{0}_{k|k,s} = \sum_{p=1}^{L}\left\{\begin{array}{l}\mu^{\varphi}_{k+1,p|q} \\ \times\left\{\begin{array}{l}\mathbf{P}^{0,0}_{k|k,p,v} + \left\{\theta^{0,0}_{k|k,p,v} - \theta^{0}_{k|k,s}\right\} \\ \qquad \times\left\{\theta^{0,0}_{k|k,p,v} - \theta^{0}_{k|k,s}\right\}^{T}\end{array}\right\}\end{array}\right\}, \tag{9.53}$$

where

$$\mathbf{P}^{0,0}_{k|k,p,v} = \sum_{u=1}^{2}\left\{\begin{array}{l}\mu^{\phi}_{k+1,u|v} \\ \times\left\{\begin{array}{l}\mathbf{P}^{0,0}_{k|k,p,u} + \left\{\theta^{0,0}_{k|k,p,u} - \theta^{0}_{k|k,p,v}\right\} \\ \qquad \times\left\{\theta^{0,0}_{k|k,p,u} - \theta^{0}_{k|k,p,v}\right\}^{T}\end{array}\right\}\end{array}\right\}. \tag{9.54}$$

Step 3. Mode update and prediction steps

Calculate $\mathcal{H}^{\alpha 1}_k,\ldots,\mathcal{H}^{\alpha 4}_k$, $\mathcal{G}^{\beta 1}_k$, $\mathcal{G}^{\beta 2}_k$, according to the Equations 9.40 through 9.45 and then update the measurement transition matrix $\mathbf{H}_k$ defined by Equation 9.39 associated with the CV technologies.

The CV-enhanced DIMM-KF gain is given by

$$\mathbf{K}_k = \mathbf{P}_{k|k-1,s}\mathbf{H}^{T}_k \times \{\mathbf{H}_k\mathbf{P}_{k|k-1,s}\mathbf{H}^{T}_k + \mathbf{R}_k\}^{-1}. \tag{9.55}$$

The update steps of the CV-enhanced DIMM-KF are given by

$$\theta_{k|k,s} = \theta_{k|k-1,s} + \mathbf{K}_k\{\mathbf{z}_k - \mathbf{H}_k\theta_{k|k-1,s}\}, \tag{9.56}$$

$$\mathbf{P}_{k|k,s} = \mathbf{P}_{k|k-1,s} - \mathbf{K}_k\{\mathbf{H}_k\mathbf{P}_{k|k-1,s}\mathbf{H}^{T}_k + \mathbf{R}_k\}\mathbf{K}^{T}_k. \tag{9.57}$$

The prediction steps of the CV-enhanced DIMM-KF are given by

$$\theta_{k+1|k,s} = \mathbf{F}\theta^0_{k|k,s} + \mathbf{G}\varphi_{k,s}, \tag{9.58}$$

$$\mathbf{P}_{k+1|k,s} = \mathbf{F}\mathbf{P}^0_{k|k,s}\mathbf{F}^T + \mathbf{G}\mathbf{Q}\mathbf{G}^T. \tag{9.59}$$

The likelihood function $\Lambda_{k,t}$ and the prediction mode probability $\mu_{k,t}$ are formulated as

$$\Lambda_{k,s} = normal(\mathbf{z}_k - \mathbf{H}_k\theta_{k|k-1,s};\ 0,\ \mathbf{H}_k\mathbf{P}_{k|k-1,s}\mathbf{H}_k^T + \mathbf{R}_k). \tag{9.60}$$

Step 4. Mode probability update

The probability at the time step k is calculated as

$$\mu_{k,s} = \Lambda_{k,s}c_s/c, \tag{9.61}$$

where c is the overall normalized constant defined as

$$c = \sum_{s=1}^{2L} \lambda_{k,s}c_s. \tag{9.62}$$

Step 5. Combination

In the final stage, the CV-enhanced DIMM-KF algorithm combines the state estimations and the covariance matrices as the following manners:

$$\theta_{k|k} = \sum_{s=1}^{2L} \mu_{k,s}\theta_{k|k,s}, \tag{9.63}$$

$$\mathbf{P}_{k|k} = \sum_{s=1}^{2L} \mu_{k,s} \times \left\{\mathbf{P}_{k|k,s} + \{\theta_{k|k,s} - \theta_{k|k}\}\{\theta_{k|k,s} - \theta_{k|k}\}^T\right\}. \tag{9.64}$$

The overall CV-enhanced DIMM-KF algorithm is described in Algorithm 9.6.

9.3.2 General performance analysis

In this section, we briefly review the information inequality, describe the framework for the designed general measurements containing the positioning-related information, and study a tight computational method of the fundamental limits on the positioning metrics, which is defined as the square position error bound (SPEB) in principle [282]. Subsequently, we transform the problem of estimation of the theoretical bound to that of analyzing the bound for the trace of the inverse matrix, propose a novel lower bound limiting the positioning estimation performance, named the mSPEB, which reduces the computation complexity compared to the calculation of the SPEB, and finally formulate the FIM of the system model for studying the estimated covariance lower bound of the CV-enhanced positioning.

Throughout this section, Ξ denotes an $\mathcal{N}$-by-$\mathcal{N}$ symmetric positive definite matrix with eigenvalues

$$\lambda_1 \le \lambda_2 \le, \ldots, \le \lambda_{\mathcal{N}}.$$

$\lambda(\Xi)$ is the set of all eigenvalues. The parameters $\mathcal{S}$ and $\mathcal{T}$ denote the bounds for the lowest and largest eigenvalues λ_1 and $\lambda_{\mathcal{N}}$ of Ξ.

$$0 \le \mathcal{S} \le \lambda_1,\ \lambda_{\mathcal{N}} \le \mathcal{T}.$$

$\mathbb{TR}(\cdot)$ is the trace operator and $\|\cdot\|_F^2$ is the F-norm operator.

Algorithm 9.6 One Trial of the CV-enhanced DIMM-KF Algorithm

Require: $\theta^{0,0}_{k|k,p,u}$, $\theta^{0,0}_{k|k,p,v}$, $\mathbf{P}^{0,0}_{k|k,p,u}$, $\mu^{\varphi}_{k+1,p|q}$, $\mu^{\phi}_{k+1,u|v}$, π^{φ}_{pq}, π^{ϕ}_{uv}, the GPS, the DFS, and the RSSI measurements $\mathbf{z}^{GPS}_k$, $\mathbf{z}^{DFS}_k$, $\mathbf{z}^{RSSI}_k$

Initial c^{φ}_q, c^{ϕ}_v, $\mu^{\varphi}_{k+1,p|q}$, $\mu^{\phi}_{k+1,u|v}$, $\theta^{0}_{k|k,s}$, and $\mathbf{P}^{0}_{k|k,s}$
for $p = 1,\dots,L$ **do**
  Calculate c^{φ}_q and $\mu^{\varphi}_{k+1,p|q}$ via (9.51) and (9.49), respectively
end for
for $u = 1,2$ **do**
  Calculate c^{ϕ}_v and $\mu^{\phi}_{k+1,u|v}$ via (9.51) and (9.49), respectively
end for
for $l = 1,\dots,2L$ **do**
  Calculate $\theta^{0}_{k|k,s}$ and $\mathbf{P}^{0}_{k|k,s}$ via (9.52)-(9.54)
end for
Calculate $\mathcal{H}^{\alpha 1}_k,\dots,\mathcal{H}^{\alpha 4}_k$, $\mathcal{G}^{\beta 1}_k$, $\mathcal{G}^{\beta 2}_k$ via (9.40)-(9.45)
Set $\mathbf{H}_k$ via (9.39) and set $\mathbf{R}_k$ according to the number of neighboring vehicles in $N_{DFS} \cup N_{RSSI}$
for $s = 1,\dots,2L$ **do**
  Update $\theta_{k|k,s}$ and $\mathbf{P}_{k|k,s}$ via (9.55)-(9.57)
  Predict $\theta_{k+1|k,s}$ and $\mathbf{P}_{k+1|k,s}$ via (9.58)-(9.59)
end for
Combine $\theta_{k|k}$ and $\mathbf{P}_{k|k}$ via (9.63)-(9.64).

9.3.2.1 CRLB

To analyze the optimal theoretical performance of an unbiased estimator, the Cramér-Rao lower bound (CRLB) is commonly regarded as the evaluation benchmark [91,180]. Note that the variance's equality with the mean squared error (MSE) for the estimator $\hat{\mathbf{\Phi}}$ strictly satisfies the information inequality [234],

$$\mathbb{E}\{(\hat{\mathbf{\Phi}} - \mathbf{\Phi})(\hat{\mathbf{\Phi}} - \mathbf{\Phi})^T\} \geq \mathbf{I}(\mathbf{\Phi})^{-1}, \tag{9.65}$$

where $\mathbf{I}(\mathbf{\Phi})$ is the FIM for the parameter vector $\mathbf{\Phi}$. However, the parameters we are interested in are merely the positioning-related error variance, which indicates that only the upper left 2×2 submatrix of $\mathbf{I}(\mathbf{\Phi})^{-1}$ is of interest in a 2D localization problem.

9.3.2.2 SPEB

The SPEB, a measure to bound the average squared position error, is commonly defined to evaluate the performance of localization accuracy on wireless collaboration networks [282]. Determining the SPEB requires to obtain the inversion of the FIM as follows:

$$SPEB = \mathbb{TR}\{[\mathbf{I}(\mathbf{\Phi})^{-1}]_{2\times 2}\} = \mathbf{I}(\mathbf{\Phi})^{-1}(1,1) + \mathbf{I}(\mathbf{\Phi})^{-1}(2,2). \tag{9.66}$$

However, by reason of $\mathbf{I}(\mathbf{\Phi})$ usually being a high-dimension matrix, the inversion of $\mathbf{I}(\mathbf{\Phi})$ is quite complex to calculate, which results in a trade-off between computation complexity and performance evaluation. In fact, only the submatrix $[\mathbf{I}(\mathbf{\Phi})^{-1}]_{2\times 2}$ can contribute the unique insights into the bounding laws on the localization problems.

9.3.2.3 EFIM

To circumvent the calculation of the matrix inversion, we first introduce the notions of the EFIM [282].

Given a parameter vector $\boldsymbol{\Phi} = [\boldsymbol{\Phi}_{\boldsymbol{\Omega}}^T, \boldsymbol{\Phi}_{\boldsymbol{\Upsilon}}^T]^T$ and let the FIM $\mathbf{I}(\boldsymbol{\Phi})$ be written as a 2×2 block matrix

$$\mathbf{I}(\boldsymbol{\Phi}) = \begin{pmatrix} \mathbf{I}_{\boldsymbol{\Omega}} & \mathbf{I}_{\boldsymbol{\Omega}\boldsymbol{\Upsilon}} \\ \mathbf{I}_{\boldsymbol{\Omega}\boldsymbol{\Upsilon}}^T & \mathbf{I}_{\boldsymbol{\Upsilon}} \end{pmatrix}, \tag{9.67}$$

where $\boldsymbol{\Phi} \in \mathbb{R}^{\mathcal{N}}$ and $\boldsymbol{\Phi}_{\boldsymbol{\Omega}} \in \mathbb{R}^{\mathcal{M}}$. $\mathbf{I}_{\boldsymbol{\Omega}} \in \mathbb{R}^{\mathcal{M}\times\mathcal{M}}$ represents the partial information of $\mathbf{I}(\boldsymbol{\Phi})$ only pertaining to $\boldsymbol{\Omega}$, $\mathbf{I}_{\Upsilon} \in \mathbb{R}^{(\mathcal{N}-\mathcal{M})\times(\mathcal{N}-\mathcal{M})}$ represents the partial information of $\mathbf{I}(\boldsymbol{\Phi})$ only pertaining to $\boldsymbol{\Upsilon}$, and $\boldsymbol{I}_{\boldsymbol{\Omega}\Upsilon} \in \mathbb{R}^{\mathcal{M}\times(\mathcal{N}-\mathcal{M})}$ represents the coupled information between $\boldsymbol{\Omega}$ and $\boldsymbol{\Upsilon}$, while the notions corresponding to the dimension meet the conditions that $1 \leq \mathcal{M} \leq \mathcal{N}$. Consequently, we obtain the EFIM of $\boldsymbol{\Omega}$ as follows:

$$\mathbf{I}(\boldsymbol{\Phi}_{\boldsymbol{\Omega}}) = \mathbf{I}_{\boldsymbol{\Omega}} - \mathbf{I}_{\boldsymbol{\Omega}\Upsilon}\mathbf{I}_{\Upsilon}^{-1}\mathbf{I}_{\boldsymbol{\Omega}\Upsilon}^T. \tag{9.68}$$

The right-hand side of Equation 9.68 is also known as the Schur complement of the sub-block I_{Υ} in $\mathbf{I}(\boldsymbol{\Phi})$ [1], which is equivalent to $\mathbf{I}(\boldsymbol{\Phi})$ for the parameters $\boldsymbol{\Phi}_{\Omega}$ in the sense that it retains all the necessary information to deduce the CRLB of $\boldsymbol{\Omega}$:

$$\left[\mathbf{I}(\boldsymbol{\Phi})^{-1}\right]_{\Omega} = [\mathbf{I}(\boldsymbol{\Phi}_{\Omega})]^{-1}. \tag{9.69}$$

9.3.2.4 The mSPEB

A novel theoretical lower bound of the MSE matrix of an unbiased positioning-related estimator is derived through studying the properties from the bounds for the trace of the inverse of a symmetric positive definite matrix.

Theorem 9.1
Given a cooperative localization network with parameter vectors $\boldsymbol{\Phi}$ *and* $\boldsymbol{\Phi} = [\boldsymbol{\Phi}_{\boldsymbol{\Omega}}^T, \boldsymbol{\Phi}_{\boldsymbol{\Upsilon}}^T]^T$ *where* $\boldsymbol{\Phi}_{\Omega}$ *is the position information vector and* $\boldsymbol{\Phi}_{\boldsymbol{\Upsilon}}$ *is the other parameter vector independent of the position information. In a positioning estimation problem, if the corresponding FIM for the parameter vector* $\boldsymbol{\Phi}$, $\boldsymbol{I}(\boldsymbol{\Phi})$, *is a positive definite matrix, then a lower bound of SPEB is given by*

$$SPEB \geq \frac{-\mathcal{T}^2\mathcal{M}^2 + (\bar{\mu}_1\mathcal{T} + \bar{\mu}_2)\mathcal{M} - \bar{\mu}_1^2}{\bar{\mu}_2\mathcal{T} - \bar{\mu}_1\mathcal{T}^2}, \tag{9.70}$$

where $\bar{\mu}_1 = \mathbb{TR}(\boldsymbol{I}(\boldsymbol{\Phi}_{\Omega}))$, $\bar{\mu}_2 = \| \boldsymbol{I}(\boldsymbol{\Phi}_{\Omega}) \|_F^2$, $\mathcal{T}$ *is the upper bound of eigenvalues of* $\boldsymbol{I}(\boldsymbol{\Phi}_{\Omega})$, *and* $\mathcal{M}$ *is the dimension of* $\boldsymbol{I}(\boldsymbol{\Phi}_{\Omega})$. $\boldsymbol{I}(\boldsymbol{\Phi}_{\Omega})$ *is the corresponding EFIM.*

Proof 9.1 The Schur complement condition on positive definite matrix states that for any symmetric matrix ■ of the form

$$\blacksquare = \begin{pmatrix} \mathbf{A} & \mathbf{B} \\ \mathbf{B}^T & \mathbf{C} \end{pmatrix},$$

if $\mathbf{C}$ is invertible, the following property will be obtained:

$$\blacksquare \succ 0 \Longleftrightarrow \mathbf{A} - \mathbf{B}\mathbf{C}^{-1}\mathbf{B}^T \succ 0 \textit{ and } \mathbf{C} \succ 0 \tag{9.71}$$

where $\blacksquare \succ 0$ meaning that ■ is a positive definite matrix. In a specified positioning estimation problem, it is clear from Equation 9.71 that the corresponding EFIM, $\mathbf{I}(\boldsymbol{\Phi}_{\Omega})$, is a positive definite matrix as long as the corresponding FIM, $\mathbf{I}(\boldsymbol{\Phi})$, is a positive definite matrix.

Having shown that the lower bound for the SPEB is a function of the parameters of the EFIM, $\boldsymbol{I}(\boldsymbol{\Phi}_\Omega)$, including the trace, the F-norm, the upper bound of eigenvalues, and the dimension, we will explain the obtained result by introducing a lemma that derives the lower and upper bounds for the trace of the inverse of a symmetric positive definite matrix presented by Bai and Golub [17].

Lemma 9.1
Let Ξ be an $\mathcal{N}-by-\mathcal{N}$ symmetric positive definite matrix, $\mu_1 = \mathbb{TR}(\Xi)$, $\mu_2 = \| \Xi \|_F^2$ and $\lambda(\Xi) \subset [\mathcal{S}, \mathcal{T}]$ with $\mathcal{S} > 0$, then

$$\begin{aligned} &\begin{bmatrix} \mu_1 & \mathcal{N} \end{bmatrix} \begin{bmatrix} \mu_2 & \mu_1 \\ \mathcal{T}^2 & \mathcal{T} \end{bmatrix}^{-1} \begin{bmatrix} \mathcal{N} \\ 1 \end{bmatrix} \\ &\leq \mathbb{TR}(\Xi^{-1}) \leq \begin{bmatrix} \mu_1 & \mathcal{N} \end{bmatrix} \begin{bmatrix} \mu_2 & \mu_1 \\ \mathcal{S}^2 & \mathcal{S} \end{bmatrix}^{-1} \begin{bmatrix} \mathcal{N} \\ 1 \end{bmatrix}. \end{aligned} \tag{9.72}$$

In addition, it is obvious that the FIM, $\boldsymbol{I}(\boldsymbol{\Phi})$, is a symmetric matrix. Consequently, in a specified position estimation problem, if the corresponding FIM is a positive definite matrix, both the FIM, $\boldsymbol{I}(\boldsymbol{\Phi})$, and the EFIM, $\boldsymbol{I}(\boldsymbol{\Phi}_\Omega)$, are symmetric positive definite matrix, so all the preconditions of the lemma are met. Now, we can use the lemma provided earlier with $\Xi = \boldsymbol{I}(\boldsymbol{\Phi})$. Note that the SPEB of the FIM, $\boldsymbol{I}(\boldsymbol{\Phi})$, is $\mathbb{TR}\left(\boldsymbol{I}(\boldsymbol{\Phi}_\Omega)^{-1}\right)$, and the dimension of the EFIM, $\boldsymbol{I}(\boldsymbol{\Phi}_\Omega)$, is $\mathcal{M}$. Expanding the left-hand side of the inequality 9.72 in 9.73, our end result is concluded in Equation 9.70.

$$\begin{aligned} &\begin{bmatrix} \mu_1 & \mathcal{N} \end{bmatrix} \begin{bmatrix} \mu_2 & \mu_1 \\ \mathcal{T}^2 & \mathcal{T} \end{bmatrix}^{-1} \begin{bmatrix} \mathcal{N} \\ 1 \end{bmatrix} \\ &= \frac{-\mathcal{T}^2\mathcal{N}^2 + (\mu_1\mathcal{T} + \mu_2)\mathcal{N} - \mu_1^2}{\mu_2\mathcal{T} - \mu_1\mathcal{T}^2}. \end{aligned} \tag{9.73}$$

Theorem 9.2
Given a cooperative localization network with parameter vectors $\boldsymbol{\Phi}$ and $\boldsymbol{\Phi} = [\boldsymbol{\Phi}_\Omega^T, \boldsymbol{\Phi}_\Upsilon^T]^T$ where $\boldsymbol{\Phi}_\Omega$ is the position information vector and $\boldsymbol{\Phi}_\Upsilon$ is the other parameter vector independent of the position information. In a positioning estimation problem, if the corresponding FIM for the parameter vector $\boldsymbol{\Phi}$, $\boldsymbol{I}(\boldsymbol{\Phi})$, is a positive definite matrix, then a lower bound of SPEB is given by

$$SPEB \geq \frac{\mathcal{M}}{\bar{\mathcal{S}}} + \sum_{\mathcal{I}=1}^{\mathcal{M}} \frac{(\bar{\mathcal{S}} - \Gamma_{\mathcal{I}\mathcal{I}})^2}{\bar{\mathcal{S}}(\bar{\mathcal{S}}\Gamma_{\mathcal{I}\mathcal{I}} - \mathcal{W}_{\mathcal{I}\mathcal{I}})}, \tag{9.74}$$

where $\bar{\mathcal{S}}$ is the lower bound of eigenvalues of $\boldsymbol{I}(\boldsymbol{\Phi})$, $\mathcal{W}_{\mathcal{I}\mathcal{I}} = \sum_{\mathcal{J}=1}^{\mathcal{N}} \Gamma_{\mathcal{I}\mathcal{J}}^2$, $\Gamma_{\mathcal{I}\mathcal{J}}$ is the $\mathcal{I}, \mathcal{J}$th element of $\boldsymbol{I}(\boldsymbol{\Phi})$ ($\mathcal{I} = 1, 2, \ldots, \mathcal{M}$, $\mathcal{J} = 1, 2, \ldots, \mathcal{N}$), $\mathcal{N}$ is the dimension of $\boldsymbol{I}(\boldsymbol{\Phi})$, and $\mathcal{M}$ is the dimension of $\boldsymbol{I}(\boldsymbol{\Phi}_\Omega)$. $\boldsymbol{I}(\boldsymbol{\Phi}_\Omega)$ is the corresponding EFIM.

Proof 9.2 Similar to the proof procedures discussed in Theorem 9.1, it is concluded that in a specified positioning estimation problem, the corresponding FIM, $\boldsymbol{I}(\boldsymbol{\Phi})$, is a symmetric positive definite matrix when the FIM meets the conditions defined in Theorem 9.2.

Having shown that the lower bound for the SPEB is a function of the parameter of the FIM, $\boldsymbol{I}(\boldsymbol{\Phi})$, including the lower bound of eigenvalues, the entries on the main diagonal, the sum of the entries on the $\mathcal{I}$th row, and the dimension of the FIM as well as the dimension of EFIM, $\boldsymbol{I}(\boldsymbol{\Phi}_\Omega)$, we will explain the obtained result by introducing another Lemma that derives the lower and upper bounds for the entries on its main diagonal of the inverse of a symmetric positive definite matrix presented by Robinson and Wathen [187].

Lemma 9.2

Let Ξ be an $\mathcal{N}-by-\mathcal{N}$ symmetric positive definite matrix, and $\lambda(\Xi) \subset [\mathcal{S},\mathcal{T}]$ with $\mathcal{S} > 0$, then

$$\frac{1}{\mathcal{S}} + \frac{(\mathcal{S}-\Gamma_{\mathcal{I}\mathcal{I}})^2}{\mathcal{S}(\mathcal{S}\Gamma_{\mathcal{I}\mathcal{I}}-\mathcal{W}_{\mathcal{I}\mathcal{I}})} \le (\Xi^{-1})_{\mathcal{I}\mathcal{I}} \le \frac{1}{\mathcal{T}} + \frac{(\mathcal{T}-\Gamma_{\mathcal{I}\mathcal{I}})^2}{\mathcal{T}(\mathcal{T}\Gamma_{\mathcal{I}\mathcal{I}}-\mathcal{W}_{\mathcal{I}\mathcal{I}})}, \tag{9.75}$$

where $\mathcal{W}_{\mathcal{I}\mathcal{I}} = \sum_{\mathcal{J}=1}^{\mathcal{N}} \Gamma_{\mathcal{I}\mathcal{J}}^2$ and $\Gamma_{\mathcal{I}\mathcal{J}}$ is the $\mathcal{I},\mathcal{J}$th element of Ξ ($\mathcal{I},\mathcal{J} = 1,2,\ldots,\mathcal{N}$).

As we know $\mathbb{TR}(\Xi^{-1}) = \sum_{\mathcal{I}=1}^{\mathcal{N}} (\Xi^{-1})_{\mathcal{I}\mathcal{I}}$, so the lower bound of $\mathbb{TR}(\Xi^{-1})$ can be written as follows:

$$\sum_{\mathcal{I}=1}^{\mathcal{N}} (\Xi^{-1})_{\mathcal{I}\mathcal{I}} \ge \sum_{\mathcal{I}=1}^{\mathcal{N}} \left\{ \frac{1}{\mathcal{S}} + \frac{(\mathcal{S}-\Gamma_{\mathcal{I}\mathcal{I}})^2}{\mathcal{S}(\mathcal{S}\Gamma_{\mathcal{I}\mathcal{I}}-\mathcal{W}_{\mathcal{I}\mathcal{I}})} \right\}. \tag{9.76}$$

Now, setting the dimension $\mathcal{N}$ to $\mathcal{M}$, the left hand of (9.76) is the SPEB, then the end result is concluded.

It is noted that the two proposed theorems are suitable for both 2D and 3D localization scenarios in general. For a specified 2D positioning problem, the dimension of the EFIM, $\mathcal{M}$, is set to 2, while for the 3-D, $\mathcal{M} = 3$.

Lemma 9.3

Given a cooperative localization network with parameter vectors $\mathbf{\Phi}$ and $\mathbf{\Phi} = [\mathbf{\Phi}_\Omega^T, \mathbf{\Phi}_\Upsilon^T]^T$, where $\mathbf{\Phi}_\Omega$ is the position information vector and $\mathbf{\Phi}_\Upsilon$ is the other parameter vector independent of the position information. In a positioning estimation problem, if the corresponding FIM for the parameter vector $\mathbf{\Phi}$, $\boldsymbol{I}(\mathbf{\Phi})$, is a positive definite matrix, the resulting theoretical mSPEB—the modified lower bound of the MSE matrix of an unbiased estimator of $\mathbf{\Phi}_\Omega$ can be expressed as

$$mSPEB = \max \left\{ \begin{array}{c} \dfrac{-\bar{\mathcal{T}}^2\mathcal{M}^2 + (\bar{\mu}_1\bar{\mathcal{T}} + \bar{\mu}_2)\mathcal{M} - \bar{\mu}_1^2}{\bar{\mu}_2\bar{\mathcal{T}} - \bar{\mu}_1\bar{\mathcal{T}}^2}, \\ \dfrac{\mathcal{M}}{\bar{\mathcal{S}}} + \displaystyle\sum_{\mathcal{I}=1}^{\mathcal{M}} \frac{(\bar{\mathcal{S}}-\Gamma_{\mathcal{I}\mathcal{I}})^2}{\bar{\mathcal{S}}(\bar{\mathcal{S}}\Gamma_{\mathcal{I}\mathcal{I}}-\mathcal{W}_{\mathcal{I}\mathcal{I}})} \end{array} \right\}, \tag{9.77}$$

where $\bar{\mu}_1 = \mathbb{TR}(\boldsymbol{I}(\mathbf{\Phi}_\Omega))$, $\bar{\mu}_2 = \| \boldsymbol{I}(\mathbf{\Phi}_\Omega) \|_F^2$, $\bar{\mathcal{S}}$ is the lower bound of eigenvalues of $\boldsymbol{I}(\mathbf{\Phi})$, $\bar{\mathcal{T}}$ is the upper bound of eigenvalues of $\boldsymbol{I}(\mathbf{\Phi}_\Omega)$, $\mathcal{W}_{\mathcal{I}\mathcal{I}} = \sum_{\mathcal{J}=1}^{\mathcal{N}} \Gamma_{\mathcal{I}\mathcal{J}}^2$, $\Gamma_{\mathcal{I}\mathcal{J}}$ is the $\mathcal{I},\mathcal{J}$th element of $\boldsymbol{I}(\mathbf{\Phi})$ ($\mathcal{I} = 1,2,\ldots,\mathcal{M}$, $\mathcal{J} = 1,2,\ldots,\mathcal{N}$), $\mathcal{N}$ is the dimension of $\boldsymbol{I}(\mathbf{\Phi})$, and $\mathcal{M}$ is the dimension of $\boldsymbol{I}(\mathbf{\Phi}_\Omega)$. $\boldsymbol{I}(\mathbf{\Phi}_\Omega)$ is the corresponding EFIM.

Proof 9.3 According to the proof procedures discussed in Theorem 1 and Theorem 2, it is obvious that $SPEB \ge mSPEB$, so the end result is concluded.

It should be mentioned that the closed form of the mSPEB needs to process a lower dimensional EFIM and calculate the bound for the minimum eigenvalue of a high-dimensional FIM, such that it is with lower complexity, when compared to the SPEB used in current literature [8,10,91] that has to calculate the inverse of the high-dimensional FIM directly.

9.3.2.5 Insights into factors affecting the CP performance

Here, the FIM can be calculated at each time instant k as

$$\begin{aligned}\mathbf{I}(\theta) &= \mathbb{E}\left\{\left[\frac{\partial \ln(f(\mathbf{z}|\theta))}{\partial \theta}\right]\left[\frac{\partial \ln(f(\mathbf{z}|\theta))}{\partial \theta}\right]^T\right\},\\ &= -\mathbb{E}\left\{\frac{\partial^2 \ln(f(\mathbf{z}|\theta))}{\partial \theta^2}\right\},\end{aligned} \tag{9.78}$$

where $\mathbf{z}$ is the measurement vector in Equation 9.38, θ the state vector in Equation 9.30, $\mathbb{E}\{\cdot\}$ is the expectation operator, and $f\{\cdot\}$ is the conditional Probability Distribution Function (PDF) of $\mathbf{z}$ on condition of the value of θ. Assume that $f(\mathbf{z}|\theta)$ follows a Gaussian distribution $normal(\mathbf{z};\ \mathbf{z}_{\text{mean}},\ \mathbf{R})$ as

$$f(\mathbf{z}|\theta) = \frac{\mathbb{EXP}\{-\frac{1}{2}(\mathbf{z}-\mathbf{z}_{\text{mean}})^T\mathbf{R}^{-1}(\mathbf{z}-\mathbf{z}_{\text{mean}})\}}{(2\pi)^{\frac{4+i+j}{2}}\sqrt{\mathbb{DET}(\mathbf{R})}}, \tag{9.79}$$

where the variable $\mathbf{z}$ is normally distributed with the mean $\mathbf{z}_{\text{mean}}$ and the covariance matrix $\mathbf{R}$, i is the total number of the neighboring vehicles associated with the DFS measurements, and j is the total number of the neighboring vehicles associated with the RSSI measurements. After deploying the natural logarithm on both sides of Equation 9.79, the formula can be rewritten as

$$\begin{aligned}\ln(f(\mathbf{z}|\theta)) = &-\frac{1}{2}\ln(|\mathbf{R}|) - \frac{1}{2}(\mathbf{z}-\mathbf{z}_{\text{mean}})^T\mathbf{R}^{-1}(\mathbf{z}-\mathbf{z}_{\text{mean}})\\ &-\frac{4+i+j}{2}\ln(2\pi).\end{aligned} \tag{9.80}$$

Then, substituting Equation 9.38 into 9.80, the result of the second-order partial derivative of the state vector θ can be written as

$$\frac{\partial^2 \ln(f(\mathbf{z}|\theta))}{\partial \theta^2} = -\mathbf{H}^T\mathbf{R}^{-1}\mathbf{H}. \tag{9.81}$$

Subsequently, substituting Equation 9.81 into 9.78, the form of the FIM $\mathbf{I}(\theta)$ can be simplified as follows:

$$\mathbf{I}(\theta) = \mathbf{H}^T\mathbf{R}^{-1}\mathbf{H}, \tag{9.82}$$

which can be formulated with the form of a block matrix as follows:

$$\mathbf{I}(\theta) = \begin{bmatrix} \mathbf{I_A} & \mathbf{I_B} \\ \mathbf{I_B^T} & \mathbf{I_C} \end{bmatrix}. \tag{9.83}$$

The elements of $\mathbf{I}(\theta)$ are given by Equations 9.84 through 9.86, where σ_{m_x}, σ_{m_y}, $\sigma_{\dot{m}_x}$, $\sigma_{\dot{m}_y}$, σ_ρ, σ_r are the elements in the covariance matrix of the measurements defined in Equations 9.87 and 9.88.

$$\Xi_{\mathbf{A}} = \begin{bmatrix} \frac{1}{\sigma_{m_x}^2} + \frac{1}{\sigma_\rho^2}\sum_{\alpha=1}^{i}(\mathcal{H}_k^{\alpha 1})^2 + \frac{1}{\sigma_r^2}\sum_{\beta=1}^{j}(\mathcal{G}_k^{\beta 1})^2 & \frac{1}{\sigma_\rho^2}\sum_{\alpha=1}^{i}(\mathcal{H}_k^{\alpha 1}\mathcal{H}_k^{\alpha 2}) + \frac{1}{\sigma_r^2}\sum_{\beta=1}^{j}(\mathcal{G}_k^{\beta 1}\mathcal{G}_k^{\beta 2}) \\ \frac{1}{\sigma_\rho^2}\sum_{\alpha=1}^{i}(\mathcal{H}_k^{\alpha 1}\mathcal{H}_k^{\alpha 2}) + \frac{1}{\sigma_r^2}\sum_{\beta=1}^{j}(\mathcal{G}_k^{\beta 1}\mathcal{G}_k^{\beta 2}) & \frac{1}{\sigma_{m_y}^2} + \frac{1}{\sigma_\rho^2}\sum_{\alpha=1}^{i}(\mathcal{H}_k^{\alpha 2})^2 + \frac{1}{\sigma_r^2}\sum_{\beta=1}^{j}(\mathcal{G}_k^{\beta 2})^2 \end{bmatrix}, \tag{9.84}$$

$$\Xi_{\mathbf{B}} = \begin{bmatrix} \frac{1}{\sigma_\rho^2}\sum_{\alpha=1}^{i}(\mathcal{H}_k^{\alpha 1}\mathcal{H}_k^{\alpha 3}) & \frac{1}{\sigma_\rho^2}\sum_{\alpha=1}^{i}(\mathcal{H}_k^{\alpha 1}\mathcal{H}_k^{\alpha 4}) \\ \frac{1}{\sigma_\rho^2}\sum_{\alpha=1}^{i}(\mathcal{H}_k^{\alpha 2}\mathcal{H}_k^{\alpha 3}) & \frac{1}{\sigma_\rho^2}\sum_{\alpha=1}^{i}(\mathcal{H}_k^{\alpha 2}\mathcal{H}_k^{\alpha 4}) \end{bmatrix}, \tag{9.85}$$

$$\Xi_{\mathbf{C}} = \begin{bmatrix} \frac{1}{\sigma_{\dot{m}_x}^2} + \frac{1}{\sigma_{\dot{\rho}}^2}\sum_{\alpha=1}^{i}(\mathcal{H}_k^{\alpha 3})^2 & \frac{1}{\sigma_{\dot{\rho}}^2}\sum_{\alpha=1}^{i}(\mathcal{H}_k^{\alpha 3}\mathcal{H}_k^{\alpha 4}) \\ \frac{1}{\sigma_{\dot{\rho}}^2}\sum_{\alpha=1}^{i}(\mathcal{H}_k^{\alpha 3}\mathcal{H}_k^{\alpha 4}) & \frac{1}{\sigma_{\dot{m}_y}^2} + \frac{1}{\sigma_{\dot{\rho}}^2}\sum_{\alpha=1}^{i}(\mathcal{H}_k^{\alpha 4})^2 \end{bmatrix}. \tag{9.86}$$

Consequently, the mSPEB of $\mathbf{I}(\theta)$ can be obtained from (9.77). $\mathbf{I_A}$ characterizes the localization information corresponding to the cooperation via intervehicle measurements using the GPS and the RSSI-related data, while $\mathbf{I_B}$ and $\mathbf{I_C}$ characterize the same type of measurements using only the GPS data. As the components of each element derived in the FIM, $\mathbf{I}(\theta)$, we can conclude that each intervehicle measurement or metrics will contribute to positioning enhancements from the point of view of the CRLB.

9.3.3 Numerical results

This section discusses a series of computer simulations used to evaluate the performance of the proposed CP method. Meanwhile, the insightful data analyses were conducted to interpret the inherent relationship between traffic incidents and positioning enhancements for mobile vehicle localization.

9.3.3.1 Simulation setup

Consider a section of urban roads with a width of four lanes (each lane with a width of 3.5 m) and a length of 1 km. It is assumed that the basic traffic setting is subject to the following conditions: (1) The traffic intensity of the road is 20 vehicles/km, (2) The CV penetration is 100%, and (3) The average velocity of the traffic flow is 90 km/h. The vehicles' dynamics are described by model (9.30), and the sampling period $\Delta t = 0.2$ s. The distribution of the system noise ζ_k takes with covariance $\mathbf{Q}_k = \text{diag}(\sigma_{a_{x,k}}^2, \sigma_{a_{y,k}}^2)$, where the elements $\sigma_{a_{x,k}} = \sqrt{0.99/2}$ m/s^2 and $\sigma_{a_{y,k}} = \sqrt{0.01/2}$ m/s^2 are the acceleration noise along the E and the N directions, respectively. The settings of $\sigma_{a_{x,k}}$ and $\sigma_{a_{y,k}}$ reveal the dynamic behavior of non-abnormal vehicles moving on the road of which driving actions on the acceleration are along the X-axis. Three dynamic models corresponding to the different accelerations of $0, -2, 5\,\text{m/s}^2$ are used, and the switching between any two of these three models is described by the first-order Markov chain φ with the transition probability $\pi_{pp}^{\varphi} = 0.8$ $(p = 1,2,3)$, and $\pi_{pq}^{\varphi} = 0.1$ $(p \neq q;\ p,q = 1,2,3)$. As shown in Figure 9.4, the TV starts at the position (300,10) in m, and then travels on the second lane from the West (W) to the East (E). The initial velocity is set as (90,0) in km/h, and the TV starts to make an approximated uniform motion between 0 and 35 step, a slow-down movement with a deceleration of -2 m/s^2 between 36 and 40 step, a straight movement with a constant velocity between 41 and 65 step, a speed-up movement with an acceleration of 5 m/s^2 between 66 and 70 step, and finally another uniform movement between 71 and 100 step.

It is assumed that the initial locations of all the neighboring vehicles are practically distributed within the road section according to the uniform distribution. The dynamics of the neighboring vehicles are subject to model 9.30. However, they keep a near stable movement during the entire 100 steps. The DSRC communication range of each vehicle is set as 300 m, and only the neighboring vehicles under the DSRC coverage of the TV could beneficially contribute to the positioning enhancements for mobile vehicle localization. It is noted that a vehicle who holds that communication capability combined with the coarse measurements obtained from the GPS and the DSRC physical layer is defined as the vehicle who holds the CV technologies. The measurement process is represented by model 9.38, and the measurement noise $\vartheta_k(\phi_k)$ is described by the Markov chain ϕ_k, which takes with the switching covariance matrix associated with the DFS and the RSSI measurements under the LOS and the NLOS conditions with the following formulations:

$$\begin{aligned}\mathbf{R}_k^{\text{LOS}} = \text{diag}(&\sigma_{m_{x,k}}^{\text{GPS}^2}, \sigma_{m_{y,k}}^{\text{GPS}^2}, \sigma_{\dot{m}_{x,k}}^{\text{GPS}^2}, \sigma_{\dot{m}_{y,k}}^{\text{GPS}^2}, \sigma_{\rho_k^1}^{\text{LOS,DFS}^2},\\ &\ldots, \sigma_{\rho_k^i}^{\text{LOS,DFS}^2}, \sigma_{r_k^1}^{\text{LOS,RSSI}^2},\\ &\ldots, \sigma_{r_k^j}^{\text{LOS,RSSI}^2})\end{aligned} \tag{9.87}$$

and

$$\begin{aligned}\mathbf{R}_k^{\text{NLOS}} = \text{diag}(&\sigma_{m_{x,k}}^{\text{GPS}^2}, \sigma_{m_{y,k}}^{\text{GPS}^2}, \sigma_{\dot{m}_{x,k}}^{\text{GPS}^2}, \sigma_{\dot{m}_{y,k}}^{\text{GPS}^2}, \sigma_{\rho_k^1}^{\text{NLOS,DFS}^2},\\ &\ldots, \sigma_{\rho_k^i}^{\text{NLOS,DFS}^2}, \sigma_{r_k^1}^{\text{NLOS,RSSI}^2},\\ &\ldots, \sigma_{r_k^j}^{\text{NLOS,RSSI}^2}),\end{aligned} \tag{9.88}$$

respectively. For the GPS measurements, assume that the variance of the position and the velocity in the LOS and the NLOS conditions are fixed. In the 2D localization problem, the LOS and the NLOS conditions are used for depicting the different situation of the propagation channel on the DSRC signals that is paralleling to the road plane, whereas the propagation of the GPS signals is not in that plane, so as to take the variances of the GPS measurements as $\sigma_{m_{x,k}}^{\text{GPS}} = \sqrt{200/2}$ m, $\sigma_{m_{y,k}}^{\text{GPS}} = \sqrt{200/2}$ m, $\sigma_{\dot{m}_{x,k}}^{\text{GPS}} = \sqrt{15/2}$ m/s, $\sigma_{\dot{m}_{y,k}}^{\text{GPS}} = \sqrt{15/2}$ m/s, respectively. For the DFS measurements, the noise variance under the LOS condition is set as $\sigma_{\rho_k^\alpha}^{\text{LOS,DFS}} = 100$ Hz, and that variance under the NLOS condition is set as $\sigma_{\rho_k^\alpha}^{\text{NLOS,DFS}} = 120$ Hz. For the RSSI measurements, the transmission power–related constant $C = 20$, and the path-loss exponent $\gamma = 3.5$. The noise variance under the LOS condition is set as $\sigma_{r_k^\beta}^{\text{LOS,RSSI}} = 5$ dBm, and that variance under the NLOS condition is set as $\sigma_{r_k^\beta}^{\text{NLOS,RSSI}} = 30$ dBm. The transition probability to describe the switching states between the LOS and the NLOS conditions is set as $\pi_{uu}^{\phi} = 0.9(u = 1,2)$, and $\pi_{uv}^{\phi} = 0.1(u \neq v; u, v = 1,2)$. Assume that the TV travels under the LOS condition at the beginning, entering into the NLOS condition between the 30th and 80th steps, and makes another transition into the LOS condition between the 81st and 100th step.

9.3.3.2 *Single-trial analysis*

To verify the effectiveness of the proposed method, a single-trial test is used to clearly demonstrate the entire scenario in which the settings follow the parameters that are described in the Section 9.3.3.1. The true trajectory of the TV (i.e., the black circle and the black dot represent the initial and the ending positions of the TV, respectively), one trail of the estimated trajectory of the TV using the CV-enhanced DIMM-KF method, and the original GPS measurements of the TV are collectively shown in Figure 9.6. Compared to the GPS-based positioning, the proposed CV-enhanced DIMM-KF method could provide much better performance on the vehicle localization throughout the entire process.

9.3.3.3 *Monte Carlo results*

To evaluate the closeness from the estimated to the true trajectories, the RMSE metrics in position is deployed at each time step k. The definitions of the RMSE are formulated as Equation 9.89, the root-mean-SPEB is $R.SPEB_k = \sqrt{\frac{1}{N_T}\sum_{T=1}^{N_T} SPEB_k(T)}$, and the root-mean-mSPEB is

$R.mSPEB_k = \sqrt{\frac{1}{N_T}\sum_{T=1}^{N_T} mSPEB_k(T)}.$

$$RMSE_k = \sqrt{\frac{1}{N_T}\sum_{T=1}^{N_T}\left\{\left(\hat{m}_{x,k}(T) - m_{x,k}(T)\right)^2 + \left(\hat{m}_{y,k}(T) - m_{y,k}(T)\right)^2\right\}}, \qquad (9.89)$$

In Equation 9.89, $(\hat{m}_{x,k}(T), \hat{m}_{y,k}(T))$ denotes the estimated position vector in the Tth Monte Carlo simulation. In Figure 9.7, the comparison between the GPS-based positioning and the proposed CP method is conducted over $N_T = 1000$ Monte Carlo runs. Each of them follows the basic traffic settings that are described in Section IV-A. Meanwhile, for the TV's position, the R.SPEB and the R.mSPEB obtained from the FIM and the EFIM at each time step k are illustrated as well. The results testify the achieved performance that is enhanced by the proposed CP method, and also indicate that the R.mSPEB is at least incredibly close to the R.SPEB in this specific 2D case.

To analyze the enhanced performance of the proposed CP method on different traffic intensities and CV penetrations, the statistical simulations are created. Each case is imitated $N_T = 500$ times, and the average achievable performance for the combinations between the

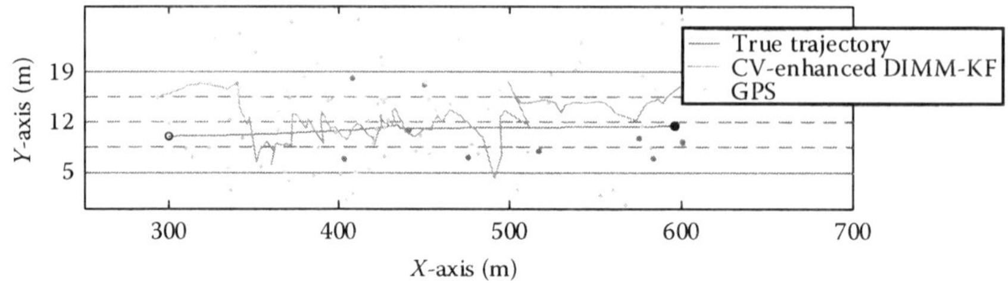

Figure 9.6 One trial demo of vehicular positioning estimation with both the DFS and the RSSI measurements.

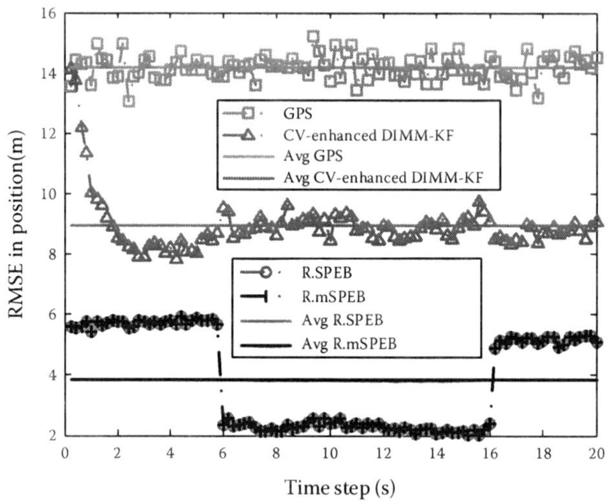

Figure 9.7 The performance of the GPS and the proposed method, and the fundamental limits bounded by the R.SPEB and the R.mSPEB.

traffic intensity and the CV penetration was evaluated. Correspondingly, the traffic intensity is set ranging from 20 to 200 vehicle/km, and the CV penetration is set ranging from 25% to 100%. Both Figures 9.8 and 9.9 are under consideration of the neighboring vehicles who can provide the DFS measurements to benefit the positioning performance on the TV. Figure 9.8 shows that the enhancements on the vehicle localization system generally increases with the increase in traffic intensity and the CV penetration, respectively. Additionally, in Figure 9.9, it should be noted that a few outliers adversely affect the achieved performance when the traffic intensity is at a relative high level. Indeed, regardless of the traffic intensity, the number of the participated neighboring vehicles is a key factor to the CP method on the vehicle localization. Figure 9.9 shows that the enhancement rate sternly increases with increase in the number of the participated neighboring vehicles. With regard to the performance metric defined as

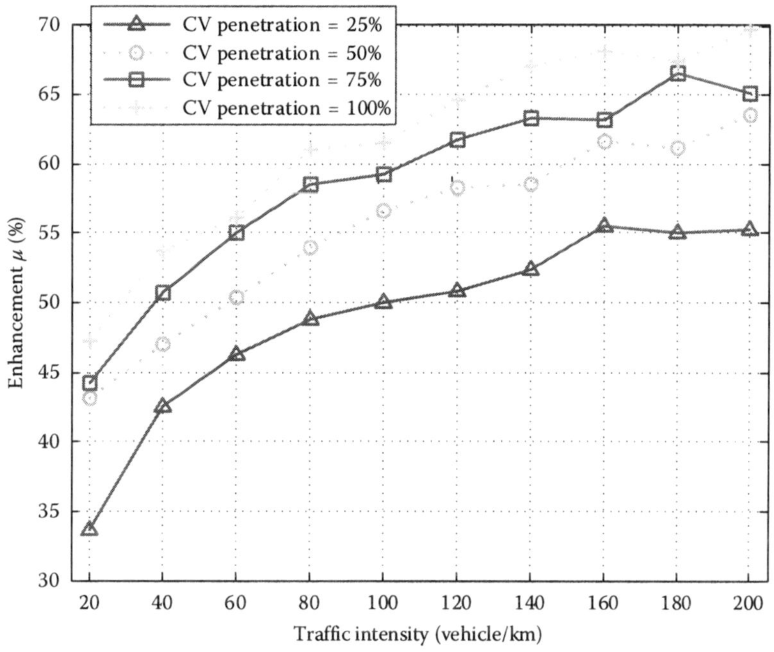

Figure 9.8 The enhancements for different CV penetrations—DFS only.

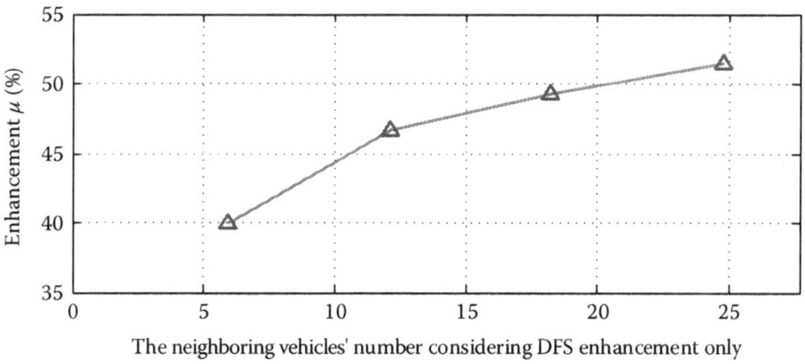

Figure 9.9 The neighboring vehicles' number and achieved performance—DFS only.

$\mu\% = [1 - RMSE_{CP}/RMSE_{GPS} \times 100\%]$, the enhancement rate over the GPS-based positioning reaches at about 35% to about 70%.

Both Figures 9.10 and 9.11 are under consideration of the neighboring vehicles who can provide the RSSI measurements to benefit the positioning performance on the TV. Figures 9.10 and 9.11 show that the trend of the related increments is similar to Figures 9.8 and 9.9, respectively. Correspondingly, the CV-enhancement method reaches at about 25% to about 60% over the GPS-based positioning.

Figure 9.12 compares the CP method enhanced by using the DFS and the RSSI measurements with the other enhancement by only using the DFS measurements, showing that the

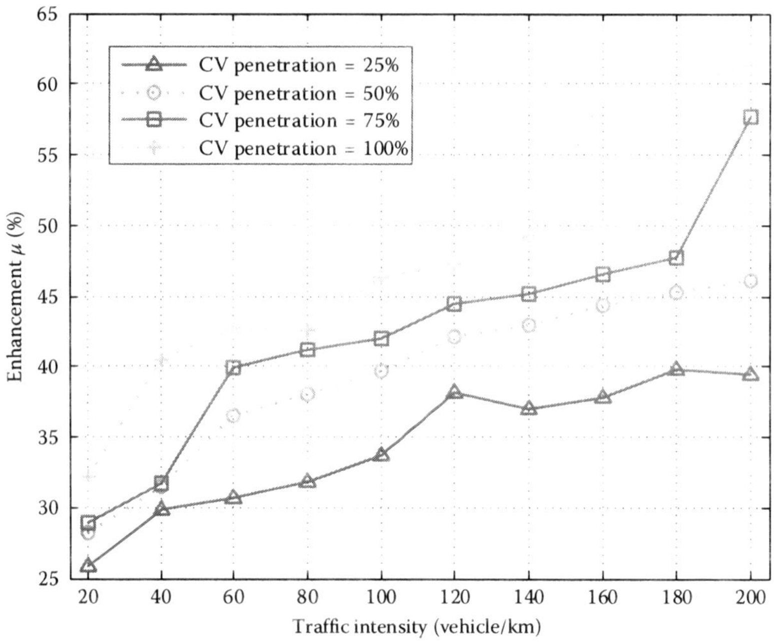

Figure 9.10 The enhancements for different CV penetrations—RSSI only.

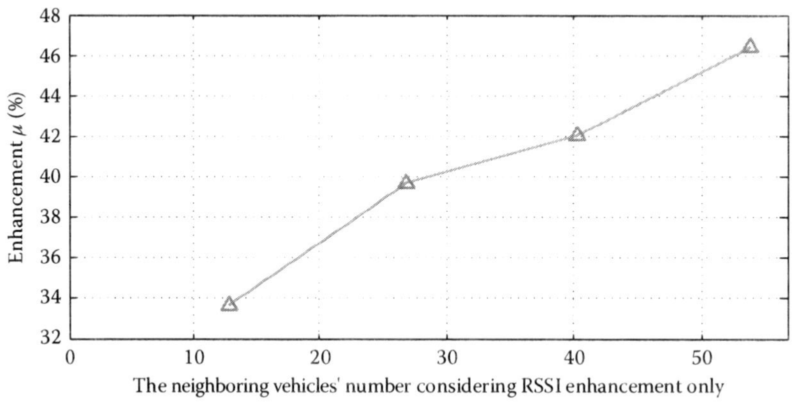

Figure 9.11 The neighboring vehicles' number and achieved performance—RSSI only.

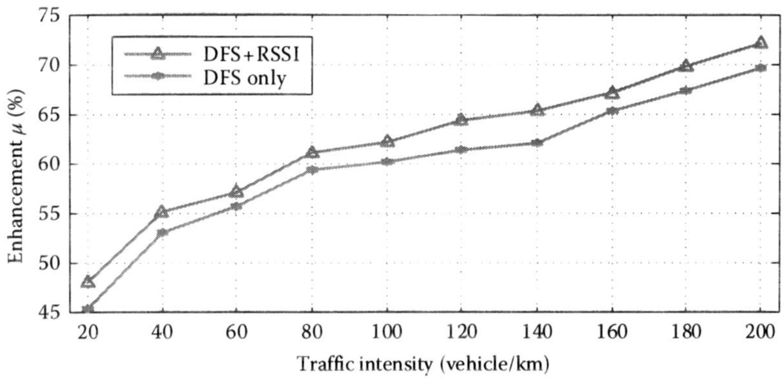

Figure 9.12 The enhancements for the DFS and the RSSI measurements combination, and for the DFS measurements only.

proposed CP enhancement approach using more measurements' data can better improve the positioning performance for mobile vehicle localization. Significantly, the achieved enhancement rate is up to 72.10% when the traffic intensity is 50 vehicle/km/lane.

9.4 Conclusion

We proposed a vehicular positioning algorithm in Section 9.2. With the assistance of CV, the observation transition matrix and the covariance matrix of observation noise are updated by the fusion data at each time instant, which provides additional useful information compared to the traditional filtering approach. Simulation results show that the proposed approach outperforms the GPS-based localization. Additionally, in Section 9.3, we proposed a novel method that combined both the DFS and the RSSI measurements extracted from the DSRC physical layer to enhance the positioning accuracy for the vehicle localization system. Avoiding some range-based methods, the proposed CP method is designed to leverage both the range-rate (DFS) and the ranging (RSSI) measurements shared in the V2V communication environments. The feasibility and the performance of the method have been investigated through the following two types of simulations: (1) the single-trial analysis and (2) the Monte Carlo results. The achieved enhancement rate on the TV localization can be increased from about 35% to about 72% compared with the stand-alone GPS method, according to different traffic intensities and the CV penetrations. The proposed mSPEB is verified to bound the fundamental limits for localization systems with less computational complexity compared to the conventional SPEB. Additional insight that all intervehicle measurements can improve the CP estimation accuracy is provided from the point of view of the CRLB.

Chapter 10

A Lightweight, Adaptive Security Mechanism

Jian Wang and Zemin Sun

Jilin University

CONTENTS

10.1	Introduction	187
10.2	Background	189
	10.2.1 Problem statement	189
	10.2.2 Related work	190
10.3	Method	191
	10.3.1 Channel capacity model	191
	10.3.2 Security cost model	193
	10.3.3 Utility model	195
	10.3.4 The optimal transmit power	195
10.4	Numerical Results	197

10.1 Introduction

Vehicular ad hoc networks (VANETs) allow vehicles to exchange information with other vehicles [vehicle to vehicle (V2V)] and with nearby roadside infrastructures [vehicle to infrastructure (V2I)] in real time, thus providing a number of value-added services to create a safer and more efficient traffic environment [59]. Due to factors closely related to driving safety and even potential life risk, VANETs impose stringent conditions on communication quality and security protection for timely and reliable information dissemination. Unlike the traditional wired network, VANETs are characterized by numerous new features, e.g., quick movement and unstable channel. These characteristics cause many challenging issues and limitations, e.g., bandwidth shortage, latency sensitivity, and serious Doppler effects, all of which make simultaneously harmonizing quality of service (QoS) and security difficult. To address these technical issues in an organizational way, European Telecommunications

Standards Institute/International Organization for Standardization (ETSI/ISO) jointly proposed a horizontal/vertical mixture as the intelligent transport systems (ITS) station reference architecture. Such elaborate design structure enables the joint optimization of two seemingly conflicting topics of the communication quality and security and facilitates the redesign of the related protocols and controlling algorithms to optimally utilize the restrictive resources.

For harvesting long-distance coverage, vehicles are intended to send packets using a strong power level, which, however, raises the serious risk of exposure to threats and attacks, such as eavesdropping and trace tracking. Security mechanisms might provide confidentiality, integrity, authenticity, and more protections, but would have to sacrifice certain communication performance [266]. Such degradation is not acceptable in vehicular networks due to the nature of restrictive network resources. Therefore, it is essential to construct a joint optimization model incorporating communication quality and security together. This methodology seems to be the obvious way of coping with the contradictions between quality of service (QoS) and security, but it is challenged with elaborately bridging the gap between the respective concerns of communication and security considerations through suitable quantitative metrics.

The term QoS is used to express the level of performance provided to users [283]. Several metrics may quantify network services at extent from different viewpoints, e.g., throughput, transmission delay, packet delivery ratio, and others. These metrics are necessary but are not the core elements needed to guarantee QoS in ad hoc environments. Vehicular networks inherently lack consistent infrastructure, and the topological structure changes rapidly, resulting in channels that are quite unstable and unreliable. Additionally, vehicular communications are heavily sensitive to the situated environmental context. For example, the varying interdistance between vehicles is apt to cause the large- and small-scale signal attenuation. In particular, fluctuating channel noise leads to random variation of signal amplitude. Regarding these factors, we adopt the channel capacity that theoretically bounds the transmission speed over a channel [207] to quantify the QoS concern.

Security is another critical issue in vehicular networks, where the information is propagated in an open access environment [196]. While it is easy to evaluate the vehicular QoS using channel capacity, the security is difficult to measure quantitatively. Life-critical information could be subject to many security threats (e.g., falsification and eavesdropping) during transmission. Risk assessment is generally performed by using a brute force model, and thus the security level is quantified relative to the attacker's capability. We consider the security cost in evaluating vehicular security by introducing the concept of information security strength.

Since the two metrics to quantify vehicular QoS and security are already available, the next attention is devoted to finding a suitable bridge to link them both. Recently, the new emerging paradigm of Decentralized Congestion Control (DCC) has aimed to activate context-awareness ability to effectively and efficiently react to vehicular dynamic ability. The key point in the DCC algorithm is exactly the transmit power. Increasing the transmit power could significantly improve the channel capacity; however, at a cost; e.g., serious channel contention. The advancement to the optimization strategies of transmit power shall be performed from various perspectives; most of the previous attempts are intended to improve communication quality [15,104]. However, communication quality and security strength are obviously inversely proportional to each other. For example, the long encryption key and the complicated cryptography algorithm can undoubtedly offer high-level security strength, but at the cost of consuming numerous network bandwidth and computing resources, resulting in a fundamental contradiction between QoS and security [266]. We are therefore exploring power adaptive control from a new viewpoint of joint optimization of QoS and security.

A power adaptive control decision should be made by the trade-off of channel capacity and security cost. The main contributions can be summarized by (1) formulating a node utility by bridging channel capacity and security cost using the transmit power, (2) introducing the concept of information security strength and providing its concrete resolving algorithm regarding Rijndael (Advanced Encryption Standard [AES]) encryption, (3) calculating the optimal transmit power for maximizing the node utility in various vehicular contexts, and (4) investigating the effects of various combinations of the concerned parameters on the proposed method.

10.2 Background

10.2.1 Problem statement

Figure 10.1 shows the V2V communication scenario, where n communication nodes are randomly placed over the $(D \times D)m^2$ area, and the eavesdropper nodes are uniformly distributed. For unicast as an example, we denote the sender and the receiver as j and k, respectively. Node j sends data packets to node k via the wireless channel that is assumed as an ideal additive white Gaussian distribution channel in which the probability density function (PDF) is constant and follows the Gaussian distribution [214]. $h_{jk}(t)$ is the channel gain between node j and node k at time t. As a prerequisite, QoS is guaranteed by the channel capacity between j and k. The reception process is probably affected by other nodes that are situated near the receiver and fall into two categories: neighbor nodes and interference nodes. On the one hand, node l is denoted as a neighbor node if the receiver k can receive node l's signal within the range of $d_{l\max}$. This indicates the maximum transmission radius within which the data packet can be accurately identified by receiver k. On the other hand, node l is considered an interference node if the receiver k can receive l's signal within the range of interference radius $d_{l\text{infer}}$ that is situated between $d_{l\max}$ and d_δ. d_δ depends on the receiver sensitivity. In practice, there are some malicious nodes that try to eavesdrop on the information transmitted by node j and

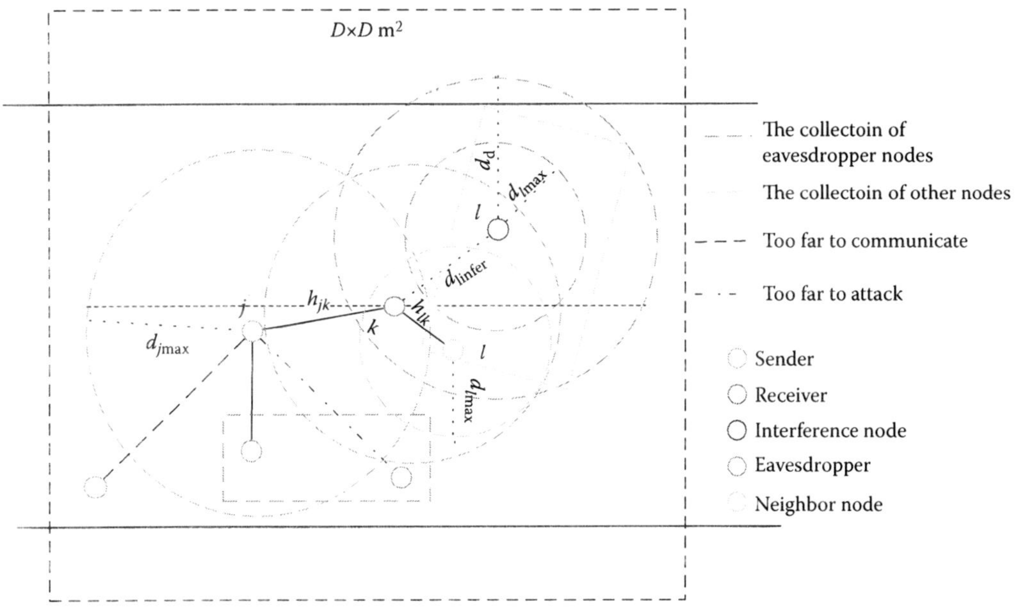

Figure 10.1 The communication scenario.

attempt to crack the ciphertext. Among all the eavesdropper nodes, only those within the circle with radius $d_{j\max}$ are considered to be eavesdropping successfully.

10.2.2 Related work

Jointly optimizing the QoS and security has attracted much attention from academia and industry, and some progresses has already been made available. Chen et al. [33] presented a model to measure the interaction between time delay and security in different application scenarios, with heterogeneous users, and with disparate services. Jindal and Singh [103] proposed a link encryption scheme that adapts to channel variations and enhances the security level of wireless local area networks (WLANs) without compromising any network performance. Zhou et al. [296] studied the throughput of large-scale decentralized wireless networks with physical layer security constraints and used the transmission capacity framework to characterize the area spectral efficiency of secure transmissions with constraints on both the QoS and the level of security.

To cope with the dynamic balance between QoS and security in vehicular networks, we first need to develop metrics to quantitatively measure them. We measure the vehicular QoS using the channel capacity devised by Shannon [206,207]. Large numbers of studies are conducted to optimize the channel capacity through adjusting and controlling the transmit power. Sengupta et al. [205] applied game theory to solve the power control problem in a code division multiple access (CDMA)-based distributed sensor network. They devised a distributed algorithm for optimal power control and proved that the system is power stable only if the nodes comply with certain transmit power thresholds. Kim [118] proposed a new online power control scheme based on evolutionary game theory. To converge a desirable network equilibrium, the proposed scheme adaptively adjusts a transmit power level in a distributed online manner. Niyato et al. [169] focused on adaptive power management for a wireless base station under various uncertainties, including renewable power generation, power price, and wireless traffic load. Gengzhong et al. [72] proposed a noncooperative game power control model based on the wireless sensor networks (WSNs) model of CDMA and proposed a distributed power control algorithm based on the game model. Shih et al. [210] analyzed the relationships among the transmission range, the carrier-sensing range, and the interference range in a case where power control was adopted, and based on the analyzed results, proposed four mechanisms to prevent the POINT problem from occurring in wireless ad hoc networks and further analyzed the superiority of each mechanism under certain situations.

Another main concern to this work is the protection and quantification of vehicular security. Some efforts have been made to secure vehicular networks. Sun et al. [220] proposed a security system for VANETs to achieve the privacy desired in vehicles and the traceability required by law enforcement authorities, in addition to satisfying fundamental security requirements, including authentication, nonrepudiation, message integrity, and confidentiality. Wasef et al. [251] introduced complementary security mechanisms to complement the security services provided by public key infrastructure (PKI) Yeh et al. [280] proposed a novel, portable, privacy-preserving authentication and access control protocol (PAACP), for nonsafety applications in VANETs. Some progress has been available in quantifying security. Madan et al. [151] modeled a security intrusion and the response of an intrusion tolerant system to the attack as a random process, which was used to analyze and quantify the security attributes of the system. Ortalo et al. [173] presented the results of an experiment in security evaluation. The system was modeled as a privilege graph that exhibits its security vulnerabilities.

In summary, most of the previous adaptive power control literature aims to improve communication quality through heuristically resolving channel contention and at the same time maintaining the effective dissemination range. These power control schemes all originated from the endless pursuit of better and faster physical- and media access control-layer metrics. However, such an isolated target does not fit the vehicular networks, where security is really another matter. Despite much progress toward the joint optimization of communication quality and security, there remains reticence of a particular advance to the trade-off between the channel capacity and security cost from the perspective of transmit power, especially in the vehicular networks. Therefore, we devote to designing a joint optimization utility model that delicately accommodates the two seemingly separate quantities of channel capacity and security cost through employing the transmit power as a bridge. We also theoretically deduce the optimal transmit power that corresponds to maximize the proposed utility, i.e., a best balance point between the enjoyed channel capacity and the paid security cost. Moreover, we perform extensive numerical calculations to collect the effects of various combinations of the parameters concerned on such balance point.

10.3 Method

Here, we first present the trade-off model, including the channel capacity model, security cost model, and utility model, all of which allow us to quantitatively characterize the communication quality and security. Then, we deduce the theoretical optimal transmit power according to the proposed trade-off model.

10.3.1 Channel capacity model

As shown in Figure 10.1, the channel capacity $I_{jk}(t)$ between nodes j and k is expressed as follows [206]:

$$I_{jk}(t) = A\log_2(1+SNR_{jk}(t)) \tag{10.1}$$

$$\mathrm{SNR}_{jk}(t) = \frac{h_{jk}(t)p_j(t)}{\sigma(t)^2 + p_{l\mathrm{sum}}} \tag{10.2}$$

$$p_{l\mathrm{sum}} = \sum_{l\in L, l\neq j} h_{lk}(t)p_l(t), \tag{10.3}$$

where the notation A denotes the channel bandwidth in unit of HZ; $\mathrm{SNR}_{jk}(t)$ is the signal-to-noise ratio at receiver k; $\sigma(t)^2$ is the white Gaussian noise; $h_{jk}(t)$ denotes the channel gain between sender j and receiver k at time t; $p_j(t)$ is sender j's transmit power; the *total interference* $p_{l\mathrm{sum}}$ suffered by receiver k is the sum of the interference caused by the other nodes (denoted by collection L), including neighbor nodes and interference nodes; $h_{lk}(t)$ represents the channel gain between nodes l and k at time t; and $p_l(t)$ is node l's transmit power. Substituting Equation 10.3 into 10.2, $\mathrm{SNR}_{jk}(t)$ can be defined by:

$$\mathrm{SNR}_{jk}(t) = \frac{h_{jk}(t)p_j(t)}{\sigma(t)^2 + \sum_{l\in L, l\neq j} h_{lk}(t)p_l(t)} \tag{10.4}$$

The power attenuation heavily depends on the distance between the sender and the receiver, and thus the received power $p_k(t)$ at a given distance d_{jk} apart from sender j can be expressed as [14]

$$p_k(t) = (1+d_{jk})^{-\beta}p_j(t), \tag{10.5}$$

where *path loss exponent* β is a constant dependent on the situated context, e.g., indoor or outdoor [23]. The relation between the transmit power and the received power is formulated as [76]

$$p_k(t) = h_{jk}(t)p_j(t). \tag{10.6}$$

From Equations 10.5 and 10.6, $h_{jk}(t)$ is expressed as:

$$h_{jk}(t) = (1 + d_{jk})^{-\beta}. \tag{10.7}$$

The channel gain $h_{lk}(t)$ between nodes l and k can be obtained in the same way and expressed as:

$$h_{lk}(t) = (1 + d_{lk})^{-\beta}. \tag{10.8}$$

Suppose each other node (i.e., neighbor nodes and interference nodes) adopts the same transmit power p_l. Thus, the total interference $p_{l\text{sum}}$ can be rewritten as:

$$p_{l\text{sum}} = p_l(t) \sum_{l \in L, l \neq j} h_{lk}(t) = p_l(t) \sum_{l \in L, l \neq j} (1 + d_{lk})^{-\beta} \tag{10.9}$$

Substituting Equations 10.7 and 10.9 into 10.4, then $\text{SNR}_{jk}(t)$ is rewritten as:

$$\text{SNR}_{jk}(t) = \frac{(1 + d_{jk})^{-\beta} p_j(t)}{\sigma(t)^2 + p_l(t) \sum\limits_{l \in L, l \neq j} (1 + d_{lk})^{-\beta}} \tag{10.10}$$

Then the channel capacity $I_{jk}(t)$ can be obtained by substituting Equation 10.10 into 10.1:

$$I_{jk}(t) = A \log_2 \left(1 + \frac{(1 + d_{jk})^{-\beta} p_j(t)}{\sigma(t)^2 + p_l(t) \sum\limits_{l \in L, l \neq j} (1 + d_{lk})^{-\beta}} \right). \tag{10.11}$$

Next, we explain how to identify receiver's *neighbor nodes* and *interference nodes* within the range of $(D \times D)m^2$. The receiver k can correctly receive the packet sent by node l if and only if [23]:

$$\text{SNR}_{lk}(t) = \frac{(1 + d_{lk})^{-\beta} p_l(t)}{\sigma(t)^2 + p_f(t) \sum\limits_{f \in L, f \neq l} (1 + d_{fk})^{-\beta}} \geq \gamma_0, \gamma_0 > 0, \tag{10.12}$$

where L is the collection of receiver k's other nodes, and γ_0 is the threshold of *SNR* that indicates the required minimum *SNR* for a successful message reception [76]. The value of γ_0 depends on the desired data rate, modulation scheme, etc. [25].

Although receiving the packet, the receiver may not accurately decode the information due to other nodes' interference, which undoubtedly degrades the communication quality. If the receiver can correctly decode the packet, the total interference shall be zero, so:

$$\sum_{f \in L, f \neq l} h_{fk}(t) p_f(t) = 0. \tag{10.13}$$

On the one hand, node l is identified as the neighbor node if receiver k receives l's signal within the range of $d_{l\max}$ that indicates the maximum transmission radius within which the information can be accurately decoded by k. For the sake of deriving the maximum transmit radius, it is assumed that node l sends packets using its maximum power, i.e.,:

$$p_l(t) = p_{l\max} \tag{10.14}$$

Substituting Equations 10.13 and 10.14 into Equation 10.12, we have $(1 + d_{lk})^{-\beta} p_{l\max}/\sigma(t)^2 \geq \gamma_0$, i.e., $d_{lk} \leq \left(p_{l\max}/\sigma(t)^2\gamma_0\right)^{\frac{1}{\beta}} - 1$, hence node l's maximum transmit radius $d_{l\max}$ is expressed as:

$$d_{l\max} = \left(\frac{p_{l\max}}{\sigma(t)^2 \gamma_0} \right)^{\frac{1}{\beta}} - 1. \tag{10.15}$$

On the other hand, node l is judged as the interference node if its transmit power meets the inequation:

$$\delta \leq h_{lk} p_l(t) \leq (1 + d_{l\max})^{-\beta}, \tag{10.16}$$

where $d_{l\max}$ is node l's maximum transmit radius defined in Equation 10.15, and δ denotes the *receiver sensitivity*. Due to $h_{lk}(t) = (1 + d_{l\text{infer}})^{-\beta}$, we get:

$$d_{l\max} < d_{l\text{infer}} < d_\delta \tag{10.17}$$

$$d_\delta = \left(\frac{p_l(t)}{\delta} \right)^{\frac{1}{\beta}} - 1 \tag{10.18}$$

10.3.2 Security cost model

The security cost model $C_j(p_j(t))$ at time t is specified as:

$$C_j(p_j(t)) = k(b_j) I_{jk}(t), \tag{10.19}$$

where $k(b_j)$ is a function that represents the *insecurity level of information* sent by node j and depends on the security strength of information itself and the number of eavesdroppers. A large number of eavesdroppers leads to a high insecurity level of information, and high information security strength contributes to a low insecurity level of information. Thus, $k(b_j)$ can be formulated as:

$$k(b_j) = -\alpha^{b_j} + 1, 0 < \alpha < 1, \tag{10.20}$$

where the parameter α denotes the newly introduced concept of information security strength. The security cost is obtained by substituting Equations 10.11 and 10.20 into Equation 10.19:

$$C_j(p_j(t)) = (-\alpha^{b_j} + 1) A \log_2 \left(1 + \frac{(1 + d_{jk})^{-\beta} p_j(t)}{\sigma(t)^2 + p_l(t) \sum\limits_{l \in L, l \neq j} (1 + d_{lk})^{-\beta}} \right), 0 < \alpha < 1. \tag{10.21}$$

We next focus on the derivation of b_j under the assumption that the total node density is μ vehicles/m^2 and the relative density of eavesdropper nodes is ρ vehichles/vehicle. The maximum transmit radius of sender j can be defined as $d_{j\max} = \left(\frac{p_{j\max}}{\sigma(t)^2 \gamma_0} \right)^{\frac{1}{\beta}} - 1$ from Equation 10.15, which holds at j employing its maximum power $p_{j\max}$. The sender j's transmit power is thus bounded by $0 < p_j(t) < p_{j\max}$, which governs the maximum coverage area using transmit power $p_j(t)$, i.e.,

$$d_{j\max}(t) = \left(\frac{p_j(t)}{\sigma(t)^2 \gamma_0} \right)^{\frac{1}{\beta}} - 1. \tag{10.22}$$

From Equation 10.22 the number of eavesdropper nodes can be roughly derived as:

$$b_j = \pi d_{j\max}(t)^2 \mu \rho = \pi \left(\left(\frac{p_j(t)}{\sigma(t)^2 \gamma_0} \right)^{\frac{1}{\beta}} - 1 \right)^2 \mu \rho. \tag{10.23}$$

Heretofore, all the variables in the function $C_j(p_j)$ have already been discussed, except for the information security strength α, which is quantified from the perspective of cryptography. The information security strength α depends not only on the complexity of the employed encryption algorithm but also on the eavesdropper's cracking capability. As for the capability of the encryption algorithm against cryptanalysis, Jakimoski et al. [99] introduced the general types of cryptanalytic attacks, such as ciphertext-only attack, known-plaintext attack, chosen-plaintext attack, adaptive-chosen-plaintext attack, and chosen-ciphertext attack, of which the known-plaintext attack and chosen-ciphertext attack are considered to be highly practical in the modern cryptosystem [21,166]. *Linear cryptanalysis* and *differential cryptanalysis* are two typical examples of known-plaintext attack and chosen-ciphertext, respectively. Consequently, we apply the *wide trail strategy* [49] that gives the bound on the resistance against differential and linear cryptanalysis to measure the strength of the employed encryption algorithm. Additionally, we use the required total number of computation operations for cracking the ciphertext to quantify the eavesdropper's ability.

Assuming that the sent packet is encrypted by the Rijndael (AES) algorithm, the information security strength α can be calculated as:

$$\alpha = \alpha_1 \alpha_2, \tag{10.24}$$

$$\alpha_1 = P_d(1-b_d) + (1-P_d)(1-b_l), \tag{10.25}$$

$$\alpha_2 = \begin{cases} \frac{t}{t_s} & t < t_s, \\ 1 & t \geq t_s, \end{cases} \tag{10.26}$$

where α_1 is the strength of the employed encryption algorithm. The eavesdropper cracks the ciphertext using differential cryptanalysis and linear cryptanalysis with the probability P_d and $1 - P_d$, respectively. b_d and b_l denote the *best differential characteristic* and the *best linear approximation*. $P_d(1-b_d)$ and $(1-P_d)(1-b_l)$ imply the ability of the encryption algorithm against differential cryptanalysis attack and linear cryptanalysis attack, respectively. α_2 means the security level hidden behind the conveyed information against the eavesdropper's brutal force attack. t is the time in unit of microseconds consumed by the eavesdropper attempting all the possible keys to crack the ciphertext, which is calculated by Equation 10.27. Suppose the key length of AES is λ *bits*, so the attacker tries on average at least half of all the possible keys in conducting a successful crack. Concretely speaking, for a given ciphertext, the attacker tries at least $2^{\lambda-1}$ times until achieving a valid decryption n implies the attacker's capability, i.e., the number of decrypting operations executed by the attacker per microsecond. t_s represents the universal lifetime and is expressed as Equation 10.28.

$$t = \frac{2^{\lambda-1}}{n}\,\mu\text{s} = \frac{2^{\lambda-1}}{n \times 10^6 \times 3600 \times 24 \times 365}\ \text{years}, \tag{10.27}$$

$$t_s = 1.4 \times 10^{10}\ \text{years}. \tag{10.28}$$

Substituting Equations 10.27 and 10.28 into 10.26, we obtain:

$$\alpha_2 = \begin{cases} \frac{2^{\lambda-1}}{n \times 10^6 \times 3600 \times 24 \times 365 \times 1.4 \times 10^{10}} & t < t_s, \\ 1 & t \geq t_s. \end{cases} \tag{10.29}$$

Hence, the information security strength α can be calculated by combining Equations 10.29, 10.25, and 10.24 and expressed as:

$$\alpha = \begin{cases} \frac{2^{\lambda-1}(P_d(1-b_d)+(1-P_d)(1-b_l))}{n \times 10^6 \times 3600 \times 24 \times 365 \times 1.4 \times 10^{10}} = \frac{2^{\lambda-1}(P_d(1-b_d)+(1-P_d)(1-b_l))}{n \times 4.41504 \times 10^{23}} & t < t_s, \\ \alpha_1 = P_d(1-b_d) + (1-P_d)(1-b_l) & t \geq t_s. \end{cases} \tag{10.30}$$

Furthermore, n can be bounded by

$$\infty+ > n > \frac{2^{\lambda-1}\left(P_d(1-b_d) + (1-P_d)(1-b_l)\right)}{4.41504 \times 10^{23}}, \tag{10.31}$$

which follows the fact that, from Equation (10.20), the information security strength α satisfies the constraint $0 < \alpha < 1$, i.e., $0 < \frac{2^{\lambda-1}(P_d(1-b_d)+(1-P_d)(1-b_l))}{n \times 4.41504 \times 10^{23}} < 1$. Substituting Equations 10.23 and 10.30 into Equation (10.21), the security cost is thus expressed as:

$$C_j(p_j(t)) = \left(-\alpha^{\frac{\pi\left(\left(\frac{p_j(t)}{\sigma(t)^2\gamma_0}\right)^{\frac{1}{\beta}}-1\right)^2}{\mu\rho}} + 1 \right) A \times \log_2 \left(1 + \frac{(1+d_{jk})^{-\beta} p_j(t)}{\sigma(t)^2 + p_l(t) \sum\limits_{l \in L, l \neq j} (1+d_{lk})^{-\beta}} \right) (0 < \alpha < 1), \tag{10.32}$$

where the information security strength α is defined by Equation 10.30.

10.3.3 Utility model

We present a node utility model to jointly optimize QoS and security. From Equations 10.19 and 10.20, we can deduce that $C_j(p_j)$ and $I_{jk}(t)$ are in the same unit bit/s because the insecurity level of information $k(b_j)$ is dimensionless. Additionally, the value range of $C_j(p_j)$ is the same as that of $I_{jk}(t)$, which can be easily explained by Equation (10.20), where $k(b_j)$ is bounded by $0 < k(b_j) < 1$, due to the constraint of $0 < \alpha < 1$. Given the communication scenario in Figure 10.1, the node utility of sender j can be defined by:

$$U_j(p_j(t)) = I_{jk}(t) - C_j(p_j(t)), \tag{10.33}$$

where $I_{jk}(t)$ is the channel capacity in unit *bit/s* between sender j and receiver k at time t, and $C_j(p_j)$ denotes the *security cost* of sender j.

At this point, substituting Equations 10.11 and 10.32 into Equation 10.33, the node utility function $U_j(p_j(t))$ is defined by:

$$U_j(p_j(t)) = A \log_2 \left(1 + \frac{(1+d_{jk})^{-\beta} p_j(t)}{\sigma(t)^2 + p_l(t) \sum\limits_{l \in L, l \neq j} (1+d_{lk})^{-\beta}} \right) \times \left(-\alpha^{\frac{\pi\left(\left(\frac{p_j(t)}{\sigma(t)^2\gamma_0}\right)^{\frac{1}{\beta}}-1\right)^2}{\mu\rho}} \right), 0 < \alpha < 1 \tag{10.34}$$

10.3.4 The optimal transmit power

Suppose p_j^* is the optimal transmit power that maximizes the node utility and set $\beta = 2$. By setting the partial derivative of Equation 10.34 with respect to $p_j(t)$ as 0, we obtain:

$$\frac{\partial U_j(p_j(t))}{\partial p_j(t)} = \left(\frac{M}{Mp_j^* + 1} + \frac{\pi\mu\rho\sqrt{\frac{1}{\sigma(t)^2\gamma_0}}\left(\sqrt{\frac{p_j^*}{\sigma(t)^2\gamma_0}} - 1\right)\ln\left(Mp_j^* + 1\right)\ln(\alpha)}{\sqrt{p_j^*}} \right) \times \frac{A \times W}{\ln(2)} = 0 \tag{10.35}$$

$$M = \frac{(1+d_{jk})^{-\beta}}{\sigma(t)^2 + p_l(t) \sum\limits_{l \in L, l \neq j} (1+d_{lk})^{-\beta}}, \tag{10.36}$$

$$W = \alpha^{\pi\mu\rho} \left(\sqrt{\frac{p_j^*}{\sigma(t)^2 \gamma_0}} - 1 \right)^2. \tag{10.37}$$

For the solving purpose, we set:

$$\hat{p} = \frac{p_j^*}{100}, 0 < \hat{p} < 1, \tag{10.38}$$

where the constraint $0 < \hat{p} < 1$ follows the fact that p_j^* should satisfy the well-known power constraint $0 < p_j^* < 100\,\text{mw}$.

$$\begin{aligned} \frac{\partial U_j(p_j(t))}{\partial p_j(t)} &= \frac{M}{100M\hat{p}+1} + \frac{\pi\mu\rho\sqrt{\frac{1}{\sigma(t)^2\gamma_0}}\left(10\sqrt{\frac{\hat{p}}{\sigma(t)^2\gamma_0}} - 1\right)\ln(100M\hat{p}+1)\ln(\alpha)}{10\sqrt{\hat{p}}} \\ &= 0. \end{aligned} \tag{10.39}$$

Here, $100M\hat{p}$ is in 10^{-5} order of magnitude. The functions $y = 10^{-5}x$ and $y = \ln(1+10^{-5}x)$ are almost completely overlapping at $0 < x < 1$. Thus, (10.39) can be approximated by $\ln(1+100M\hat{p}) \approx 100M\hat{p}$ into:

$$\frac{\partial U_j(p_j(t))}{\partial p_j(t)} = \frac{M}{100M\hat{p}+1} + 10\pi\mu\rho\sqrt{\frac{1}{\sigma(t)^2\gamma_0}}\left(10\hat{p}\sqrt{\frac{1}{\sigma(t)^2\gamma_0}} - \sqrt{\hat{p}}\right)M\ln(\alpha) = 0. \tag{10.40}$$

The functions $y = \sqrt{x}$ and $y = x$ are also approximately the same at $0 < x < 1$, so (10.40) can be further simplified to:

$$\frac{M}{100M\hat{p}+1} + 10\pi\mu\rho\sqrt{\frac{1}{\sigma(t)^2\gamma_0}}\left(10\sqrt{\frac{1}{\sigma(t)^2\gamma_0}} - 1\right)\hat{p}M\ln(\alpha) = 0. \tag{10.41}$$

The value of $\hat{p}$ is obtained by directly solving Equation 10.41.

$$\hat{p} = -\frac{S + \sqrt{S^2 - 400M^2S}}{200MS} \tag{10.42}$$

$$S = 10\pi\mu\rho M\sqrt{\frac{1}{\sigma(t)^2\gamma_0}}\left(10\sqrt{\frac{1}{\sigma(t)^2\gamma_0}} - 1\right)\ln(\alpha). \tag{10.43}$$

According to Equation 10.38, the approximate optimal transmit power $p_j^{*\sim}$ of node j can be finally obtained by:

$$p_j^{*\sim} = 100\hat{p} = -\frac{S + \sqrt{S^2 - 400M^2S}}{2MS}, \tag{10.44}$$

where the information security strength α is defined by Equation 10.30, M is defined by Equation 10.36, and S is defined by Equation 10.43. Herein, all the variables are expressed by the known elements.

10.4 Numerical Results

Table 10.1 lists the parameters in numerical calculations. In this section, we comprehensively analyze the effects of various parameters on the node utility and the optimal transmit power.

Figure 10.2 shows the effect of information security strength α on the node utility U_j and the optimal transmit power $p_j^{*\sim}$, both of which grow as the parameter α increases. This is because, for a given number of the eavesdropper nodes b_j, increasing the information security strength α undoubtedly reduces the insecurity level of information $k(b_j)$ according to Equation 10.20, and thus weakens the security cost according to Equation 10.19. Additionally, the value of $p_j^{*\sim}$ increases with the growth of α in terms of Equation 10.44. From Figure 10.2, one can know that a high security level of the sent information contributes to better QoS and secure communication, which however needs a relative high transmit power to achieve the optimal utility.

Figure 10.3 provides the effect of the relative density of eavesdropper nodes ρ on the node utility U_j and the optimal transmit power $p_j^{*\sim}$. The curves show that ρ poses a negative

Table 10.1 Parameters and Values

Parameters	*Meaning*	*Default Value*
$D \times D$	Scenario range	6000 m × 6000 m
A	Wireless channel bandwidth	10 MHz
σ^2	White Gaussian noise	1
γ_0	Threshold of SNR	5×10^{-5}
β	Pass loss exponent	2
p_l	The transmit power of other nodes	30 MW
ρ	Relative density of eavesdropper nodes	0.001 vehicles/vehicle
μ	Total node density	0.004 vehicles/m^2
δ	Receiver sensitivity	8×10^{-9}
k	Key length	128, 192, 256 bits
P_d	Probability of eavesdropper choosing differential cryptanalysis	0.5
b_d	The best differential characteristic	2×10^{-150}
b_l	The best linear approximation	2×10^{-76}

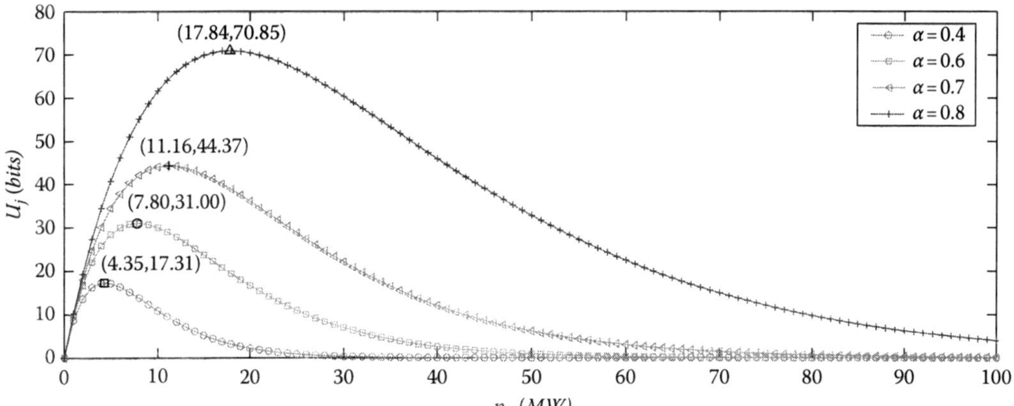

Figure 10.2 Effect of the information security strength on the node utility and the optimal transmit power.

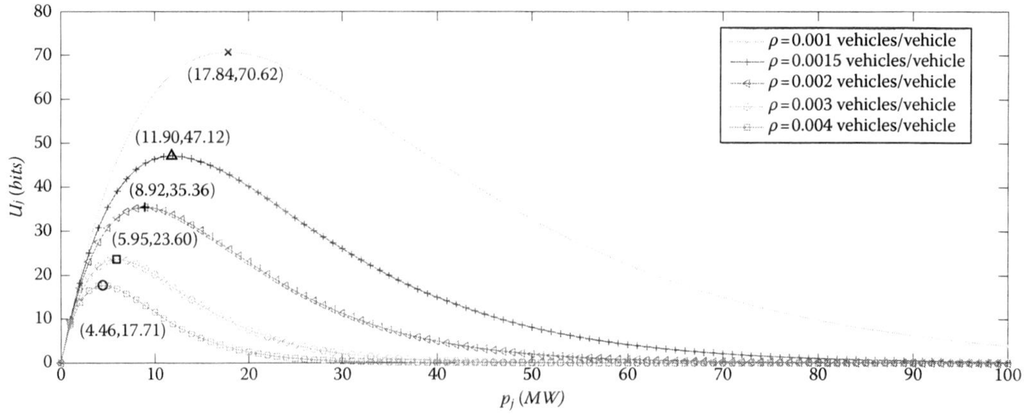

Figure 10.3 Effect of the relative density of eavesdropper nodes on the node utility and the optimal transmit power, where $\alpha = 0.8$.

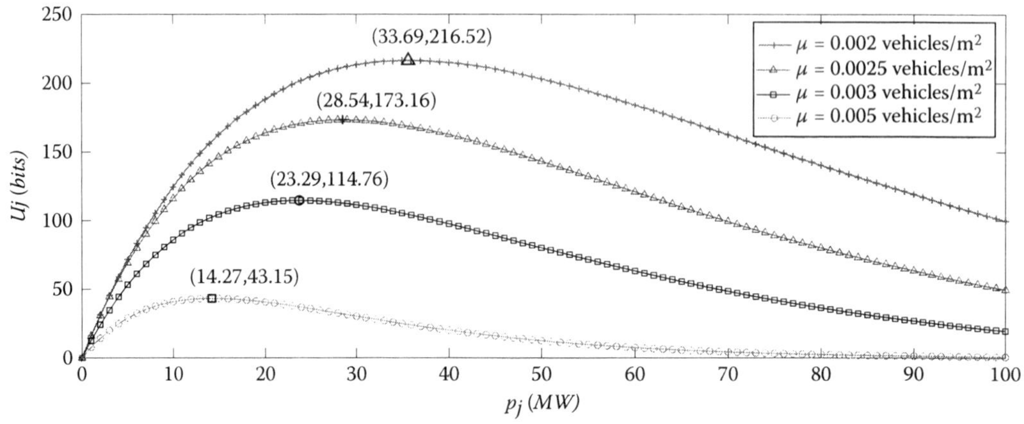

Figure 10.4 Effect of the total node density on the node utility and the optimal transmit power, where $\alpha = 0.8$.

effect on the node utility U_j and the optimal transmit power $p_j^{*\sim}$, which is because, according to Equation 10.23, increasing eavesdropper nodes density ρ apparently raises the number of eavesdropper nodes and further burdens the security cost regarding Equation 10.21. Moreover, from Equation 10.44 one can deduce that the value of $p_j^{*\sim}$ also decreases as ρ increases. Thus, the sender needs to adaptively control its transmit power according to the situated context for efficient and secure communication. To be precise, the sender should employ low power in the case of high-density eavesdropper nodes. On the other hand, the high transmit power can improve the node utility if there are few malicious nodes.

Figure 10.4 restricts attention to the effect of the total node density μ on the node utility U_j and the optimal transmit power $p_j^{*\sim}$, in which the two concerned indicators drop down as the total node density μ increases. This is because the high value of the total node density tends to increase the other nodes and eavesdropper nodes. On the one hand, increasing μ can rise up the total interference number $p_{I\text{sum}}$, which can be explained by Equation 10.3. On the

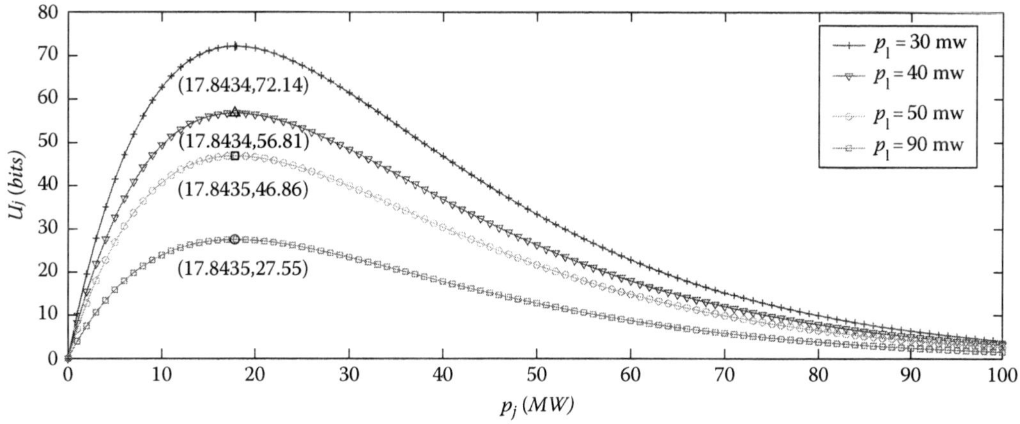

Figure 10.5 Effect of the transmit power of other nodes on the node utility and the optimal transmit power, where α = 0.8.

other hand, according to the definition of eavesdropper nodes b_j in Equation 10.23, large μ contributes to a high value of b_j. Generally speaking, according to Equations 10.34 and 10.44, the total node density μ exerts a negative impact on both U_j and $p_j^{*\sim}$, which corresponds to the positive correlations between μ and b_j as well as between μ and $p_{l\text{sum}}$. Consequently, the proposed method can effectively improve the communication quality but without compromising any security through controlling the transmit power adaptively to the vehicle density.

Figure 10.5 displays the effect of the transmit power of other nodes (i.e., neighbor nodes and interference nodes) on the node utility U_j and the optimal transmit power $p_j^{*\sim}$. We observe that when p_l moves from 30 to 90 MW, the node utility U_j goes down. The downtrend of U_j is reasonable, since increasing the transmit power of the other nodes worsens the interference suffered by the receiver, which can be easily explained by Equation 10.3. Unlike the correlation between p_l and U_j, the other nodes' power hardly imposes any effect on the optimal transmit power. This is because increasing p_l does not dominate the positive effect of p_l on $p_j^{*\sim}$ according to Equation 10.44, and thus the resulting tiny influence can be ignored. In a nutshell, the sender's utility decreases as the other nodes increase their transmit power, but the utility only can be maximized by the same transmit power, i.e., about 17.84 MW, regardless of the other nodes' power.

Figure 10.6 shows the effect of the wireless channel bandwidth A on the node utility U_j and the optimal transmit power $p_j^{*\sim}$. The node utility U_j gradually declines as the bandwidth A varies from 6 to 2 MHz. The bandwidth indicates the transmission capacity of the wireless channel; thus the wide bandwidth enables a better communication service and poses a positive effect on the node utility U_j, according to Equation 10.34. Interestingly, the bandwidth does not bring any effect on the optimal transmit power. This is because the bandwidth only affects the amplitude of the node utility curve, which can be inferred from Equation 10.1. Moreover, it can be learned from Equation 10.44 that the bandwidth A doesn't play any role in deciding the optimal transmission power $p_j^{*\sim}$. The wide bandwidth can enhance the node utility and thus improve the communication quality and security. The utility with different bandwidths is, however, maximized by the same transmit power, i.e., about 17.85 MW.

Figure 10.7 shows the effect of the key length λ on the node utility U_j and the optimal transmit power $p_j^{*\sim}$, where the key length λ exerts a positive effect on the two metrics concerned. As presented in Equation 10.27, the attacker tries at least $2^{\lambda-1}$ to crack the ciphertext

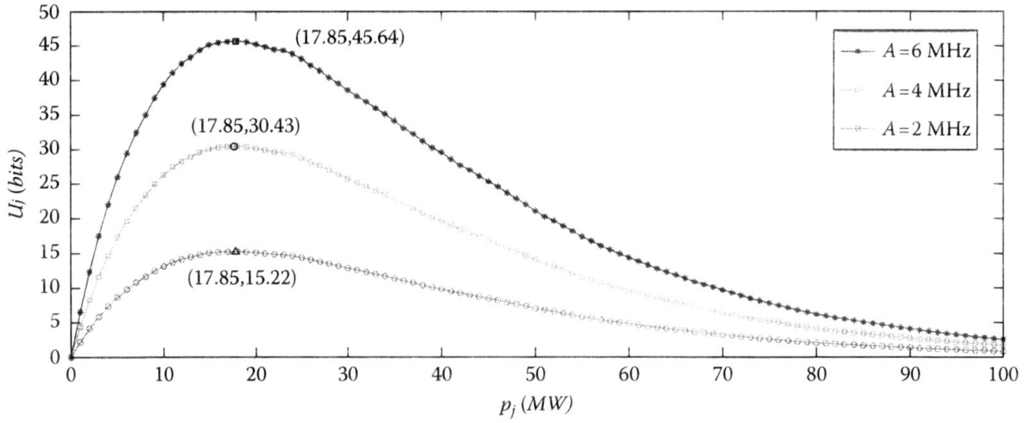

Figure 10.6 Effect of the wireless channel bandwidth on the node utility and the optimal transmit power, where α = 0.8.

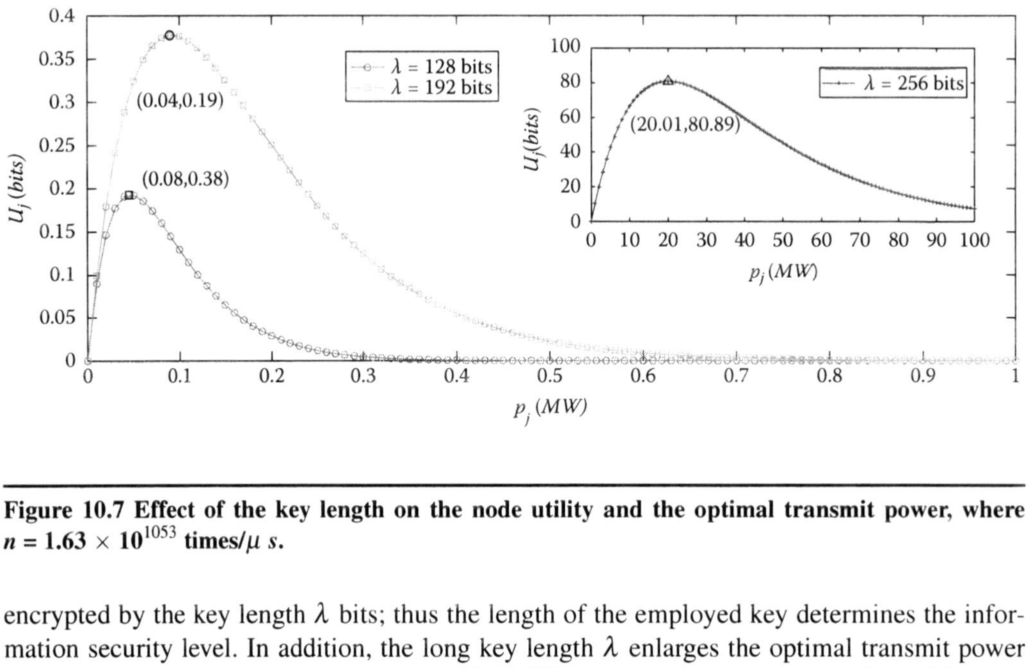

Figure 10.7 Effect of the key length on the node utility and the optimal transmit power, where $n = 1.63 \times 10^{1053}$ times/μ s.

encrypted by the key length λ bits; thus the length of the employed key determines the information security level. In addition, the long key length λ enlarges the optimal transmit power $p_j^{*\sim}$, which can be explained by Equation 10.44. We also can observe that the effect of λ on U_j and $p_j^{*\sim}$ is gradually reinforced as the key length increases. For the given cracking ability of eavesdropper nodes, employing a long key length can certainly improve the QoS and security together and increasing the transmit power can reach to the optimal utility without worrying about being eavesdropped upon.

Figure 10.8 plots the effect of the attacker's capability n on the node utility U_j and the optimal transmit power $p_j^{*\sim}$ against three key lengths. The attacker's capability reveals that both U_j and $p_j^{*\sim}$ slip down as n increases. This is because, from Equation 10.30 it can be inferred that a negative correlation occurs between the information security strength α and the attacker's capability n, which accordingly exerts negative effects on U_j and $p_j^{*\sim}$, according to Equations 10.1 and 10.44. So the proposed method facilitates the adaptive power control to balance the QoS and security against different attacker's capabilities.

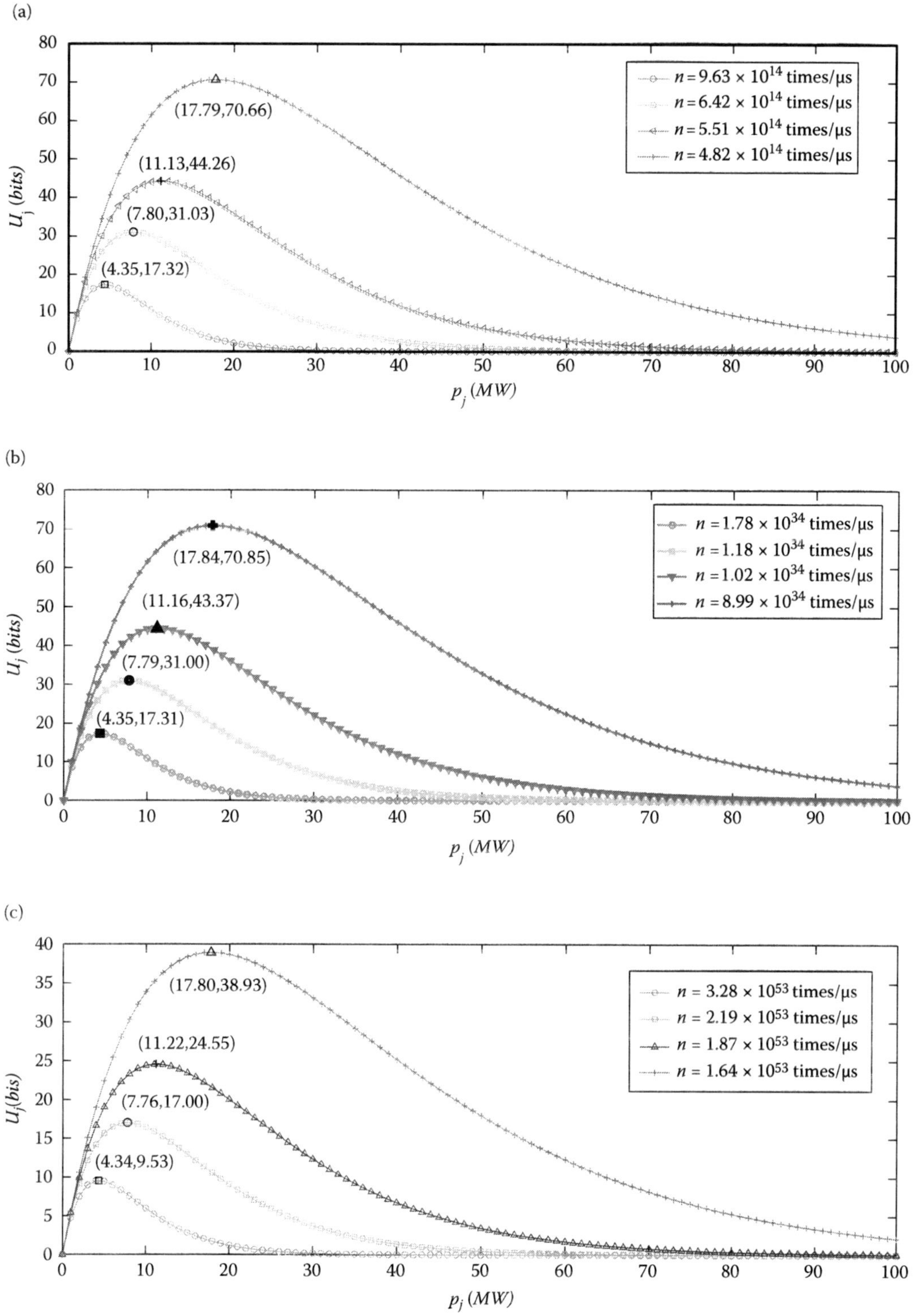

Figure 10.8 Effect of the attacker's capability on the node utility and the optimal transmit power, where (a) λ = 128 bits, (b) λ = 192 bits, and (c) λ = 256 bits.

Chapter 11

A Social-Network-Enabled Green Transportation System Framework Driven by Connected Vehicle Innovations

Wei Shu
University of New Mexico

Guohui Zhang
University of Hawaii

Min-You Wu and Jia-Liang Lu
Shanghai Jiaotong University

CONTENTS

11.1 Introduction 203
11.2 Problem Definition and Related Work 204
11.3 Approach 206
11.4 Cyberspace, Social-Network-Enabled Architecture 207
11.5 Incentives and Pricing Formulation 209
11.6 PV Assignment and Hyper Simulator 210
11.7 Conclusion 211

11.1 Introduction

With increasing economic and social activities, travel demand has increased significantly over the past few decades, overloading many existing roadway systems. Public transportation in concentrated urban areas has helped reduce energy consumption but has suffered from last-mileI problem and inconvenience. In this chapter, we propose a social network approach for future transportation systems. This approach combines social networks, public-owned vehicles, mechanism design and pricing incentives, optimal vehicle assignments, and hyper simulators to provide an ultra-green transportation system as a whole. Here, social networks minimize transportation needs, and in cases where transportation is necessary, they minimize the energy consumption of transportation by optimal assignment and scheduling of vehicles. Publicly-owned vehicles can be shared by many people thus minimizing Vehicle Miles Traveled (VMT) and the total number of trips made. This approach is more energy saving than the use of privately-owned vehicles, more flexible than the use of buses, and less expensive and more efficient than the use of taxis. Mechanism design and pricing incentives provide the theoretical basis of energy efficiency by introducing a marketing mechanism into transportation assignment and scheduling. A hyper simulation enables accurate scheduling and routing of vehicles, thus reducing the overall energy consumption.

This chapter aims to create a service-oriented transportation system that targets significant reductions in energy consumption, pollution impact, and traffic congestion and provides solutions with affordable costs from the perspective of both individuals and transportation agencies. In this chapter, we develop and demonstrate a prototype for a social-network-enabled transportation system as a platform for vehicle users, municipalities, transportation authorities and industries, and internet inventors to develop products and services and to explore monetization.

In this chapter a social-network-enabled cyberspace is created virtually which would allow people to interact, generate schedules, and make reservations. Scenario-driven traffic operations are simulated to provide a prediction of arrival time with the best routing and to enable optimal vehicle assignment, subject to change with dynamic traffic information. To optimize traffic system operations, various congestion pricing strategies are performed to enhance network-wide traffic assignment. Various pricing mechanisms, such as pricing for advanced reservation and real-time request, for sharing and nonsharing, for special requests such as must share with friends, fastest possible, lowest cost, etc., are designed and deployed. Global optimal pricing strategies are identified and executed. These pricing incentives facilitate prevailing traffic control and management applications, such as navigation (routing), lane scheduling, intersection control, accident response, etc. Eventually, a traffic condition–based scenario-driven network-wide congestion pricing mechanism is developed to guide real-world traffic system operations.

11.2 Problem Definition and Related Work

In the United States, during the last three decades, annual VMTs increased by 94.2%, while roadway lane miles increased by only 8.3% [229]. Transportation consumes 27% of the country's energy resources and accounts for about 11% of all expenses in the economy. Vehicle emission have become one of the major contributors to excessive air pollution. More than 50% of all carbon monoxide produced in the United States comes from transportation systems. The relevant public health costs of traffic-related pollution is huge—between $40 billion and $64 billion per year [229]. Public transportation in concentrated urban areas helps reducte energy consumption but experiences the last-mile problem and can be inconveniet.

Substantial technology advances have been made in the areas of autonomous automobiles and connected vehicles, which open a wide landscape for future traffic system operations. We posit a situation in which the current system of privately owned cars, licensed taxis, and public mass transportation is replaced by a sharing-dominated system. In our proposed system, vehicles could be autonomous or human driven, vehicles, sizes could vary, and so on. Unlike the current Dial-a-Ride system, we could create configured routes integrated with cyberspaces hobby, daily needs, as well as social-connected favorite partners. Previous studies on vehicle sharing, congestion pricing, and traffic simulation model development provide a solid foundation for this development.

Over the past few years, vehicle sharing has gained in popularity for providing enhanced mobility services. The concept of vehicle sharing was initially practiced as carpooling, which has been widely used in the United States since the 1960s. Today, the vehicle-sharing process is simplified in support of the various wireless communication and mobile devices. Wide-range Internet availability facilitates car-sharing processes by allowing private car owners to share their cars with others. Demand-based car-sharing mechanisms cannot be fully established because of unbalanced supply–demand distribution. Vehicles may possibly get stuck in low-mobility-demand zones but become unavailable in high-demand zones. Similar problems have existed in bike sharing areas and greatly motivated the bike relocation method development in the recent years [45,157]. Car-sharing mechanisms must actively address their reallocation issues to enhance their applicability. The system we propose has ideas in common with a Dial-a-Ride scheme [63], where both approaches have the advantages of efficiency and flexibility. The main difference between the typical Dial-a-Ride system and the system described here is that in our system, a public vehicle (PV) is a driverless system that significantly reduces costs, thus becoming a realistic daily transportation tool. Our system also provides the information support and user-friendly platform for the PV system.

To reduce traffic congestion and raise revenues, congestion pricing has been proposed and studied for several decades since, Pigou [177], and Knight [122] initially explored congestion pricing theory. In the 1960s, research interest in congestion pricing was resurrected by the work of Walters [239], Beckmann [136], and Vickrey [236]. Vickrey developed a dynamic vehicle congestion model to derive socially optimal tolls featured with flexible departure and arrival time. Since then, substantial research has been conducted by transportation researchers and practitioners [138]. The well-known first-best pricing theory has attracted much research attention [89,136,274]. However, the first-best pricing theory has limited practical value. The second-best pricing principles have been proposed as a practical solution to determining tolls considering physical and economic constraints [235]. Additionally, many researchers such as Ferrari [297], Larsson and Patriksson [127], and Inouye [62] have derived link tolls under capacity constraints based on Wardropian traffic equilibrium [250]. Vickrey [236] and Downs [44] proposed that congestion pricing should be determined based on trial-and-error efforts to enhance its applicability. Li [131], Yang et al. [275], and Meng et al. [156] proposed the iterative toll adjustment mechanisms according to the single and networkwide link flows without demand information. Arnott et al. [143] compared the four distinct pricing strategies and concluded that considerable benefits can be achieved under congestion pricing. In practice, congestion pricing strategies have been implemented worldwide, for instance in Singapore [172]. The cases of congestion pricing in the United States include the HOT and express toll lane systems of State Route (SR) 91 and Intercontinental Highway (I) 15 and I-15 in California, I-10 and US-290 in Texas, and I-394 in Minnesota. [60].

Microscopic traffic simulation is commonly utilized in transportation engineering fields, including transportation system design, traffic operations, and management alternative evaluation. Hence, simulation-based investigations of toll-based traffic operations is of practical importance for traffic engineers to quantify toll impacts, optimize tolling strategies, and identify potential problems prior to implementation. Verkehr In Stdten - SIMulationsmodell (VISSIM) is one of the most powerful microscopic simulation tools developed to model urban traffic operations. This software can simulate and analyze traffic operations under a broad range of scenarios. Many simulation studies have been conducted using VISSIM. Gomes et al. [61] developed and calibrated a VISSIM model for simulating a congested freeway operation. Lelewski et al. [130] built up a VISSIM simulation model to analyze express toll plaza operations. Zhang et al. [286] conducted simulation-based investigations of HOT lane operations on Washington SR-167. Many studies [199,213] indicate that the output data from a simulation run are inherently correlated.

11.3 Approach

The current social networks focus on the virtual world, even with physically connected bodies also presented as an entity in virtual world. It can be envisioned that the partial and selective cyber space can be mapped into the real physical system to further impact physical transportation system operations. We deal with this new type of social network as a social-network-enabled transportation cyber physical system (CPS) architecture, which consists of the cyber space and the physical layer, as shown in Figure 11.1. The cyber space is reflected from entities in social networks, where a stationary node (s-node) represents a Point of Interest (PoI) that can be mapped onto a physical place; an edge represents connectivity between two stationary nodes (mapped to multiple transportation routes). A mobile node (m-node) represents an avatar (mapped to a person), where one or more avatars can be grouped as a mobile squad; an activity link (a-link) represents association from a PoI to one more s-nodes, and a friend link (f-link) represents a binding from a friend group name onto m-nodes.

For each avatar, his/her daily life is defined by many activities associated with various PoIs and certain time windows. Execution of these activities is carried by moving avatars to corresponding s-nodes to meet the specified timing requirements. Planning of activities in daily life can be made in advance, dynamically adjusted, spontaneously created, or changed. Social networks provide a natural platform to integrate s-nodes, links, avatars, and events.

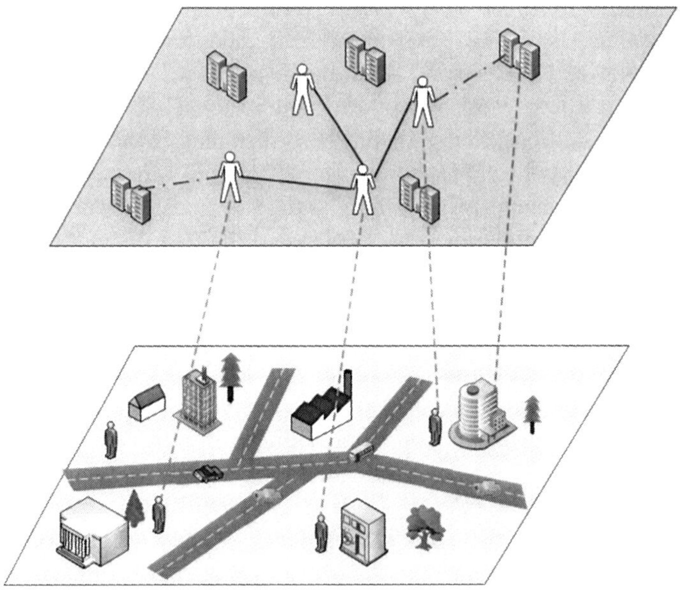

Figure 11.1 Cyberspace and physical space.

In the physical plane, people are notated as p-entities and locations as l-entities. A road system connects the l-entities. A p-entity can be located at an l-entity for a time interval, or in transit from an origination l-entity to a destination l-entity via a route and by a mover. Movers can be of various types, such as publicly-owned vehicles, community car pools, public transportation, including subways and buses, and so on. In this section, we concentrate on a particular type of mover, the PV, a publicly-owned vehicle that can be shared by many people. A PV is classified as an autonomous car (driverless or self-driving). Upon request, relying on the traffic assignment system and simulator, the most suitable PV (with some empty space, close to the location, and being able to move the person to the destination) will be scheduled to serve the need.

With the introduction of cyberspace and physical space, the social-network-enabled component mainly lies in cyberspace to manage all m-nodes and s-nodes as well as edges, activities, a-link, and f-links, while the transportation component mainly exists in physical space to manage p-entities and l-entities as well as roads/routes and movers. Thereafter, the cyber-physical system is required to establish a connection between two components.

This system creates a cyber space that runs in parallel with the physical world and assists in decision making in the physical world. Tools are designed in the cyberspace to easily and effectively schedule daily activities for people. The transportation component works with activity demands to achieve a service-oriented coordination. To achieve this goal, we need to address the following key issues:

- Establishment of a connection between cyberspace and the physical world to enable an efficient scheduling service in terms of energy consumption and service quality.
- Provision of incentives to both users and transportation providers that will enable the creation of an efficient transportation service.
- Creation of a fast and accurate information dissemination system that provides optimal routes for every mover.

11.4 Cyberspace, Social-Network-Enabled Architecture

The social-network-enabled component mainly lies in the cyber space to manage all m-nodes and s-nodes as well as edges, activities, a-links, and f-links. Every m-node is characterized by its interest profile, friendship profile, and preference profile. As an initial design, an interest profile is a set of 4-tuples, which is in the format of (m-node ID, a-link ID, f-link ID, time). An activity link (a-link) represents association from PoI to one or more s-nodes. To establish the complete associations from an m-node to its corresponding s-nodes, there exists an interest binding principle to clarify that the activities specified by a-links can be performed based on one of s-nodes in the s-node list. For example, activity coffee and activity breakfast can be associated with a same s-node satellite coffee. Such an interest binding can be manually specified or automatically generated via matching. The binding can be changed from time to time. An f-link points to a special group of m-nodes, who are friends/avatars to stay with as activity partners or to share resources. Thus, there is a friendship profile in cyberspace for every avatar, consisting of multiple triplets (m-node, f-link, m-node, m-node).

Mapping of m-nodes in the cyber plane onto a p-entity on a physical plane can be a many-to-one relation that implies that a person as a p-entity can have multiple avatars in cyberspace. It can use two avatars to handle weekday and weekend activities, respectively; or a special avatar for business trip and another for family vacations since a specialized m-node should have its well-defined interest profile, friend profile, and preference profile. A restriction is that at any time, no more than one m-node can be mapped onto a p-entity.

Mapping of s-node in the cyber plane onto an l-entity in physical plane is either one-to-one or one-to-zero relation. For a one-to-zero case, no physical location is required for some PoI, such as online Web conferencing. Otherwise, an s-node is mapped onto an l-entity based on its location. Mapping of edge onto routes in physical plane is a one-to-many relation since the route from origin l-entity to destination l-entity is not necessarily unique.

Mapping of edges in cyberspace usually comes with transportation requests to transport a mobile node or a mobile squad between s-nodes. Such requests need to be physically evaluated by PV assignment and routing. The edge transition can be assigned with different service classes based on individual preferences, such as single segment (end-to-end), multiple segment, exclusive, shared small, shared large, or sharing with favorites, etc. Connection hubs are social function enabled, such as coffee shops, grocery markets, day cares, schools, gyms, etc. For example, a weekday working avatar would like to schedule a coffee stop over on the way from home to office. Meanwhile, its favorite friend has not been met for this week. A satisfactory activity schedule will be an exclusive last-mile pick up from home to coffee shop as a transportation hub while its friend will also join him for breakfast (a PoI serving both coffee and breakfast) at the same hub. Thereafter, both of them will share a mover from the hub to office.

An avatar coordinates with Avatars interactively and iteratively in the cyber space to manipulate aggregation of clauses to produce activity schedule through the activity manager described later. When an edge in the schedule needs to be mapped into transportation in the physical space, a request is generated and then processed by the transportation manager. It requires several system modules to accomplish the interaction between the cyber space and physical space, proposed as follows:

1. Activity manager. The activity manager is responsible for manipulating all activity clauses submitted from the cyber space to automatically produce feasible schedules for every avatar. Transition between two consecutive activities changes binding to different s-nodes that may result in generation of transportation needs. Then, the activity manager needs to communicate with the transportation manager to fulfill edge transition (from an s-node to another). The activity manager also needs to evaluate many feasible schedules, if they exist, and to negotiate iteratively with m-node agents to make the optimal schedules. In general, the daily activity clause can be planned ahead of days, hours, or minutes. But changes in activity, time window, or personal preference can happen dynamically any time before the activities taking place. Therefore, the activity manager has to deal with dynamics and is able to accommodate the changes incrementally and dynamically. Overall, specifications coming with edge transition requests may include but not be limited to origination, destination, departure time window, arrival time window, sharing, fastest time or travel time limit, lowest price or price limit. Among these specifications, origination and destination are necessary; one of departure time or arrival time must be provided; other specs are optional.

2. Transportation manager. The transportation manager takes requests from the activity manger to fulfill edge transition (from an s-node to another) based on its interaction with both transportation system in physical space and simulator in cyber space. Every edge transition is mapped onto physical road system and proper PVs to serve the request. Two major issues need to be addressed. First, the transportation manager makes assignment of PVs for transition requests. Meanwhile, PVs themselves need to consume the road system in the physical space to fulfill transportation requirements. In this case, a simulator will run a number of scenario simulations to find out possible options to these

transportation requests. The simulator must be superfast, so enough number of alternatives can be generated in a short time period. It must be accurate enough, so a precise assignment and schedule can be provided. It has to be powerful enough to handle a massive amount of transportation requests simultaneously, especially in a large-scale metropolitan area. Second, a pricing system is necessary to provide the incentive to both the transportation consumers and the transportation providers. To do so, technologies such as mechanism design, stable assignment, etc. are applied to the incentive pricing system. The transportation manager executes above tasks and negotiates with the activity manger iteratively. The negotiation between the activity manager and the transportation manager utilizes the behavior described in the preference profile to reach the goal with different priorities, such as better performance and higher price, or less price and compromised performance.

11.5 Incentives and Pricing Formulation

To maximize the system operation efficacy, considerable service requests should be planned ahead and have longer departure window. Planning ahead will enable efficient assignment of PVs and reduce pickup overhead. More people can share the PV when the departure window is longer. What will motivate people to have a nonimmediate service and to wait longer for the service? Why won't they always request no-sharing, fastest time, and immediate service like a taxi? Creating appropriate incentives will be critical to answer these questions and ensure the success of PV paradigm implementation. A proper incentive mechanism can minimize unnecessary requests for the best service and waste system resources. Balancing utilization disciplines and pricing strategies can satisfy the real service needs and optimize the utilization efficiency of system-wide resources. With such incentives and flexibility, the PV system may provide even lower price than a bus system can do.

A price is presented from the assignment system with pickup time, travel time, number of persons in the vehicle, and so on. A number of options may also be presented so that the negotiation system can make a choice of trade-off of price and performance.

Price can be determined according to the cost of transportation. Normally, people asking for no-sharing, fastest time, and immediate service will be charged the highest price. On the other hand, those with sharing and no travel time limit will get the favorite price. The price presented to a person can be determined by

$$\text{Price} = \Delta C + E + P, \tag{11.1}$$

where ΔC is the increased cost to pick up the person; E is the compensation to the existing passengers already in the vehicle to compensate for their loss in longer travel time and in facing discomfort due to more people in the vehicle; and $P = p \times l$ is the profit to the PV company, where p is the profit per man-mile and l is the travel length.

The price is reverse proportional to the request-ahead time. Longer the request-ahead time, the lower the price. With this pricing scheme, people have incentives to plan ahead and have more flexible travel schedule to pay lower price. However, the scheme is not optimal for all partners who join the game. Each partner will try to maximize profit or has lower price for a trip. The PV company wants to make a higher price than its real cost. The existing passengers want to obtain more compensation than their loss. They all want to make a higher price but the person ordering the service wants to lower the price. Thus, the pricing scheme is modified to achieve a truthful mechanism. The approach of strategy-proof mechanism is applied for this purpose [260,289].

11.6 PV Assignment and Hyper Simulator

Efficient utilization of the PV system has two major issues, assignment and routing. First, we have to find optimal assignment of PVs to the requests. Then, we have to find the best route for the PV. These two issues interact with each other and must be dealt with at the same time. Optimal PV assignment and routing are to be studied to minimize some metrics to serve a set of requests, such as the total traveling time of all PVs, minimal energy consumption, and so on. First, a problem of static assignment of a set of requests is defined and a solution is obtained. For a large city, a distributed algorithm must be provided. A mechanism is to be researched to find the best match between many requests and PVs when multiple requests exist. The well-known Gale–Shapley algorithm is modified to obtain a stable solution. In real life, the requests will arrive at runtime, so dynamic scheduling and routing algorithms are necessary for PVs. Different from previous works, we propose a routing algorithm for global assignments. This algorithm balances the load among the roads, achieving the best global performance. The problem can be defined as follows. Given a road system and a set of PVs, each of them has an origin–destination (OD), the schedule for the minimal total travel time of all PVs is to be generated. The similar problem can be finding schedule for minimal energy consumption, minimal pollution, etc.

The optimal scheduling and assignment problem is an nondeterministic polynomial time (NP)-complete problem; thus we propose a heuristic algorithm. This results in a near-optimal route for each PV. The more difficult problem is to find optimal solutions for a system where there exist heterogonous types of vehicles. In this situation, PVs are scheduled to balance the existing road traffic, benefiting other vehicles, while the PVs have the best route, resulting in a win–win solution. A key issue is to have accurate traffic information and traffic prediction on time. We propose a hyper simulator that uses the best methods to create accurate simulation and use high performance computers along with the fastest algorithm to provide a 100-fold faster simulation. With such a super faster simulator, we may have a high-quality global routing. A scenario simulation of many different situations can be simulated to find the best one or a number of alternatives of routing solutions. In the following, we will describe design issues of the hyper simulator.

To form up a close loop cyber transportation system, a hyper simulator is developed to emulate social network activity–driven traffic system operations. The cyber-space-based traffic system control and management strategies, vehicle departure and scheduling optimization, and congestion pricing mechanism are implemented in the hyper simulator. Traffic simulation platform, VISSIM, is used to enable multiagent-based individual vehicle and urban infrastructure interactions. The hyper simulator system architecture and modulated components are illustrated in Figure 11.2. To implement connected-vehicle (CV)-based two-way communications between vehicle to vehicle (V2V) and vehicle to infrastructure (V2I) in VISSIM, an external module is developed. This external module is essential for simulating individual vehicle behavior and for supporting the CV-based vehicle routing functions. Besides its standard built-in modules, VISSIM offers component object model (COM) interfaces for executing COM commands from external programs. Such customer-specific COM applications provide extensive simulation capacities needed for satisfying various requirements from users. After the VISSIM COM server is registered in the computer operating system, communications between the external program and the VISSIM model are set up. The COM objects, such as individual vehicles and roadway segments, can be utilized and controlled by external programs. Through COM interfaces, an external program can emulate social network activities and transport needs and access the VISSIM simulation model to retrieve traffic data and logic decisions in real time. In this study, Microsoft Visual C# is used as the computer language to implement the module.

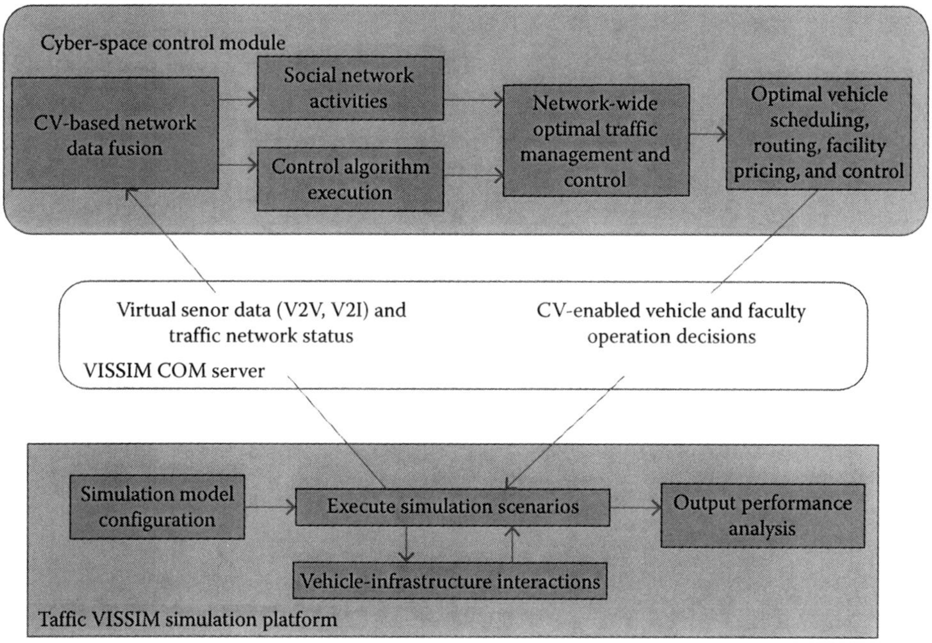

Figure 11.2 Hyper simulator platform system architecture and modulated components. V2V, vehicle-to-vehicle; V2I, vehicle-to-infrastructure.

With the developed simulation platform, the effectiveness of the proposed traffic control and vehicle routing system can be evaluated and the benefits achievable for individual travelers and transportation agencies can be quantified. On the basis of the hyper simulator, we may compute many different routing alternatives, such as end-to-end (single segment), x-segment, exclusive, shared small, shared large, and so on.

11.7 Conclusion

Transportation-related pollutants are some of the largest contributors to air pollution. Exposure to traffic emissions has been linked to many adverse health effects. Substantial technology advances have been made in area of autonomous automobiles and connected vehicles, which opens a wide landscape for future traffic system operations. We have proposed a social network approach for future green transportation system in this paper. Vehicle sharing with public-owned vehicles can significantly reduce the number of vehicles on road. A set of new technologies including cyber space interaction, mechanism design and pricing incentives, optimal vehicle assignment, and hyper simulators enables efficient vehicle sharing. The proposed framework provides a solid theoretical foundation for large-scale, network-wide PV sharing exploration and provides services that can directly impact everyone's daily life and make significant societal impacts, such as reducing air and noise pollution and fossil fuel consumption and optimizing traffic control and management strategy.

Chapter 12

A Mobility Model for Connected Vehicles Induced by the Fish School

Daxin Tian, Keyi Zhu, Jianshan Zhou, and Yunpeng Wang

Beihang University

CONTENTS

12.1 Introduction 214
12.2 Model of Cooperative Vehicles 215
12.2.1 Cooperative multivehicle mobility model 215
12.2.1.1 Cooperative multivehicle mobility model 215
12.2.1.2 Stability analysis of connected vehicles 222
12.2.2 Improved cooperative multivehicle mobility model 226
12.3 Application and Analysis of Cooperative Multivehicle Mobility Model 231
12.3.1 Simulation and verification 231
12.3.1.1 Simulation and verification of cooperative multivehicle mobility model 231
12.3.1.2 Simulation and verification of improved cooperative multivehicle mobility model 232
12.3.2 Influence of intervehicle communication on mobility of multiple vehicles ... 235
12.3.2.1 Under the traffic flow scenario of low density and high mobility 235
12.3.2.2 Under the traffic flow scenario of high density and low mobility 238
12.3.3 Influence of intervehicular communication on vehicle safety and traffic efficiency 238
12.3.3.1 Analysis of collision probability 239
12.3.3.2 Analysis of flow rate 242
12.4 Conclusion 246

12.1 Introduction

With the rapid growth of wireless communication technologies, a large number of vehicles have formed a vehicular ad hoc network (VANET) and are cooperating with each other; thus, collision warning, road obstacle warning, intersection collision warning, and lane-change assistance can be improved [88,247]. These connected vehicles move in a cooperative manner, which is similar to some cooperative behaviors of animal flocking in nature. Therefore, it is convenient for researchers to model the connected vehicles as a group by drawing an analogy to animals flocking. The goals of this paper are twofold: (1) to mathematically model the cooperative movements of connected vehicles with wireless communications based on the fish school (2) to provide comprehensive evaluation of the influence of connected vehicles on safety and efficiency.

Wireless communication technologies expand the sensing range of vehicles and enhance the information interplays among vehicles moving on the same road section. For example, one popular application of these wireless communication technologies in vehicular system is the vehicular ad hoc network (VANET), which is supported through vehicle-to-roadside (V2R) and vehicle-to-vehicle (V2V) communications. There are large numbers of studies on signal propagation in V2R and V2V communications, such as Refs. [243,261], which focus on the mechanism of communication under various traffic situations. In addition there are other studies that address the impact of connected vehicles on the transportation system, where communication technologies are widely deployed. For instance, Ref. [37] investigates the impact of the number of cooperative vehicles on network performance under Nakagami fading channel. Moreover, wireless communications can also be used to assist in developing a driving-assistance system [244]. There are few studies paying attention to the cooperative behavior of the connected vehicles moving in the same road section as a group. It is a large challenge to model the cooperative behaviors of connected vehicles, because the interplays among vehicles, as well as some environmental factors influencing vehicular movement, should first be carefully identified, and then mathematically formulated. Nevertheless, this chapter aims at addressing this issue by following the bio-inspired modeling approach. Since cooperation is one characteristic of animals in flocks, such as fish school, we draw on the mechanism of flocking as it is present in the biosphere (i.e., in the fish school) to model the cooperative behaviors of connected vehicles.

Inspired by the aforementioned behaviors of fish school, we can draw an analogy between the connected vehicles and the fish school. With the assistance of advanced sensors, such as a velocity sensor, vehicular positioning system, or a navigation system, a vehicle can be provided with the real-time information such as velocity, acceleration, position, and other basic parameters, and it can provide the collected information to neighboring vehicles through wireless communications. That is, the wireless communication systems can make each vehicle able to sense environmental information relevant to itself and to others as well as to share the collected information with its neighbors, which is similar to the environment-sensing behavior of fish school. Furthermore, with wireless communication, vehicles are able to coordinate their movements (velocity and direction) according to the environmental situation and the overall mobility of the vehicle group. Each vehicle can interact with its neighbors in real time via a wireless communication system, so that they can move in a cooperative manner. This is similar to the cooperative behavior of fish school that is guided by the interplay of different fish. In this sense, some rules characterizing the behavior of fish school such as cohesion separation, and alignment can be adopted to present the cooperation of those connected vehicles to some extent. In addition, another important characteristic of fish school behavior is danger avoidance (such as avoiding obstacles or predators), which can be analogous to collision avoidance by car drivers.

Moving fish that have formed a coordinated school can shift back to being an amorphous shoal within seconds of facing the emergence of obstacles [161]. Obstacle avoidance has been studied in research on flocks [137,197,225]. However, these studies have not considered the real-world applications of these models. Hence, it is worthwhile to provide mathematical models for describing the mobility of connected vehicles when obstacles/avoidance is taken into account. Some important factors should be additionally considered, defined, and formulated in to the modeling. These include the constraints of the road and, of traffic rules, which must be taken into consideration when drawing an analogy between groups of vehicles and the fish school.

In this chapter, a novel model is proposed to formulate the movement of connected vehicles by considering some real-world traffic situations where the constraints of the road and road obstacles exist. By analogy, connected vehicles also follow rules similar to those governing the fish school. The interplay among connected vehicles is mathematically modeled by introducing the potential field functions that are similar to fish school behavior. In addition, a theoretical analysis framework is provided to verify the rationality of the proposed model. Finally, some numerical experiments are also described to demonstrate the model as well as to provide a better understanding of the improvements to safety and to traffic efficiency that might result from by wireless communication.

The rest of the chapter is as follows: Section 12.2 proposes a mobility model to describe connected vehicles and presents a stability analysis. In Section 12.3, extensive simulations are performed to verify the proposed model. Section 12.4 concludes the chapter.

12.2 Model of Cooperative Vehicles

12.2.1 Cooperative multivehicle mobility model

Moving as a group on the same section of road each vehicle needs to maintain a different velocity and accelerate under the effects of forces. In the model, we address the attraction of the goal, the repulsion of obstacles, the constraints of the road, and the interplay among vehicles in the group, including both attractive and repulsive forces. Moreover, we assume that vehicles can communicate with other neighbor vehicles via wireless communication technologies, which reflect in sensing range that will be provided later.

In this model, each vehicle i belongs to the set of vehicles denoted by $N = \{1,2,\ldots,n\}$ with the total number of vehicles $n(n>2)$. In a vehicle group, we concentrate on the target vehicle i and its neighbor j, $\forall i \neq j \in N$. The vector $p_i(t) = [x_i(t), y_i(t)]^T$ signifies the location of vehicle i at time t. In addition we describe the force between mobility vehicles due to the distance changes among them, so that the interplay ranges of each vehicle are significant. In this chapter, each vehicle is treated as a virtual sphere with the radius r_i in order to correct the drawbacks of overlap (see in Figure 12.1). The concept of the "local warning scope" is defined as the local circle zone with radius r_i^W of the vehicle to warning that the situation may be dangerous when other vehicles or obstacles enter this area. In this chapter, the vehicles described can get information on the location, velocity, and behaviors on neighboring vehicles by communicating with other vehicles in its sensing range with the radius as r_i (see in Figure 12.1).

12.2.1.1 Cooperative multivehicle mobility model

In order to develop a mobility model that can reflect an accurate traffic situation, the model not only considers the interactive forces among vehicles in the group but also takes the repulsive

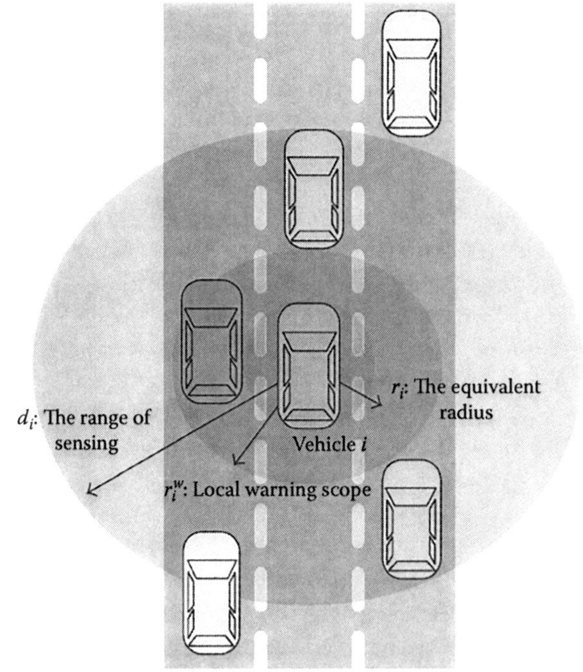

Figure 12.1 The interplay range of vehicle i.

forces of obstacles and road boundaries into account. These forces are described by a vector at every location in the force field. Following these considerations, we define the effects produced by accumulated forces exerted on vehicle i as follows:

$$m_i v_i(t) = F_i^{(1)}(t) + F_i^{(2)}(t) + F_i^{(3)}(t) \tag{12.1}$$

in which $v_i(t)$ is the actual velocity at time t of vehicle i with mass m_i. In formula (12.1), $F_i^{(1)}(t)$ is the attractive force by the goal of vehicle i, $F_i^{(2)}(t)$ is the interactive force among connected vehicles, and $F_i^{(3)}(t)$ is the environmental effects of vehicle i. Without special notation, the letters in bold represent the term is a vector within a two-dimensional plane.

The vehicles are expected to move in a certain direction at the desired speed. Therefore, $F_i^{(1)}(t)$ stimulates vehicle i with a certain desired velocity v_i^0 in the expected direction $e_i(t)$. Hence, the expected velocity can be formulated as:

$$v_i^0(t) = v_i^0 e_i(t). \tag{12.2}$$

However, in reality, the velocity $v_i(t)$ at time t of a vehicle may deviate from the desired velocity $v_i^0(t)$ due to the necessary deceleration, acceleration, or other unknowns. Thus, $F_i^{(1)}(t)$ as a stimulus to force the vehicle back to the $v_i^0(t)$ again with a relaxation time τ_i can be given by

$$F_i^{(1)}(t) = m_i \frac{v_i^0(t) - v_i(t)}{\tau_i}. \tag{12.3}$$

One of the characteristics of connected vehicles is cooperative behavior. The vehicles benefit from wireless communication so that they can cooperate moving as a group. The vehicles aggregate together in the same movement pattern while keeping a certain safe distance from each other. For the purpose of formulating the group forces upon vehicles, we set the interactive forces into two categories, including attractive forces and repulsive forces, which can make the vehicles move as a group while maintaining a certain distance and avoiding collisions. We can combine the attractive and repulsive forces in $F_i^{(2)}(t)$ as follows:

$$F_i^{(2)}(t) = \sum_{\forall j \in N_i(t)} (F_{ij}^{\mathrm{A}}(t) + F_{ij}^{\mathrm{R}}(t)). \tag{12.4}$$

$F_i^{(2)}(t)$ is the accumulated force upon target vehicle i and all other neighboring vehicles j in its sensing range. $F_{ij}^{\mathrm{A}}(t)$ is the attractive force on vehicle i originating from its neighboring vehicles. Under the influence of the attractive force, the vehicles will accelerate toward each other to keep together as a group. Meanwhile, $F_{ij}^{\mathrm{R}}(t)$ is the repulsive force on vehicle i issuing from neighboring vehicles, which keeps them at a safe distance and avoids collisions among connected vehicles.

In short distances, like the warning scope, the repulsion increases with a decrease in the distances between vehicle i and its neighbors. Likewise, attraction will play a leading role in making vehicles stay together as a group when the distances between vehicle i and its neighbors are larger than the warning scope. If the neighboring vehicle is beyond the bounds of the sensing range d_i, the attraction will not exist. Similarly, the repulsive force will not contribute to the acceleration of a vehicle if it is not in the sensing range of the target vehicle. To depict this situation, we use the concept of potential functions to formulate the attractive and repulsive forces. In addition, according to Refs. [26,126], potential functions are used to describe the interplays among individuals in both fish school and vehicles, which also can control individuals and help them avoid an obstacle through the repulsive potential field.

The effects of two kinds of interplays among vehicles, i.e., attraction and repulsion, vary along with the changing relative distance between vehicles. In addition, the comprehensive impact of interplays can be appropriately modeled upon the potential field functions. A bump function model in Ref. [198] is adopted to formulate the smooth potential functions with finite cutoffs. In this chapter, for the sake of simplicity, the bump function used to indicate the proximity between the vehicles i and j is denoted as $\varphi_{ij}(t) = \psi_l(p_i(t), p_j(t))$. By combining $\varphi_{ij}(t)$ and the attractive/repulsive potential functions $U_{\mathrm{A}}(p_i(t), p_j(t))$ and $U_{\mathrm{R}}(p_i(t), p_j(t))$, the attractive/repulsive forces are defined as:

$$F_{ij}^{\mathrm{A}}(t) = -m_i \nabla U_{\mathrm{A}}(p_i(t), p_j(t)) - \rho_i \varphi_{ij}(t) v_i(t), \tag{12.5}$$

$$F_{ij}^{\mathrm{R}}(t) = -m_i \nabla U_{\mathrm{R}}(p_i(t), p_j(t)) - \mu_i \varphi_{ij}(t) v_i(t), \tag{12.6}$$

where ρ_i and μ_i are two positive real coefficients used to scale the magnitude of the attractive/repulsive forces, and they are limited as $\rho_i, \mu_i(0,1)$ in this chapter. According to the form of the bump function, $\varphi_{ij}(t)$ can be expressed as:

$$\varphi_{ij}(t) = \begin{cases} 1, \text{when } \frac{\Delta d_i j(t) - (r_i + r_j)}{(d_i - (r_i + r_j))} \in [0, l]; \\ k_{\mathrm{v}} \left(1 + \cos\left(\pi \left(\frac{\Delta d_i j(t) + (l-1)(r_i + r_j) - l \times d_i}{(1-l)(d_i - (r_i + r_j))} \right) \right) \right), \text{when } \frac{\Delta d_{ij}(t) - (r_i + r_j)}{d_i - (r_i + r_j)} \in (l, 1]; \\ 0, \text{otherwise}, \end{cases} \tag{12.7}$$

where k_{v} is similarly a positive parameter within $(0,1)$ and the variable $\Delta d_i j(t)$ denotes the intervehicular distance between the vehicles i and j at the time t. To explore the influence of the different parameters k_{v} and l on the adopted bump function, we conduct the

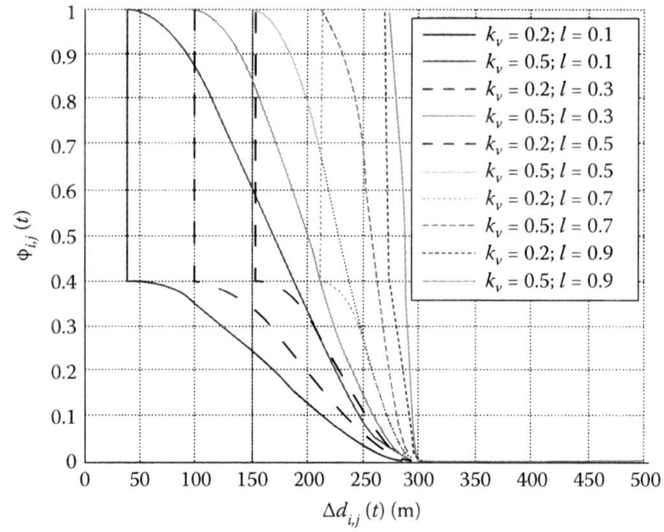

Figure 12.2 Bump function with different parameters.

numerical experiment with the settings of $d_i = 300\,\text{m}$, $r_i = r_j = 5\,\text{m}$, $k_\text{v} = \{0.2, 0.5\}$, and $l = \{0.1, 0.3, 0.5, 0.7, 0.9\}$. The results are shown in Figure 12.2. It can be found that this bump function is a scalar function that ranges between 0 and 1 and has different cutoffs with different values of k_v and l. It also can be seen in Figure 12.2 that when $k_v = 0.2$, the function has a sudden jump at a certain point. To guarantee the smoothness of the potential function, we fix $k_\text{o} = k_\text{L} = k_\text{v} = 0.5$ and $l = 0.3$ in the following experiments.

The potential function of attractive force is designed as the form of a tangent function:

$$U_\text{A}(p_i(t), p_j(t)) = \begin{cases} \frac{1}{2}k_\text{A} \tan\left(\frac{\pi\left(\|p_i(t)-p_j(t)\|-(r_i+r_j)\right)}{2(d_i-(r_i+r_j))}\right), \text{when} & \|p_i(t)-p_j(t)\| \in [(r_i+r_j), d_i) \\ 0, \text{otherwise.} \end{cases} \tag{12.8}$$

and the repulsive potential function is given as follows:

$$U_\text{R}(p_i(t), p_j(t)) = \begin{cases} \frac{1}{2}k_\text{R}\left(\frac{1}{\left(\|p_i(t)-p_j(t)\|-(r_i+r_j)\right)^2} - \frac{1}{\left(d_i-(r_i+r_j)\right)^2}\right)^2, \text{when} & \|p_i(t)-p_j(t)\| \in ((r_i+r_j), d_i] \\ 0, \text{otherwise,} \end{cases} \tag{12.9}$$

where the vector $p_i = [x_i, y_i]^T$ and $p_j = [x_j, y_j]^T$ denote the position of vehicles i, and j, respectively. k_A and k_R are two scaling positive real coefficients. By introducing the potential functions $U_\text{A}(p_i(t), p_j(t))$ and $U_\text{R}(p_i(t), p_j(t))$, we can further assume that the magnitudes of the attractive and repulsive effects existing between two neighboring vehicles i and j are equal when distance $\Delta d_i j(t) = \|p_i(t) - p_j(t)\|$ between them stays at a certain desired value. Let this desired distance be $\Delta d_\text{desire} \in ((r_i + r_j), d_i)$. Note that the attractive and repulsive effects are represented by the gradients of the potential functions $U_\text{A}(p_i(t), p_j(t))$ and $U_\text{R}(p_i(t), p_j(t))$

with respect to the relative distance between the vehicles i and j, respectively. Accordingly, k_A and k_R should be set to satisfy the partial differential equation $\|\frac{\partial U_A(p_i(t),p_j(t))}{\partial \Delta d_{ij}(t)}\| = \|\frac{\partial U_R(p_i(t),p_j(t))}{\partial \Delta d_{ij}(t)}\|$ under the condition of $\Delta d_{ij}(t) = \Delta d_{desire}$. By substituting $\Delta d_{ij}(t) = \Delta d_{desire}$ into Equations 12.8 and 12.9 and solving the equation $\|\frac{\partial U_A(p_i(t),p_j(t))}{\partial \Delta d_{ij}(t)}\| = \|\frac{\partial U_R(p_i(t),p_j(t))}{\partial \Delta d_{ij}(t)}\|$, we can find the relationship of the magnitudes of k_A and k_R as follows:

$$\frac{k_R}{k_A} = \frac{\frac{\pi}{4\left(d_i-(r_i+r_j)\right)}\left(\cos^{-2}\left(\frac{\pi}{2}\left(\frac{\Delta d_{desire}-(r_i+r_j)}{d_i-(r_i+r_j)}\right)\right)\right)}{2\left(\Delta d_{desire}-(r_i+r_j)\right)^{-3}\left[\left(\Delta d_{desire}-(r_i+r_j)\right)^{-2}-\left(d_i-(r_i+r_j)\right)^{-2}\right]}. \qquad (12.10)$$

Based on Equation 12.10, it can be seen that, once the desired distance Δd_{desire} between vehicles and the parameters d_i, r_i, and r_j are given, the settings on k_A and k_R are expected to satisfy the relationship $k_R = \xi \times k_A$, where ξ is set to be the result of the right side of Equation 12.10. To explore the impact of the parameters r_i and r_j on the potential functions $U_A(p_i(t),p_j(t))$ and $U_R(p_i(t),p_j(t))$, we conduct the following numerical experiment where the sensing range d_i is fixed at 300 m and r_i has the same value with r_j for simplicity, i.e., $r_i = r_j = 5$ m. In addition, the desired distance Δd_{desire} is set to be 50 m, and the parameter k_A discretely varies from 0.01 to 0.5. According to Equation 12.10, the ratio of k_R and k_A can be calculated as $\xi = 1.482 \times 10^5$. Hence, in this experiment, we can set the other parameter k_R as $k_R = \xi \times k_A = 1.482 \times 10^5 k_A$. The relevant results are given in Figure 12.3.

From Figure 12.3, it can be found that the potential functions are both monotonically varying, along with the value of the distance $\|p_i(t) - p_j(t)\|$ between vehicle i and vehicle j. The definitions of the functions $U_A(p_i(t),p_j(t))$ and $U_R(p_i(t),p_j(t))$ guarantee that the attractive and repulsive potential fields have important properties—the smaller the distance $\|p_i(t) - p_j(t)\|$ the smaller the effect of the attractive potential field becomes, while the effect of the repulsive potential field increases along with decreasing the distance $\|p_i(t) - p_j(t)\|$. As shown in Figure 12.3, when $\|p_i(t) - p_j(t)\|$ decreasingly approaches the value of $(r_i + r_j)$, $U_A(p_i(t),p_j(t))$ is reduced to zero while $U_R(p_i(t),p_j(t))$ dramatically increases. This indicates that when a vehicle tends to collide with another vehicle, the attractive potential effect should be weakened and the repulsive effect strengthened at the same time. Hence, the comprehensive effect resulting from both the attractive and repulsive potential fields will force the vehicle to avoid the collisions. Additionally, from Figure 12.3, it can be found that the larger the parameters k_R and k_A, the steeper the slopes of $U_A(p_i(t),p_j(t))$ and $U_R(p_i(t),p_j(t))$ become. At this point, large parameter settings for k_R and k_A would make the potential fields more sensitive to the time-related changes in the intervehicular distance. Hence, an appropriate value of k_R and k_A should be chosen carefully. According to the results presented by Figure 12.3, we fix $k_A = 0.1$ and $k_R = \xi k_A$ for our model in other subsequent experiments for the sake of controlling the sensitivity of the potential field functions $U_A(p_i(t),p_j(t))$ and $U_R(p_i(t),p_j(t))$.

Some other environmental factors influencing the movement of the connected vehicles should be taken into account as well, such as road obstacles and road constraints. That is, the vehicles on the road might face obstacles like accidents, road maintenances, and other emergencies. In addition, vehicles are also constrained by road boundaries, a situation which is different from that of fish school. In this chapter, the effects arising from road obstacles and road boundaries are modeled as:

$$F_i^{(3)}(t) = \sum_{\forall k \in M} F_{ik}^{o}(t) + F_i^{L}(t), \qquad (12.11)$$

where $F_{ik}^{o}(t)$ is the repulsive force of the obstacle k, $k = \{1,2,\ldots,M\}$ (M is the number of road obstacles), and $F_i^{L}(t)$ is the virtual repulsive force that forces a vehicle to be away from the

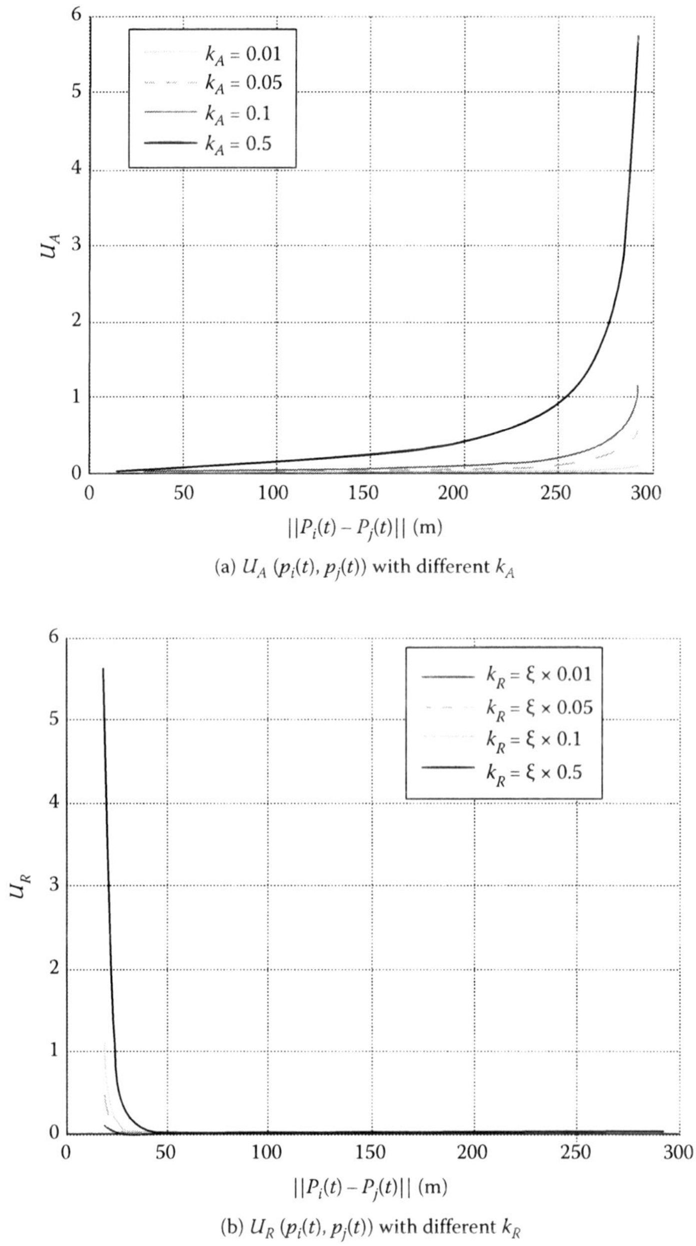

Figure 12.3 The variation of potential functions with different parameter settings.

boundary of the road and stay on the road when the vehicle is too close to one boundary and would drive off the road. Since the decrease in the relative distance between a vehicle and an obstacle or between a vehicle and a road boundary could lead to an increase in the repulsive effects on the vehicle, the repulsive forces of road obstacles and road boundaries should be formulated as decreasing functions of such a relative distance. By referring to the bump function given in Equation 12.7, we propose potential field functions to model the repulsive effect arising from road obstacles and road boundaries as follows.

We define the vector $p_k^O = [x_k^O(t), y_k^O(t)]^T$ as the position of a certain obstacle denoted by O_k. In this chapter, the obstacle is treated as a ball with the radius r_k^O. When the vehicle i approaches the obstacle O_k in proximity, and the distance $\|p_i(t) - p_k^O\|$ becomes smaller than the sensing range d_i, the vehicle i will detect it and prepare to avoid it. Furthermore, $F_{ik}^O(t)$ should be a monotonically decreasing function of the distance between the obstacle O_k and the vehicle i in the sensing range, and the vehicle i should keep a certain distance from the obstacles so as to avoid a collision. Then, the function of the repulsive force induced by a bump function defined in Ref. [198] is given as

$$F_{ik}^O(t) = F_{ik}^O \times \varphi_l(p_i(t), p_k^O(t)) n_{ik}(t), \tag{12.12}$$

where F_{ik}^O is the maximum magnitude of the obstacle repulsive force $F_{ik}^O(t)$, and $n_{ik}(t)$ is a unit vector used to indicate the repulsive acting direction at the time instant t

$$n_{ik}(t) = \frac{(p_i(t) - p_k^O(t))}{\|p_i(t) - p_k^O(t)\|} \tag{12.13}$$

and the bump function $\varphi_l(p_i(t), p_k^O(t))$ is represented as

$$\varphi_l(p_i(t), p_k^O(t)) = \begin{cases} 1, \text{when} \quad \frac{\Delta d_{ik}(t) - (r_i + r_k^O)}{d_i - (r_i + r_k^O)} \in [0, l]; \\ k_O\left(1 + \cos\left(\pi\left(\frac{\Delta d_{ik}(t) + (l-1)(r_i + r_k^O) - l \times d_i}{(1-l)(d_i - (r_i + r_k^O))}\right)\right)\right), \text{when} \quad \frac{\Delta d_{ik}(t) - (r_i + r_k^O)}{d_i - (r_i + r_k^O)} \in (l, 1]; \\ 0, \text{otherwise}, \end{cases} \tag{12.14}$$

where $\Delta d_{ik}(t) = \|p_i(t) - p_k^O(t)\|$ and k_O and l are positive real coefficients whose values are limited within $(0, 1)$. Considering that the constraints of the road are essential for realistic modeling of connected vehicles moving as a group, the road constraints force that is similar to the force of the obstacle is defined as

$$F_i^{\mathrm{L}}(t) = F_i^{\mathrm{L}} \psi_l(p_i(t), p_i^{\mathrm{L}}(t)) n_i^{\mathrm{L}}(t), \tag{12.15}$$

where F_i^{L} is the maximum magnitude of the force $F_i^{\mathrm{L}}(t)$, and $\psi_l(p_i(t), p_i^{\mathrm{L}}(t))$ is also designated as a bump function as follows:

$$\psi_l(p_i(t), p_i^{\mathrm{L}}(t)) = \begin{cases} 1, \text{when} \quad \frac{\Delta d_i^{\mathrm{L}}(t) - r_i}{R - r_i} \in [0, l]; \\ k_{\mathrm{L}}\left(1 + \cos\left(\pi\left(\frac{\Delta d_i^{\mathrm{L}}(t) + (l-1)r_i - lR}{(1-l)(R - r_i)}\right)\right)\right), \text{when} \quad \frac{\Delta d_i^{\mathrm{L}}(t) - r_i}{(R - r_i)} \in (l, 1]; \\ 0, \text{otherwise}, \end{cases} \tag{12.16}$$

where $k_{\mathrm{L}} \in (0, 1)$ and $p_i^{\mathrm{L}}(t)$ represents the position of the vehicle $i's$ center projection point on the nearest road boundary line at the time instant t, and $\Delta d_i^{\mathrm{L}}(t) = \|p_i(t) - p_i^{\mathrm{L}}(t)\|.n_i^{\mathrm{L}}(t)$ is the unit direction vector from $p_i^{\mathrm{L}}(t)$ to $p_i(t)$ that is perpendicular to the road boundary line, which can be calculated as:

$$n_i^{\mathrm{L}}(t) = \frac{p_i(t) - p_i^{\mathrm{L}}(t)}{\Delta d_i^{\mathrm{L}}(t)}. \tag{12.17}$$

The parameter R is the maximum range of the virtual road constraint force acting on a vehicle. From the equation above, it can be seen that when the relative distance between the vehicle i and its nearest road boundary (i.e., $\Delta d_i^{\mathrm{L}}(t)$) is larger than R, this virtual road constraint force is set to 0; otherwise, the closer this vehicle is to the road boundary, the larger the repulsive force.

In addition, in order to make the repulsive forces $F_{ik}^{O}(t)$ and $F_i^{L}(t)$ resulting from the obstacles and the road constraints dynamically coordinated with the attractive force arising from the desired velocity $v_i^{O}(t)$, we combine them with $F_i^{(1)}(t)$, defined in Equation 12.3, and then define the maximum magnitudes of these two repulsive forces as follows:

$$\begin{cases} F_{ik}^{O}(t) = \gamma_{ik} \times |F_i^{(1)}(t) \cdot n_{ik}(t)| = \gamma_{ik} \frac{m_i}{\tau_i} |(v_i^{O}(t) - v_i(t)) \cdot n_{ik}(t)| \\ F_i^{L}(t) = \beta_i \times |F_i^{(1)}(t) \cdot n_i^{L}(t)| = \beta_i \frac{m_i}{\tau_i} |(v_i^{O}(t) - v_i(t)) \cdot n_i^{L}(t)| \end{cases}, \tag{12.18}$$

where $\gamma_i k$ and β_i are positive sensitivity coefficients within $(0,1]$, which are used to scale and coordinate the magnitudes of $F_{ik}^{O}(t)$ and $F_i^{L}(t)$.

12.2.1.2 Stability analysis of connected vehicles

Since a connected vehicle are equipped with wireless communications, it is able to interact with others in its sensing range. Thus, these mobile terminals constitute a local vehicular network. The communication topology of these connected vehicles at any time instant t can be presented by a bidirectional graph, which can also be called a communication graph. Let $G(t) = (V, E(t))$ denote this communication graph, and $V = \{i | \forall i \in N\}$. As mentioned in Section 12.2.1.1, we assume that the sensing range of all the vehicles is identical, i.e., $d_i = d(\forall i \in V)$. Subsequently, an edge in this communication graph $G(t)$ is used to indicate a bidirectional wireless communication interaction between a vehicle and one of this vehicle's neighbors. That is, the set of those edges in $G(t)$ can be defined as $E(t) = \{(i,j) | \|p_i(t) - p_j(t)\| < d\}$ and $E(t) \in V \times V$. For simplicity, we also denote the neighboring nodes of any node i (i.e., the vehicle i) in the given communication graph $G(t)$ at any time instant t as a set $N_i(t) = \{j | (i,j) \in E(t), j \in V\}$. From the notation of $E(t)$, it is obvious that the topology of this communication graph varies over time due to the possibility of a vehicle entering into or out of the communication range of the host vehicle. Attempting to analyze the overall mobility of a given vehicle group moving on the same road, we assume that the initial velocity of all those vehicles is the same and that their desired velocities at any time instant t are also identical, i.e., $v_i(0) = v$ and $v_i^0(t) = v^0(t)$ for all vehicles $i \in V$. It should be noted that this assumption is reasonable, since vehicles would move with a constant velocity in the same direction when they are in an equilibrium traffic flow of the same road and encounter no emergencies or disturbances. Also, we assume that the sensing ranges of all the vehicles are equal, i.e., $d_i = d_j$ for $\forall i \neq j \in V$. We introduce the concept of the group centroid, which is represented as $p_c(t) = (1/N) \times \sum_{i=1}^{N} p_i(t)$. Now, we state the analysis results of our proposed mobility model.

Corollary 1. Consider that connected vehicles with wireless communications evolve under the mobility model defined by Equation 12.1. Accept that $\sum_{\forall k \in M} \gamma_{ik} +_i \leq 1$, $v_i(0) = v$ and $v_i^0(t) = v^0(t)$ for all vehicles $i \in V$, the overall group can asymptotically converge on the desired velocity $v^0(t)$, which is defined in Equation 12.2.

Proof: We can combine all the forces of every vehicle as follows:

$$\sum_{\forall i \in V} m_i v_i(t) = \sum_{\forall i \in V} F_i^{(1)}(t) + F_i^{(2)}(t) + F_i^{(3)}(t). \tag{12.19}$$

Since the condition is given that $d_i = d_j$ for $\forall i \neq j \in V$, and these forces $F_{ij}^{A}(t)$ and $F_{ij}^{R}(t)$ are bidirectional for all vehicles $i \in V$, we have $F_{ij}^{A}(t) = -F_{ji}^{A}(t)$ and $F_{ij}^{R}(t) = -F_{ji}^{R}(t)$ according to

Equations 12.5 and 12.6. At this point, we further have:

$$\begin{aligned}
&\sum_{\forall i \in V} F_i^{(2)}(t) \\
&= \sum_{\forall i \in V} \left(\sum_{\forall j \in N_i(t)} (F_{ij}^A(t) + F_{ij}^R(t)) \right) \\
&= \sum_{\substack{\forall i \neq j \in V \\ \forall (i,j) \in E(t)}} (F_{ij}^A(t) + F_{ij}^R(t) + F_{ji}^A(t) + F_{ji}^R(t)) = 0.
\end{aligned} \tag{12.20}$$

Therefore, Equation 12.1 can be rearranged as $\sum_{\forall i \in V} m_i v_i(t) = \sum_{\forall i \in V} (F_i^{(1)}(t) + F_i^{(3)}(t))$. Now, recalling that the bump functions $\varphi_l(p_i(t), p_k^o(t))$ and $\varphi_l(p_i(t), p_i^L(t))$ are limited within $[0,1]$, we can get the inequality as follows:

$$\begin{aligned}
\|F_i^{(3)}(t)\| &= \left\| \left(\sum_{\forall k \in M} F_{ik}^o(t) \times \varphi_l(p_i(t), p_k^o(t)) n_i k(t) \right) + F_i^{\mathrm{L}}(t) \psi_l(p_i(t), p_k^{\mathrm{L}}(t)) n_i^{\mathrm{L}}(t) \right\| \\
&\leq \left\| \left(\sum_{\forall k \in M} F_{ik}^o(t) \times \varphi_l(p_i(t), p_k^o(t)) n_i k(t) \right) \right\| + \|F_i^{\mathrm{L}}(t) \psi_l(p_i(t), p_k^{\mathrm{L}}(t)) n_i^{\mathrm{L}}(t)\| \\
&\leq \sum_{\forall k \in M} F_{ik}^o(t) + F_i^{\mathrm{L}}(t) \\
&= \sum_{\forall k \in M} \gamma_{ik} \frac{m_i}{\tau_i} |(v_i^o(t) - v_i(t)) \cdot n_{ik}(t)| + \beta_i \frac{m_i}{\tau_i} |(v_i^o(t) - v_i(t)) \cdot n_i^{\mathrm{L}}(t)| \\
&\leq \sum_{\forall k \in M} \gamma_{ik} \frac{m_i}{\tau_i} \|v_i^o(t) - v_i(t)\| + \beta_i \frac{m_i}{\tau_i} \|v_i^o(t) - v_i(t)\| \\
&= \left(\sum_{\forall k \in M} \gamma_{ik} + \beta_i \right) \frac{m_i}{\tau_i} \|v_i^o(t) - v_i(t)\| \leq \|F_i^{(1)}(t)\|
\end{aligned} \tag{12.21}$$

Hence, the magnitude of the attractive force resulting from the desired velocity $v_i^o(t)$ is larger than that of the force resulting from the obstacle and the road boundary. This implies that the mobility of the overall vehicle group is mainly dominated by the potential field that could force vehicles to keep moving with, or asymptotically converge on, the desired velocity, even after being disturbed by the repulsive potential resulting from obstacles and road constraints. At this point, we prove Corollary 1.

Now, we first provide some properties of the potential functions $U_{\mathrm{A}}(p_i(t), p_j(t))$, and $U_{\mathrm{R}}(p_i(t), p_j(t))$ related to the interactive force $F_i^{(2)}(t)$ as follows.

For simplicity, we equivalently denote $U_{\mathrm{A}}(\Delta d_{ij}(t)) = U_{\mathrm{A}}(p_i(t), p_j(t))$ and $U_{\mathrm{R}}(\Delta d_{ij}(t)) = U_{\mathrm{R}}(p_i(t), p_j(t))$ where $\Delta d_{ij}(t) = \|p_i(t) - p_j(t)\|$. It can be mathematically proven that Corollary 2 $U_A(\Delta d_{ij}(t))$ satisfies the following properties:

(1) it is continuously differentiable for $\Delta d_{ij}(t) \in [(r_i + r_j), d_i)$ and is a monotonically increasing function of $\Delta d_{ij}(t)$ within $[(r_i + r_j), d_i)$ that makes $U_{\mathrm{A}}(\Delta d_{ij}(t)) \to 0$ with $\Delta d_{ij}(t) \to (r_i + r_j)$ and $U_{\mathrm{A}}(\Delta d_{ij}(t)) \to \infty$ with $\Delta d_{ij}(t) \to d_i$; (2) $u_{ij}^A = u_{ji}^A$ when denoting $u_{ij}^A = \partial U_{\mathrm{A}}(\Delta d_{ij}(t)) / \partial \Delta d_{ij}(t)$ for $\forall (i,j) \in E(t)$; and (3) $\sum_{\forall (i,j) \in E(t)} \partial U_{\mathrm{A}}(\Delta d_{ij}(t)) / \partial p_i(t) = \sum_{\forall i \neq j \in V} \partial U_{\mathrm{A}}(\Delta d_{ij}(t)) / \partial p_i(t)$.

Proof: It is easily shown that the partial derivative of $U_{\mathrm{A}}(\Delta d_{ij}(t))$ with respect to $\Delta d_{ij}(t)$ is expressed as:

$$u_{ij}^{A} = \frac{\partial U_{\mathrm{A}}\left(\Delta d_{ij}(t)\right)}{\partial \Delta d_{ij}(t)} = \frac{\pi}{4}\left(d_i - (r_i + r_j)\right)k_{\mathrm{A}} \frac{1}{\left(cos\left(\frac{\pi\left(\Delta d_{ij}(t) - (r_i + r_j)\right)}{2(d_i - (r_i + r_j))}\right)\right)^2} > 0 \tag{12.22}$$

for $\Delta d_{ij}(t) \in [(r_i + r_j), d_i)$. At this point, $U_{\mathrm{A}}(\Delta d_{ij}(t))$ is a monotonically increasing function that is defined on $[(r_i + r_j), d_i)$. Consequently, we further get

$$\begin{cases} \lim_{\Delta d_{ij}(t) \to (r_i + r_j)} U_{\mathrm{A}}\left(\Delta d_{ij}(t)\right) = U_{\mathrm{A}}\left((r_i + r_j)\right) = 0 \\ \lim_{\Delta d_{ij}(t) \to d_i} U_{\mathrm{A}}\left(\Delta d_{ij}(t)\right) = U_{\mathrm{A}}(d_i) = \infty. \end{cases} \tag{12.23}$$

and $u_{ij}^{A} > 0$ for $\Delta d_{ij}(t) \in ((r_i + r_j), d_i)$. Thus, the property (1) is proven. Moreover, recall that we have assumed $d_i = d_j$ for $\forall i \neq j \in V$. This means that the sensing ability (sensing range) of every vehicle is identical. Hence, according to $\Delta d_{ij}(t) = \Delta d_{ji}(t)$, it can be seen that $\partial U_{\mathrm{A}}(\Delta d_{ij}(t))/\partial \Delta d_i j(t) = \partial U_{\mathrm{A}}(\Delta d_{ji}(t))/\partial \Delta d_{ji}(t)$, i.e., $u_{ij}^{A} = u_{ji}^{A}$. Thus, the property (2) is proven. Noting the definition of $U_{\mathrm{A}}(\Delta d_{ij}(t))$ and the communication graph $G(t)$, when vehicle j is not in the neighborhood $N_i(t)$ of vehicle i, the relative distance between them $\Delta d_{ij}(t)$ is larger than the sensing range d_i. Then, the potential $U_{\mathrm{A}}(\Delta d_{ij}(t))$ is set to 0 as well as $\frac{\partial U_{\mathrm{A}}(\Delta d_{ij}(t))}{\partial \Delta d_{ij}(t)} = 0$. So, we have:

$$\begin{aligned} \sum_{\forall i \neq j \in V} \frac{\partial U_{\mathrm{A}}\left(\Delta d_{ij}(t)\right)}{\partial p_i(t)} &= \sum_{\forall j \in V \setminus N_i(t)} \frac{\partial U_{\mathrm{A}}\left(\Delta d_{ij}(t)\right)}{\partial p_i(t)} + \sum_{\forall j \in N_i(t)} \frac{\partial U_{\mathrm{A}}\left(\Delta d_{ij}(t)\right)}{\partial p_i(t)} \\ &= 0 + \sum_{\forall j \in N_i(t)} \frac{\partial U_{\mathrm{A}}\left(\Delta d_{ij}(t)\right)}{\partial p_i(t)} \\ &= \sum_{\forall (i,j) \in E(t)} \frac{\partial U_{\mathrm{A}}\left(\Delta d_{ij}(t)\right)}{\partial p_i(t)}. \end{aligned} \tag{12.24}$$

This finishes the proof of the property (3).

Similarly, we present the following Corollary 3 for the potential function $U_{\mathrm{R}}(\Delta d_{ij}(t))$. Its proof can also be finished in the similar way of proving Corollary 2, so it does not need to repeat them here in consideration of space.

Corollary 3. $U_{\mathrm{R}}(\Delta d_{ij}(t))$ is (1) continuously differentiable for $\Delta d_{ij}(t) \in ((r_i + r_j), d_i]$ while it is a monotonically decreasing function defined in $((r_i + r_j), d_i]$ that satisfies $U_{\mathrm{R}}(\Delta d_{ij}(t)) \to \infty$, with $\Delta d_{ij}(t) \to (r_i + r_j)$ and $U_{\mathrm{R}}(\Delta d_{ij}(t)) \to 0$, with $\Delta d_{ij}(t) \to d_i$. (2) The partial derivative $u_{ij}^{\mathrm{R}} = \frac{\partial U_{\mathrm{R}}(\Delta d_{ij}(t))}{\partial \Delta d_{ij}(t)} (u_{ij}^{\mathrm{R}} < 0)$ also satisfies the symmetry $u_{ij}^{\mathrm{R}} = u_{ji}^{\mathrm{R}}$ for $\forall (i,j) \in E(t)$ and (3) $\sum_{\forall (i,j) \in E(t)} \frac{\partial U_{\mathrm{R}}(\Delta d_{ij}(t))}{\partial p_i(t)} = \sum_{\forall i \neq j \in V} \frac{\partial U_{\mathrm{R}}(\Delta d_{ij}(t))}{\partial p_i(t)}$.

In addition, it is significant to analyze intragroup mobility outside of the basic characteristics of the overall group. From model 12.1, it can be noted that the connected vehicle group model can be broken down into two terms: the attractive/repulsive forces resulting from the external potentials including $F_i^{(1)}(t)$ and $F_i^{(3)}(t)$, and the interactive force $F_i^{(2)}(t)$ that exists among vehicles. According to the principle of inertial frame of reference, when the group centroid $p_{\mathrm{c}}(t)$ is set as an original point of the inertial frame of reference, each vehicle's motion relative to this reference frame is mainly dependent on the interactive force $F_i^{(2)}(t)$ resulting from the attractive/repulsive potentials between vehicles instead of the external force potentials

$F_i^{(1)}(t)$ and $F_i^{(3)}(t)$. Thus, in order to analyze the intragroup mobility, we turn to focus on the influence of the interactive force $F_i^{(2)}(t)$ between vehicles. Then, the derivative of a vehicle's velocity relative to the reference frame (the relative velocity is denoted as $v_i^c(t)$) with respect to the time instant t can be expressed as

$$m_i v_i^c(t) = F_i^{(2)}(t). \tag{12.25}$$

Also, let $p_i^c(t)$ be the relative position of the vehicle i in the reference frame. The vehicle dynamic system in the reference frame is then formulated as follows:

$$\begin{cases} p_i^c(t) = v_i^c(t) \\ v_i^c(t) = \frac{1}{m_i} F_i^{(2)}(t) = -\sum_{\forall i \neq j \in V} \nabla U_A(\Delta d_{ij}(t)) + \nabla U_R(\Delta d_{ij}(t)) - \sum_{\forall i \neq j \in V} \frac{(\rho_i+\mu_i)}{m_i} \varphi_{ij}(t) v_i^c(t). \end{cases} \tag{12.26}$$

Based on a vehicle dynamic system model explained earlier, one form of the Lyapunov function $L(t)$ for the whole group can be designed by combining the potential field energy and the kinematic energy as

$$L(t) = U(t) + V(t), \tag{12.27}$$

where the potential field term is $U(t) = \sum_{\forall i \in V} \sum_{\forall (i,j) \in E(t)} (U_A(\Delta d_{ij}(t)) + U_R(\Delta d_{ij}(t)))$ and the kinematic energy is $V(t) = 1/2 \sum_{\forall i \in V} \|v_i^c(t)\|^2$. Thus, by differentiating $L(t)$ with respect to the time variable t, we further get

$$\begin{aligned} L(t) &= \sum_{\forall i \in V} \sum_{\forall (i,j) \in E(t)} (\nabla U_A(\Delta d_{ij}(t)) + \nabla U_R(\Delta d_{ij}(t)))^T \cdot p_i^c(t) + \sum_{\forall i \in V} (v_i^c(t))^T \cdot v_i^c(t) \\ &= \sum_{\forall i \in V} \sum_{\forall (i,j) \in E(t)} (\nabla U_A(\Delta d_{ij}(t)) + \nabla U_R(\Delta d_{ij}(t)))^T \cdot v_i^c(t) \\ &\quad + \sum_{\forall i \in V} (v_i^c(t))^T \cdot (-\sum_{\forall i \neq j \in V} \nabla U_A(\Delta d_{ij}(t)) + \nabla U_R(\Delta d_{ij}(t)) - \sum_{\forall i \neq j \in V} \frac{(\rho_i + \mu_i)}{m_i} \varphi_i j(t) v_i^c(t)). \end{aligned} \tag{12.28}$$

According to Corollaries 2 and 3, it should be noted that

$$\begin{cases} \sum_{\forall (i,j) \in E(t)} (\nabla U_A(\Delta d_{ij}(t)))^T \cdot v_i^c(t) - \sum_{\forall i \neq j \in V} [(v_i^c(t))^T \cdot \nabla U_A(\Delta d_{ij}(t))] = 0 \\ \sum_{\forall (i,j) \in E(t)} (\nabla U_R(\Delta d_{ij}(t)))^T \cdot v_i^c(t) - \sum_{\forall i \neq j \in V} [(v_i^c(t))^T \cdot \nabla U_R(\Delta d_{ij}(t))] = 0 \end{cases}. \tag{12.29}$$

Therefore, $L(t) = -\sum_{\forall i \in V} \sum_{\forall i \neq j \in V} (\rho_i + \mu_i)/m_i \varphi_{ij}(t)(v_i^c(t))^T \cdot v_i^c(t)$. Since $(v_i^c(t))^T \cdot v_i^c(t) \geq 0$ and $(\rho_i + \mu_i)/m_i \varphi_{ij}(t) \geq 0$, we have $L(t) \leq 0$, which means that those vehicles in the reference frame can asymptotically converge to a stable state. In the actual reference frame, this fact suggests that those vehicles having knowledge of some other vehicles' state in their local neighborhood tend to move as an overall group. Now, based on this, we declare the following result to illustrate the potential of collision avoidance among vehicles equipped with wireless communications.

Theorem 1. Consider that connected vehicles with wireless communications evolve under the mobility model (1). Given that $\sum_{\forall k \in M} \gamma_{ik} + \beta_i \leq 1$, $v_i(0) = v$ and $v_i^0(t) = v^0(t)$ for all vehicles $i \in V$. If $\Delta d_{ij}(0) > (r_i + r_j)$ for $\forall i \neq j \in V$, those vehicles will move as a group that tends to keep avoiding intrasystem collisions.

Proof: According to Corollary 1, those vehicles tend to maintain the same desired velocity $v^0(t)$. Furthermore, since we have $L(t) \leq 0$ for $t \geq 0$, $L(t) \leq L(0) < \infty$. If there were at least two vehicles i and j about to collide with each other, i.e., $\Delta d_{ij}(t) \rightarrow (r_i + r_j)$, $L(t) = (U(t) + V(t)) \rightarrow \infty$ due to $U(t) \rightarrow \infty$ under the consideration of the property (1) in Corollary 3. Hence, the contradiction occurs. At this point, any two vehicles i and j in the group do not extend to collide with each other. That is, this theorem is proven.

Theorem 2. Consider that connected vehicles with wireless communications evolve under the mobility model (1), and given that $\sum_{\forall k \in M} \gamma_{ik} + \beta_i \leq 1, v_i(0) = v$ and $v_i^0(t) = v^0(t)$ for all vehicles $i \in V$. Those vehicles that initially belong to the neighboring member of one certain vehicle i, i.e., those ones $(i, j) \in E(0)(\forall i \in V)$, tend to keep connecting to this vehicle i when all vehicles move as an overall group.

Proof: Similar to the proof of Theorem 1, since $L(t) \leq 0$ for $t \geq 0$, $L(t) \leq L(0) < \infty$. Recall property (1) in Corollary 2. When any one neighboring member j of a vehicle i tends to move out of the sensing range, i.e., $\Delta d_{ij}(t) \rightarrow d_i((i, j) \in E(0))$, $U_{\mathrm{A}}(\Delta d_{ij}(t)) \rightarrow \infty$ so that $U(t) \rightarrow \infty$. Hence, $L(t) = (U(t) + V(t)) \rightarrow \infty$ when at least one pair (i, j) satisfying $\Delta d_{ij}(t) \rightarrow d_i$. This is contradicts the fact that $L(t) \leq 0$ for $t \geq 0$. In this way, we prove Theorem 2.

12.2.2 Improved cooperative multivehicle mobility model

The vehicles in this radius are expected to interact with each other via the forces of the attractive and repulsive fields so as to keep a safe relative distance between them.

To describe the behaviors seen in cooperative multivehicle mobility, the total force effect of a different social field on vehicle i at time t is given by

$$\vec{F}_i(t) = m_i \frac{\mathrm{d}\vec{v}_i}{\mathrm{d}t} = \vec{F}_i^G(t) + \vec{F}_i^O(t) + \vec{F}_i^S(t) + \vec{F}_i^W(t). \tag{12.30}$$

The main social forces in Equation 12.30 that determine the mobility of a vehicle i with mass m_i consist of four terms, and each term in the aforementioned equation is explicitly designed as follows:

1. $\vec{F}_i^G(t)$ denotes the attractive effect that results from the desire of a vehicle's driver to reach a certain destination. If the motion of the vehicle is not interrupted, it will travel in the desired direction $\vec{e}_i(t)$ with the expected speed v_i^0. Thus, the expected velocity can be represented as

$$\vec{v}_i^0(t) = v_i^0 \vec{e}_i(t)$$

In fact, the necessary deceleration or collision avoidance processes might lead to a deviation of the actual velocity $\vec{v}_i(t)$ at time t from the desired velocity. After experiencing a certain disturbance in the motion of the vehicle i, the vehicle i should tend to return to the desired speed $\vec{v}_i^0(t)$ with a time τ_i. The time τ_i represents a vehicle system–related parameter, which is the vehicle reaction time when the vehicular velocity changes from $\vec{v}_i(t)$ to $\vec{v}_i^0(t)$. The value of τ_i depends on many factors, such as the performance of a vehicle's electromechanical system and environmental conditions including road slope, road surface, load conditions, etc. For simplicity, but without loss of generality, we consider that this parameter mainly relies on the response time of the vehicle's electromechanical system.

Therefore, the attractive term $\vec{F_i^G}(t)$ is related to the desired speed and the desired moving direction so that it can guarantee that this vehicle will approach its destination; and it is formulated as follows:

$$\vec{F}_i^G(t) = m_i \frac{\vec{v}_i^0(t) - \vec{v}_i(t)}{\tau_i} \tag{12.31}$$

2. $\vec{F}_i^O(t)$ reflects the repulsive effects of obstacles. In this research, obstacles, accidents, road maintenances, and other road emergencies are considered to be able to disturb the smooth movement of the vehicle swarm. Those dangerous entities are described as a series of virtual obstacles on roads. Thus, in the mobility model of the vehicle swarm, obstacle avoidance has to be comprehensively considered. Following this concern, $\vec{F}_i^O(t)$ is modeled as the repulsive force on the vehicle i resulting from its tendency to avoid crashing into the obstacle O_i^k. The detailed design of computing $\vec{F}_i^O(t)$ is given in the following parts.

The set of obstacles O_i^k is defined by $N = \{0, 1, 2, \ldots, m\}$, with the total amount number m of obstacles, $m \geq 0$, and $k \in N$. The obstacle O_i^k is also a closed circle with a radius r_k^O. Then, the vector $\vec{p}_k^O = [x_k^O(t), y_k^O(t)]^T$ is defined as the position of the centroid of obstacle O_i^k. R_k^O denotes the interaction range of obstacle O_i^k and the $\|\vec{p}_i(t) - \vec{p}_k^O\|$ reflects the centroid distance of vehicle i to the nearest obstacle O_i^k. Figure 12.4b schematically illustrates the interaction rang of an obstacle O_i^k related to the vehicle i. When the vehicle i approaches the obstacle O_i^k in proximity, and the distance $\|\vec{p}_i(t) - \vec{p}_k^O\|$ becomes smaller than the communication range d_i, the vehicle i will begin to notice the information broadcasted by the vehicles who first cross the section and prepare to avoid it. Furthermore, $\vec{F}_i^O(t)$ should be a monotonically increasing function of the distance between the obstacle O_i^k and the vehicle i in the interaction range, and the vehicle i should keep a certain distance from the obstacles so as to avoid collision.

Helbings exponential force decay equation [46] is adopted to depict the repulsive force exerted on the vehicle by the obstacle in the sensing range d_i. Consequently, the repulsive force induce by the obstacle O_i^k can be given by

$$\vec{F}_i^O(t) = m_i \sum_{k \in OS} A_k^O \exp - \frac{\|\vec{p}_i(t) - \vec{p}_k^O\| - (r_i + r_k^O)}{(R_k^O)} \times I(d_i - \|\vec{p}_i(t) - \vec{p}_k^O\|) \times \frac{(\vec{p}_i(t) - \vec{p}_k^O)}{\|\vec{p}_i(t) - \vec{p}_k^O\|}, \tag{12.32}$$

where A_k^O and μ_1 are real positive constants, and the term $(\vec{p}_i(t) - \vec{p}_k^O)/\|\vec{p}_i(t) - \vec{p}_k^O\|$ is a unit vector in the direction from the centroid of the obstacle O_i^k to that of the vehicle i. $\vec{F}_i^O(t)$ is the accumulated repulsive forces between i and O_i^k in the sensing range. In the model, $\|\vec{p}_i(t) - \vec{p}_k^O\| - (r_i + r_k^O)$ denotes the distance of two objects that consider the equivalent radius (such as r_i, r_k^O) of each entity. Therefore, it can avoid overlaps and collisions among individuals that are always neglected by other models.

The repulsive forces of obstacles are not effective on the vehicle until vehicles get the information about it and the distance between them gets smaller than the sensing range d_i. Therefore, $I(x)$ is an indicator function to control the individual magnitudes of the repulsive forces defined as:

$$I(d_i - \|\vec{p}_i(t) - \vec{p}_k^O\|) = \begin{cases} 0, d_i - \|\vec{p}_i(t) - \vec{p}_k^O\| < 0; \\ 1, d_i - \|\vec{p}_i(t) - \vec{p}_k^O\| \geq 0. \end{cases} \tag{12.33}$$

From Equation 12.33, it can be seen that when $d_i > \|\vec{p}_i(t) - \vec{p}_k^O\|$, indicating that the vehicle i is in the repulsive area, the repulsive force $\vec{F}_i^O(t)$ is expected to take effect. Otherwise, when $d_i < \|\vec{p}_i(t) - \vec{p}_k^O\|$, which implies that the vehicle i cannot sense the obstacle in its intervehicular communication range, $\vec{F}_i^O(t)$ is set to be zero so that it does not work on the vehicle mobility.

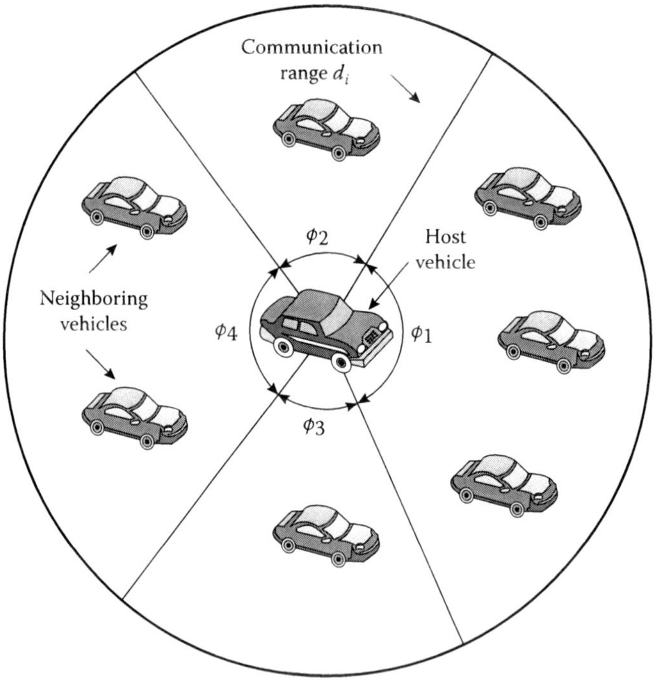

(a) Neighboring vehicles

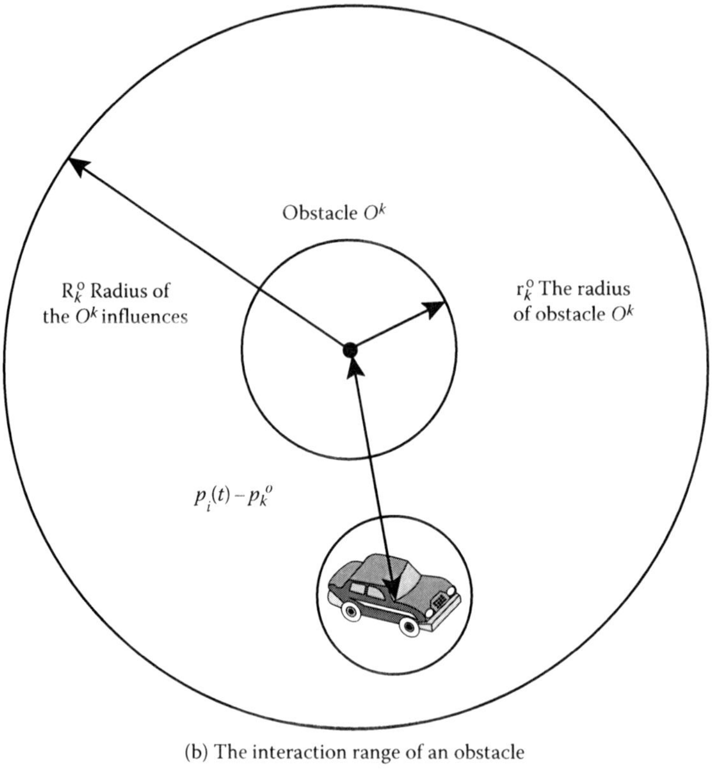

(b) The interaction range of an obstacle

Figure 12.4 The interaction range.

3. The model also takes into account those attractive and repulsive effects existing between vehicular pair interactions. Let $\vec{F}_i^S(t)$ represent the total effects of the neighboring vehicles in the vehicle swarm on the host vehicle i. Considering the fact that neighboring cars coming from different relative directions and in different relative positions could have different influences on the driving decisions made by the host vehicle i, different social forces induced by those neighbors of the host vehicle i are formulated respectively in different relative directions and in relative positions. First, the interaction circle of vehicle i with radius R_i is divided into four parts, each of which is marked by the angle $\varphi_k(k = 1,2,3,4)$ (see Figure 12.4a). Then, let $j(\varphi_k) \in N(\varphi_k)$ represent the vehicle where the $N(\varphi_k)$ is the set of those neighbors of vehicle i in the neighboring area of $\varphi_k(k = 1,2,3,4)$. Thus, the attractive force and repulsive force are taken into account in terms of different neighboring areas. For capturing the specific real traffic situation, the model recognizes that different neighboring areas have different impact factors for influencing the mobility of the vehicle i. Logically, the influence of those front neighbors of the vehicle i is stronger than that of those on both sides of this vehicle i, while the neighbors located behind the vehicle i are expected to have the weakest influence. The impact factors of each area are respectively denoted as $\omega_k(k = 1,2,3,4)$, which should satisfy $\omega_1 > \omega_2 = \omega_3 > \omega_4 > 0$.

Furthermore, the attractive potential force and the repulsive potential force among vehicles should vary with their relative distance. The model also considers that the magnitude of the repulsive effect should be larger when the vehicle pair is at a relatively short distance than that of the attractive effect for the sake of intervehicle collision avoidance. On the contrary, the attractive effect should be more enhanced than the repulsive effect when the vehicle pair is at a greater distance so that they can smoothly aggregate to form a swarm. Following these general considerations of realistic events, the sigmoid function is adopted to model those effects,

$$\lambda_i = g(p_{ij}(t), d_{ij}^s(t)) = \frac{1}{1 + \exp - c(p_{ij}(t) - d_{ij}^s(t))}, \tag{12.34}$$

where c is a system-specified positive real constant of a sigmoid function.

The formulation introduces the sigmoid function to depict the smoothness of the interaction force among vehicles. The numerical experiments shown in Figure 12.5

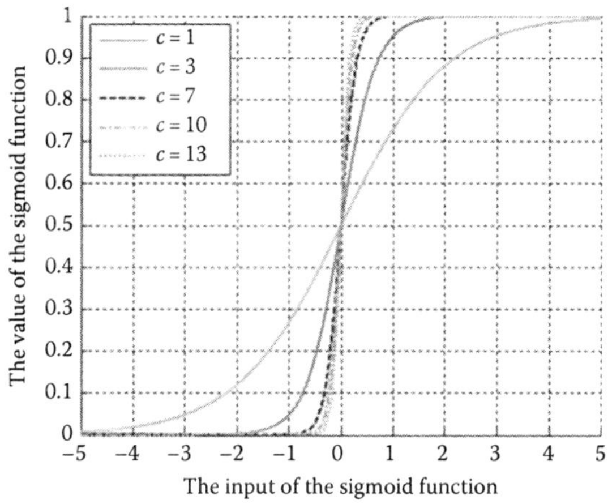

Figure 12.5 Sigmoid function with different parameters.

investigate the impacts of the different values of $c = \{1,3,7,10,13\}$ on the sigmoid function. To ensure the smoothness of the sigmoid function, in the experiments we set $c = 7$.

Moreover, the $p_{ij}(t)$ in Equation 12.34 is expressed as follows:

$$p_{ij}(t) = \|\vec{p}_i(t) - \vec{p}_j(t)\| - (r_i + r_j). \tag{12.35}$$

The variable $p_{ij}(t)$ is used to describe the position relationship of vehicle i, j, which takes the equivalent radius of each individual into account as aforementioned. In the movement of multiple vehicles, the relative distance between two adjacent vehicles is preferred to be around an ideal value, which is called as the safety distance. According to Refs. [95,226], it can be molded as follows

$$d_{ij}^{s}(t) = d_0 + \|\vec{v}_i(t)\| T_i - \frac{\|\Delta\vec{v}_{ij}(t)\|\|\Delta\vec{v}_i(t)\|}{2\sqrt{ab}}, \tag{12.36}$$

where the parameter d_0 is the stopping distance between two vehicles when the traffic flow is jammed and $\Delta\vec{v}_{ij}$ is the velocity difference between vehicle i and vehicle j. In addition parameter a is the maximum acceleration and b is the deceleration of the vehicle. T_i is the safety total reaction time between vehicle i and vehicle j given by

$$T_i = t_r + t_d \tag{12.37}$$

where t_r is the driver reaction time and t_d is the intervehicle information perception delay via vehicular communication.

Subsequently, the attractive force based on the negative exponential function is defined as follows

$$\vec{F}_i^{SA}(t) = m_i A_i^{SA} \exp\left(-\frac{\|\vec{p}_i(t) - \vec{p}_j(t)\| - (r_i + r_j)}{R_i}\right) \times \frac{\vec{p}_j(t) - \vec{p}_i(t)}{\|\vec{p}_j(t) - \vec{p}_i(t)\|}. \tag{12.38}$$

And the repulsive force is similarly given by

$$\vec{F}_i^{SR}(t) = m_i A_i^{SR} \exp\left(-\frac{\|\vec{p}_i(t) - \vec{p}_j(t)\| - (r_i + r_j)}{R_i}\right) \times \frac{\vec{p}_i(t) - \vec{p}_j(t)}{\|\vec{p}_i(t) - \vec{p}_j(t)\|}. \tag{12.39}$$

Here, it should be pointed out that the main difference between Equations 12.38 and 12.39 lies in the parameters A_i^{SA}, A_i^{SR}, μ_2, μ_3, and their directions $\vec{p}_i(t) - \vec{p}_j(t)/\|\vec{p}_i(t) - \vec{p}_j(t)\|$, $\vec{p}_j(t) - \vec{p}_i(t)/\|\vec{p}_j(t) - \vec{p}_i(t)\|$.

Then, by following those equations mentioned earlier, the attractive force is used to bring individuals to form a swarm; in addition, the repulsive force is used to avoid a collision by individual vehicles that can be summed together. Terms relating to all neighbors of vehicle i can be presented as

$$\begin{aligned} \vec{F}_i^{S}(t) = & \sum_{j(\varphi_1)\in N(\varphi_1)} (\varphi_1)[\lambda_i \vec{F}_i^{SA}(t) + (1-\lambda_i)\vec{F}_i^{SR}(t)] \\ & + \sum_{j(\varphi_2)\in N(\varphi_2)} (\varphi_2)[\lambda_i \vec{F}_i^{SA}(t) + (1-\lambda_i)\vec{F}_i^{SR}(t)] \\ & + \sum_{j(\varphi_3)\in N(\varphi_3)} (\varphi_3)[\lambda_i \vec{F}_i^{SA}(t) + (1-\lambda_i)\vec{F}_i^{SR}(t)] \\ & + \sum_{j(\varphi_4)\in N(\varphi_4)} (\varphi_4)[\lambda_i \vec{F}_i^{SA}(t) + (1-\lambda_i)\vec{F}_i^{SR}(t)] \end{aligned} \tag{12.40}$$

4. $\vec{F}_i^W(t)$ is used to reflect the influence of boundaries of the lane on the vehicle i. In the model, $\vec{F}_i^W(t)$ is formulated as follows:

$$\vec{F}_i^W(t) = f(t)\vec{e}_f, \tag{12.41}$$

where $f(t)$ is a scalar function and $\vec{e}_f$ is a column vector used to denote the target direction of the vehicle's next movement. The magnitude of the force $\vec{F}_i^W(t)$ depends on the value of $f(t)$, so it can be designed as:

$$f(t) = h\frac{\psi \text{sig}(L_0 - \Delta L_i(t))}{\|\Delta L_i(t)\|^{1/\sigma}}, \tag{12.42}$$

where L_0 is the total width of the road and $\Delta L_i(t)$ denotes the distance from the vehicle i to its nearest boundary of the roadside at time t. The parameters h and σ are both positive real constants, while the parameter ψ is a speed-related variable, which is given by

$$\psi = \|\vec{v}_i^t - \vec{v}_i(t)\| \cdot \sqrt{1 - \left(\frac{(\vec{v}_i^t - \vec{v}_i(t))^T \vec{v}_i^t}{\|\vec{v}_i^t - \vec{v}_i(t)\| \cdot \|\vec{v}_i^t\|}\right)^2}, \tag{12.43}$$

where $\vec{v}_i^t$ is the speed vector that is identical to the projection of $v_i(t)$ on the vehicle i's desired movement direction. And $\vec{e}_f$ is defined as

$$\vec{e}_f = \frac{\vec{v}_i^t - \vec{v}_i(t)}{\|\vec{v}_i^t - \vec{v}_i(t)\|}. \tag{12.44}$$

It is worth pointing out that the vehicles moving on the road should obey all traffic rules; they should go back to their lanes immediately after changing lanes to avoid obstacles. Hence, $\vec{F}_i^W(t)$ is regarded as the force that keeps vehicle i from deviating from its previous lane.

12.3 Application and Analysis of Cooperative Multivehicle Mobility Model

12.3.1 *Simulation and verification*

12.3.1.1 Simulation and verification of cooperative multivehicle mobility model

In this section, simulations achieved by MATLAB® are provided to verify the model proposed in the previous section. An analysis of the results is also given in this section in order to characterize the benefits of for safety and traffic efficiency of wireless communication among vehicles. We set up a traffic scene that includes an accident situation on a unidirectional road and where an obstacle has to be avoided. Then, we create the diagram of the trajectories of the connected vehicles to demonstrate that the model can achieve the obstacle avoidance effectively. In addition, we also create a comparative numerical experiment showing vehicles with wireless communications and those without wireless communications. The experiments show the impact of wireless communications on the connected vehicles.

The simulation scene is assumed to be on a section of a unidirectional road. The unidirectional road we concentrate on is 2000 m long and 50 m wide. We suppose that an accident occurs on the road and the location of the center of it is marked as O_1. The obstacle is a circular region with the center set to be 1000 m away from the beginning boundary of the simulation range and with radius of $r_i^O = 5$ m that needs to be detoured. A collision would occur if a the vehicle entered the circular area.

The number and the velocity of vehicles are produced randomly at the start of simulations. During the simulations, those connected vehicles are controlled by our proposed model to move as a cooperative group whose direction is consistent. In all simulations, we fix the time interval for updating the kinematic information of each vehicle at 0.1 s. In addition, the physical parameters of the vehicle are given *as* $r_i = 5$ m, $d_i = 300$ m, $m_i = 2000$ kg, $r_i^W = 7$ m. Moreover, other parameters required in the simulations are set as $\Delta d_{\text{desire}} = 50$ m, $\tau_i = 0.05$ s, $\rho_i = 0.1$, $\mu_i = 0.1$, $k_v = 0.5$, $k_A = 0.1$, $k_R = 1.482105 k_A|$, $k_O = 0.5$, $l = 0.3$, $k_L = 0.5$, $R = 3$ m, $\gamma_{ik} = 0.8(k = 1, \text{and } \forall i \in V)$, $\beta_i = 0.6$.

Figure 12.6 shows the various traveling trajectories of each vehicle when the vehicle group comes across the obstacle at different average velocities. In these simulations, the average velocities are set to 10, 15, and 20 m/s separately in order to explore how vehicles avoid obstacles at different velocities. Furthermore, the total number of vehicles is the same in these three simulations, set as $n = 20$. The axes in Figure 12.6a through c are the referenced coordinates used to measure the real-time position of vehicles moving on the road, and the unit is meter. From Figure 12.6, it can be observed that when encountering the obstacle, the vehicles change into two subgroups, similar to the behavior of fish school when the established information is locally broadcasted to the whole group. Subsequently, they are reunified into a group, having successfully avoided the obstacle. In addition, comparing the traveling trajectories at different average velocities, the conclusion can be drawn that the lower the average velocity of vehicles, the easier it is to bring the vehicles back together. Although various velocities have different traveling trajectories, all the vehicles reach their destination successfully and remain together as a group, having avoided collisions or obstacles. This result indicates that the model of connected vehicles based on the behaviors of fish school is effective to describe cooperative behaviors in avoiding obstacles.

12.3.1.2 Simulation and verification of improved cooperative multivehicle mobility model

The traffic scenario for the simulation experiments is shown in Figure 12.7; the whole region in the simulation scenario is a one-way road with three lanes. The length of the road is 2000 m and the width is 50 m. In the simulations, multiple vehicles induced by the proposed model are all expected to achieve cooperative obstacle avoidance and intervehicle collision avoidance.

The obstacles described in this chapter stand for the position of the incident such as vehicle breakdowns, car crashes, road maintenance, and spilled loads on the road. At the same time, given the reasonable ranges of each obstacle to guarantee that the vehicles will cross it safely without collision on it. The position of the incident and the location of areas of danger broadcast to other vehicles by communication technologies in advance so as to make the drivers slow down and make decisions ahead. As shown in Figure 12.7, on the road, a rear-end collision between two vehicles is assumed to occur at the location marked by O^1, and all other vehicles are required to avoid this collision point and the dangerous circle range with radius r_1^O. Additionally, two traffic signs are assumed to be set up at the locations O^2 and O^3, respectively, one of which warns of a dangerous spot and the other that informs drivers of the special local spot that does not allow entry. Vehicles moving on the road are required to avoid those dangerous circle ranges cooperatively. These circle ranges with centers marked by O^1, O^2, and O^3 and radius of r_1^O, r_2^O, r_3^O separately. The vehicle that first discovers the obstacle will send messages about it throughout the whole vehicle swarm via V2V communication so that other vehicles are warned to focus on the danger area and to prepare it avoid it. It is worth pointing

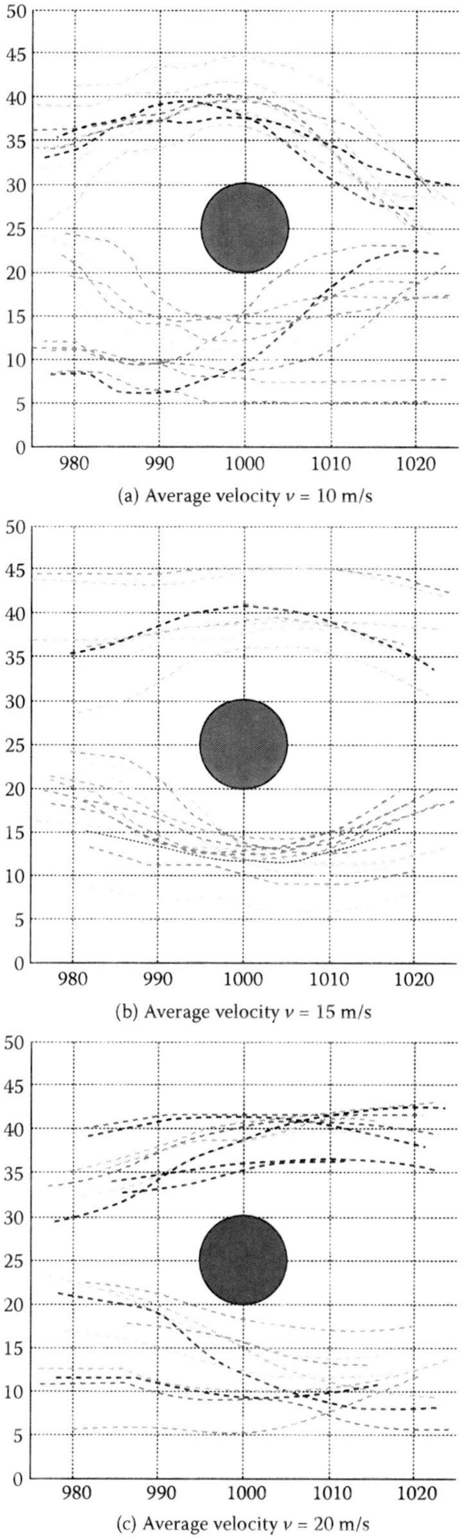

Figure 12.6 The traveling trajectory of vehicles.

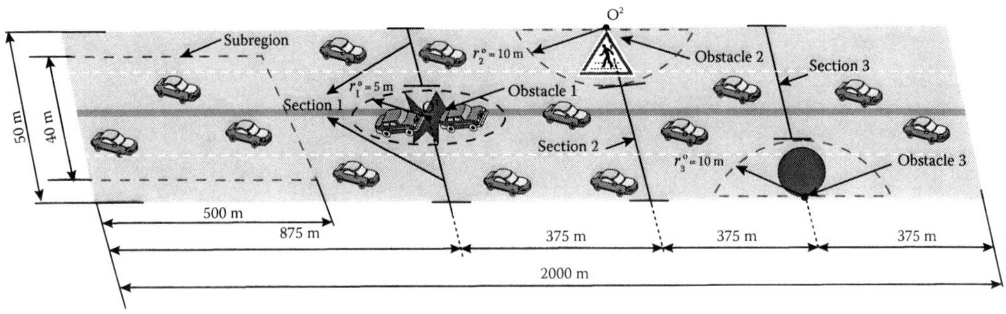

Figure 12.7 The road situation for the simulation.

out that those aforementioned local spots are uniformly treated as a series of virtual obstacles that vehicles are required to avoid. Thus, the radius of each obstacle should be the equivalent of the range of the corresponding spot into which vehicles are inhibited to move. Following this abstraction, the radiuses of those three obstacles are set to be 5, 10, and 10 m respectively in our simulations. Furthermore, Obstacle 1 is set to be 875 m away from the starting boundary of the considered region, and the horizontal distance between any two adjacent obstacles is set to be 375 m. Other detailed settings can also be referred to in Figure 12.7. The communication range of each vehicle is set to be 300 m, and the delay parameter $(t_r + t_d)$ is (0.75 + 0.054) s [224]. While in the compared conditions, the visual sensing range of each vehicle unequipped with vehicular communications is set to be 100 m, and t_d and t_r are set to be 0 and 1.5 s, respectively [67].

At the beginning of the simulation, a certain number of vehicles are randomly generated with an initial speed and randomly placed in the given subregion, whose length is set to be 500 m as shown in Figure 12.7. Each side of the sub-region is 5 m apart from the corresponding side of the road. All those cooperative multiple vehicles that can be seen in Figure 12.7 are moving in same direction during the simulation. Furthermore, in all simulations, every 0.1 s, each vehicle updates its kinematic information.

Additionally, in our simulations, the number of simulated vehicles is determined by the traffic density parameter, denoted as $\rho(\text{veh/m}^2)$. For instance, since vehicles are initially generated within a given subregion aforementioned whose length is equal to 500 m and whose width is identical to 40 m, the number of vehicles is then set to be $0.0001 \times 500 \times (50 - 5 \times 2) = 20$. Besides, it is reasonable enough to assume the initial velocity of any one vehicle following a certain stochastic distribution. According to Refs. [95,226], the velocity of vehicles moving in a certain scenario, such as in a dense traffic flow with a low mobility or in a sparse traffic flow with a relative high mobility, will satisfy a certain probability distribution. For example, vehicle velocity in a free flow or in a peak flow follows a certain normal distribution denoted as $N(\mu, \sigma)$, where μ represents the mean value of the flow speed and σ is its standard deviation. Therefore, this assumption is adopted in these simulations to initialize the velocity of any vehicle. The detailed settings on $N(\mu, \sigma)$ are given in the following subsection. Furthermore, other settings of the parameters relevant to the model used in simulations are summarized in Table 12.1. r_i is the radius of vehicle; in the experiments we set it to 3 m to represent the average value. The parameter τ_i in Equation 12.31 is used to denote the time delay of vehicle reaction; based on the results of study [224], it can be in the range of 0.1–0.5 s. R_k^O is the interaction range of obstacle O_i^k, which is given from the range $[20, 100]$ considering the real-world situation. L_0 denotes the total width of the road so that it is given by an intuitive range $[7.5, 30]$ m.

Table 12.1 Parameters

Parameters	*Values*
r_i	3 m
τ_i	0.26 s
A_k^O	600
R_k^O	$[25,50,50]$
c	7
d_0,a,b	$d_0 = 5$ m, $a = 2.5$ m/s^2, $b = 1.3$ m/s^2
A_i^{SA},R_i	$A_i^{SA} = 5$, $R_i = 5$ m
A_i^{SR}	20
$\omega_{\varphi_1},\omega_{\varphi_2},\omega_{\varphi_3},\omega_{\varphi_4}$	$\omega_{\varphi_1} = 0.7, \omega_{\varphi_2} = \omega_{\varphi_3} = 0.2, \omega_{\varphi_4} = 0.1$
h ,L_0,γ	$h = 23.04, L_0 = 20$ m$, \gamma = 0.2$

The values of these parameters are not unique, and other suitable values could also be adopted in our model. c is a chosen constant of Equation 12.34. Different values make different function curves as shown in Figure 12.6, in the simulation it is chosen 7. Moreover, d_0 can be seen as the minimal distance between two vehicles when stopped; following the study result in Ref. [106], the value is given in the range [2,5] m. The parameters a and b are the maximum acceleration, and b is given separately as the expected deceleration of the vehicle. As a result, various vehicles and situations lead the value of them differently; based on the study results in Ref. [106], the value ranges of a and b are given among the ranges $[1.0,2.5]$ and $[1.3,3.7]$. A_k^O, A_i^{SA}, A_i^{SR}, and R_i here are scaling positive real coefficients. In addition, γ_{ik} and β_i are positive sensitivity coefficients within $(0,1]$, which are used to scale and coordinate the magnitudes of $F_{ik}^o(t)$ and $F_i^L(t)$. h ,σ are constants chosen for the simulation. To make the functions smooth, these parameters are set as shown in Table 12.1.

12.3.2 Influence of intervehicle communication on mobility of multiple vehicles

Figure 12.8a shows the traveling trajectories of the connected vehicles in the simulation, where $\mu = 30.93$ m/s, $\sigma = 1.2$ m/s, and $\rho = 0.001$ veh/m^2. The spots in Figure 12.8a stand for the vehicles in the simulation and lines represent trajectories of vehicles. Besides, the black areas and light circles represent the dangerous areas and the interaction ranges of each obstacle. The vehicle will feel the repulsive force when it enters into the light circle, which is the interaction range of obstacle. If the vehicle moves into the black area which represents it collides with the obstacle. Figure 12.8a shows that *CMVM* model can successfully control the mobility of multivehicles *advoiding collisions* among vehicles and obstacles. This observation is coincident with the theoretical expectation of the model in Section 12.2.

12.3.2.1 Under the traffic flow scenario of low density and high mobility

To simulate the specific traffic flow whose density is low while speed is high on average, the paper refers to the study of [226] to adopt the specific settings on μ and σ as $\mu = 30.93$ m/s and $\sigma = 1.2$ m/s. Also, the traffic density parameter ρ is set as $\rho = 0.001$ veh/m^2. Thus, based on the parameter characterization, the velocities of multiple vehicles are initialized by following the specific normal distribution $N(30.93,1.2)$. It is worth pointing out that the chapter repeats simulations 30 times under each comparative case so as to derive the numerical results on average.

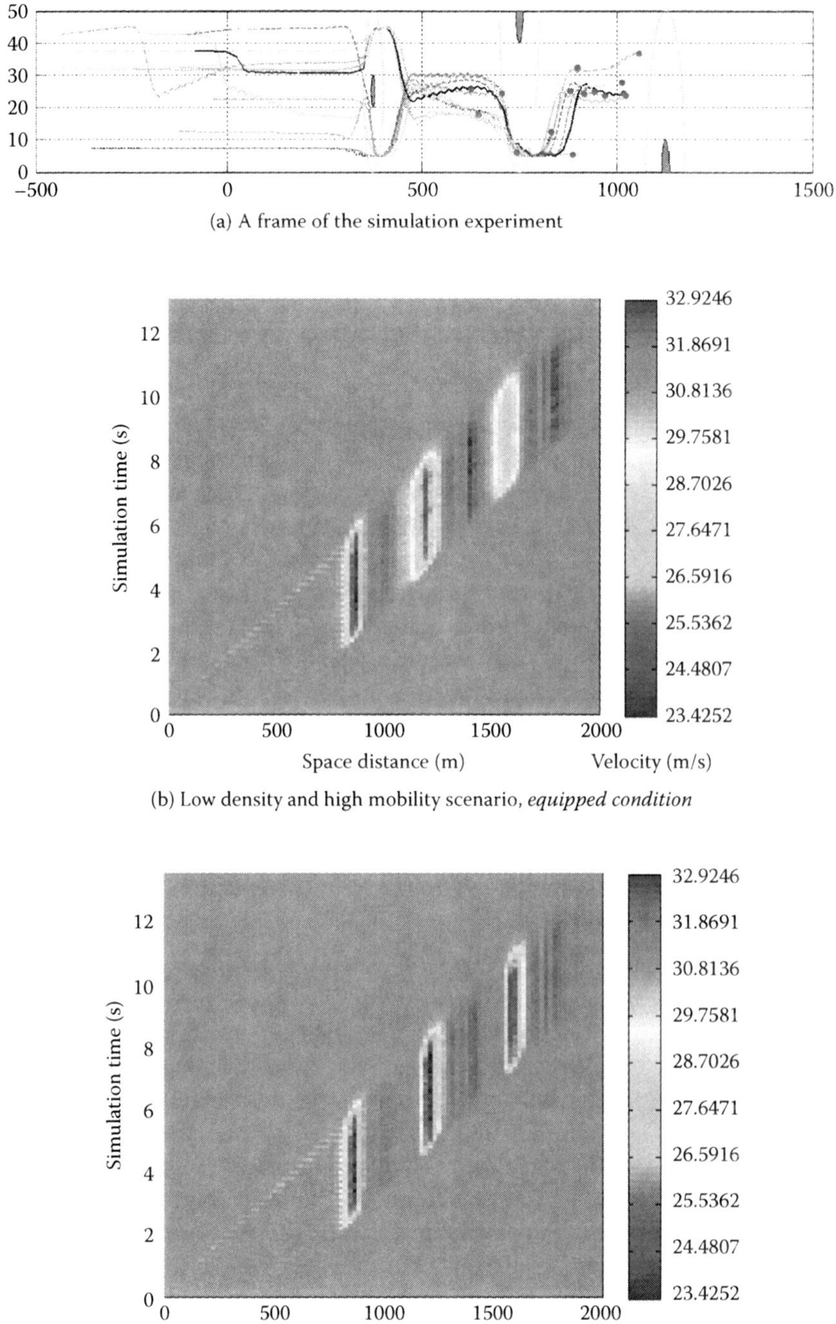

(a) A frame of the simulation experiment

(b) Low density and high mobility scenario, *equipped condition*

(c) Low density and high mobility scenario, *unequipped condition*

Figure 12.8 The mobility of multiple vehicles in different traffic scenarios. (*Continued*)

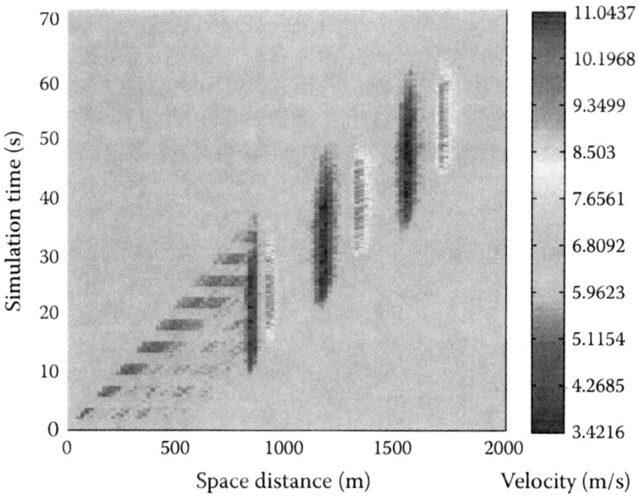

(d) High density and low mobility scenario, *equipped condition*

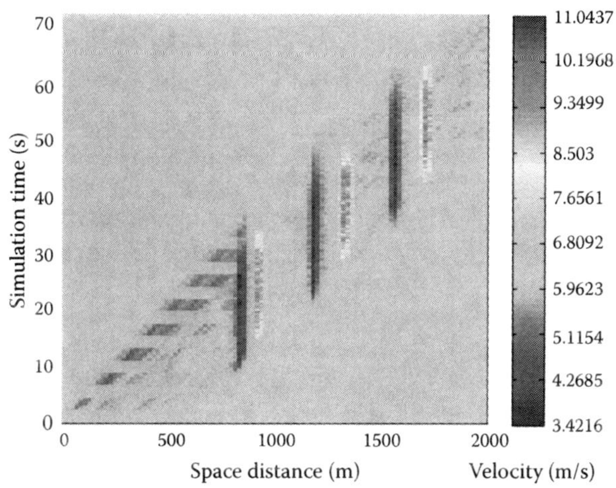

(e) High density and low mobility scenario, *unequipped condition*

Figure 12.8 (CONTINUED) The mobility of multiple vehicles in different traffic scenarios.

The space–time–velocity diagrams as illustrated in Figure 12.8b through e show the simulation traffic conditions. The horizontal and vertical coordinates represent space distance and simulation time and the different colors denote the various velocities. The color of velocity over space distance and time in respective graphs illustrates mobility of the vehicle. From Figure 12.8b and c, it is observed that there exist three local spots stained with dark or close to dark color, each of which represents the position of the section at each corresponding obstacle. By referring to the color bar that is used to map different velocities to different colors, it is obvious that the mean velocity of vehicles at the local regions close to the obstacles under both conditions is lower than the mean traffic flow speed (it should be noted that the mean traffic flow speed is identical to $\mu = 30.93$ m/s, which is represented by light color). This result implies that vehicles brake to reduce their velocity when they are achieving obstacle avoidance. Induced by the proposed model, multiple vehicles are expected to slow down when they move across each section at each obstacle. On the other hand, when comparing the result shown by

Figure 12.8b with that of Figure 12.8c, it can be seen that a dark color is attained at the three local spots in Figure 12.88c. That is, when unequipped, vehicles tend decelerate abruptly so as to achieve obstacle avoidance; the equipped vehicles slow down more gently. This is because those vehicles equipped with intervehicle communications are able to sense the obstacle earlier via intervehicular communication. These equipped vehicles can make the driving decision to slow down well ahead of time with the assistance of the intervehicular communication network so that they can efficiently avoid an abrupt deceleration. This indicates that the mobility of multiple vehicles can be guaranteed as a result of vehicular communication.

12.3.2.2 Under the traffic flow scenario of high density and low mobility

The traffic density ρ is increased to 0.0022 veh/m^2 and follows the normal distribution $N(6.083, 1.2)$ to initially generate the velocity of multiple vehicles at the beginning of simulation. Thus, the parameter characterization allows us to simulate the traffic flow scenario of relatively high traffic density and relatively low vehicle mobility. The results are given in Figure 12.8d and e, respectively.

When comparing the result in Figure 12.8d with that in Figure 12.8e, a similar conclusion can also be drawn that the multiple equipped vehicles can slow down more gently than those unequipped with vehicular communication when they are approaching to the local spots of the obstacles. Additionally, the average width of each dark spot in Figure 12.8d is larger than that in Figure 12.8e (it is worth pointing out that the width of a certain dark spot is measured by its coverage along the space axis labeled as Space distance). This means that those equipped vehicles are able to reduce their velocity in advance even when they are not so close to the obstacle. Compared to the results from those unequipped vehicles, a relatively more sufficient distance for braking can be reserved by those equipped vehicles so that they can better prepare for obstacle avoidance in a cooperative manner.

12.3.3 Influence of intervehicular communication on vehicle safety and traffic efficiency

To quantify the improvement to vehicle safety and traffic efficiency by vehicular communication, two evaluation metrics collision probabilities and flow rate are defined. The collision probability is represented as the frequency of a vehicle colliding with any one obstacle or with another vehicle. Such a collision is defined as an event in which a vehicle moves into a certain interaction range of any obstacle or vehicle. The interaction range of one obstacle is identical to its radius $r_i^O(i = 1, 2, 3)$ as shown in Figure 12.4b, and the interaction range of a vehicle is identical to radius r_i as given in Table 12.1. The collision probability at the *ith* simulation denoted by $Probability_i$ can be calculated as follows.

$$Probability_i = \frac{\sum_{k=1}^{SimEpoch_i} CollisionEvent_i(k)}{SimEpoch_i \times N_i} \tag{12.45}$$

where $SimEpochi$ represents the total epochs in the *ith* simulation and $CollisionEventi(k)$ represents the number of the collision event occurring at the *kth* epoch in the *ith* simulation. N_i denotes the total number of vehicles generated in the *ith* simulation.

On the other hand, the flow rate is used to quantify the traffic efficiency of multiple vehicles moving as a swarm. The flow rate is measured at the section of each obstacle (the section is illustrated in Figure 12.7). For example, in the *i*th simulation, the time when the first vehicle moves across the section at a certain obstacle O_i^k is denoted by $t_{i,k}^1$. The other time instant when the last vehicle moves across the same section is denoted by $t_{i,k}^2$. Thus, the flow rate at this section of the obstacle O_i^k can be calculated by the following equation:

$$FlowRate_i^{O_i^k} = \frac{N_i}{t_{i,k}^1 - t_{i,k}^2} \tag{12.46}$$

12.3.3.1 Analysis of collision probability

1. Influence of different initial vehicle velocity on collision probability

   To analyze the impact on road safety level, first we vary the initial vehicle velocity from 10 to 30 m/s and fix the total number of vehicles at $N = 30$. At each vehicle velocity point, simulation has been performed 30 times repeatedly, and the metrics P_o and P_v have then been averaged. Additionally, the standard deviations of these evaluation metrics are also calculated at each velocity. The relevant results are shown in Figure 12.9. As can be seen from these results, the danger severity indicated by P_o, P_v increases with increasing vehicle velocity. The higher the velocity at which a vehicle moves on the road, the more time this vehicle needs to decelerate to a certain lower velocity, and the larger the possibility of an obstacle or of other vehicles entering into the predefined local warning scope of vehicle i. The danger severity on the road indicated by P_o and P_v is higher with higher vehicular mobility. Nevertheless, assisted by wireless communications, decisions on braking can be made in advance in connected vehicles, and a greater distance can be reserved for changing motion trajectory. Therefore, the metrics P_o and P_v obtained in the case marked "Equipped" are lower on the overall level than those in the Unequipped cases. Especially in Figure 12.9b, P_v in the Equipped case is reduced by 57.89% on average compared with that in the Unequipped case when the vehicle velocity is larger than 10 m/s. Hence, this suggests that vehicles can benefit from wireless communications in terms of extending perception range and enhancing traffic safety level.

2. Influence of different vehicle numbers on collision probability

   Next, we turn to focus on the effects of different vehicle numbers of the group on the safety level. For comparison, the following experiment as is processed by varying the vehicle number from 10 to 70 and fixing the initial vehicle velocity at 20 m/s. Also, the simulations have been performed with 30 replications per vehicle and subsequently, the results of P_o and P_v obtained for each vehicle number were averaged. The error bars in Figure 12.10 also indicate the standard deviations of P_o and P_v. Some similar conclusions can be drawn from Figure 12.10. The safety level on the road can be enhanced even further by wireless communications. When the number of vehicles increases on the road, the traffic density is increased. Consequently, the distance between vehicles on average decreases; this leads to a lower safety level. However, the averaged value of P_o corresponding to different number of vehicles in the Equipped case is about 11.5% lower than that obtained in the Unequipped case, while the overall averaged P_v of the Equipped case is reduced by about 22.61% compared to that of the Unequipped case. The reason is that, as discussed before, wireless communications expand the sensing range of each equipped vehicle in a group so as to enable them to adjust their velocity to the overall velocity of the group ahead of time, in the meanwhile guaranteeing a relatively larger reserved distance among vehicles.

   In fact, there were no obstacle collisions or vehicle collisions during simulations, because the proposed mobility model was adopted to simulate the cooperative behaviors of connected vehicles. Furthermore, as theoretically analyzed before, collisions can be

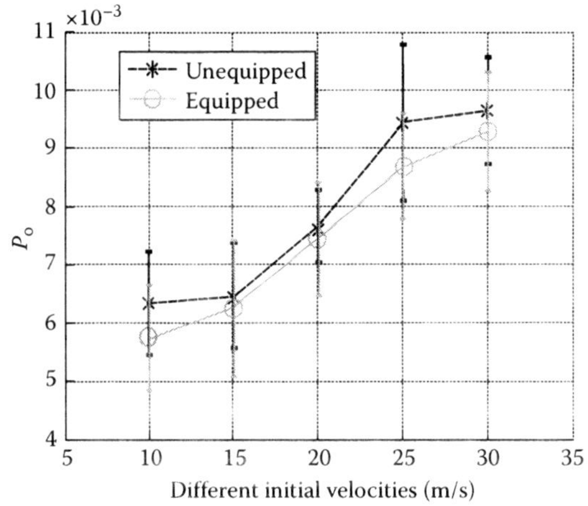

(a) P_o

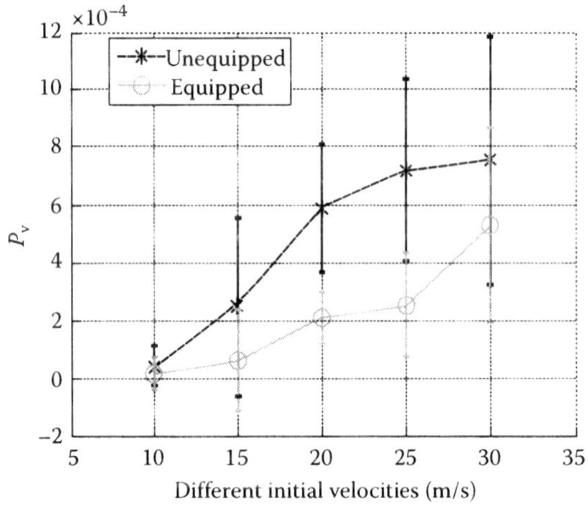

(b) P_v

Figure 12.9 The results of P_o and P_v under different initial velocity settings.

avoided between vehicles evolving with the comprehensive effects of potential forces in the proposed mobility model. Even though no collisions were caused, the potential danger on the road in actual situations may exist. Thus, considering the evaluation metrics P_o and P_v instead of collision-related statistics, we can quantify the impacts of wireless communications on the traffic safety level. This fact is shown in Figures 12.9 and 12.10 implies that road capacity could be increased by deploying wireless communications onboard when considering keeping the road danger severity under a certain level. This is also in accordance with the results in [228].

3. Influence of different communication ranges on collision probability

   To analyze the communication ranges on vehicle safety, the simulations as well as the relevant discussions are performed under two typical traffic scenarios, one of which is a free flow scenario and the other a peak-hour flow scenario. The parameter settings related to a general free flow are given in Ref. [190]. The mean value of the normal distribution for initializing vehicle velocity is set as $\mu = 29.15$ m/s, and the standard deviation is

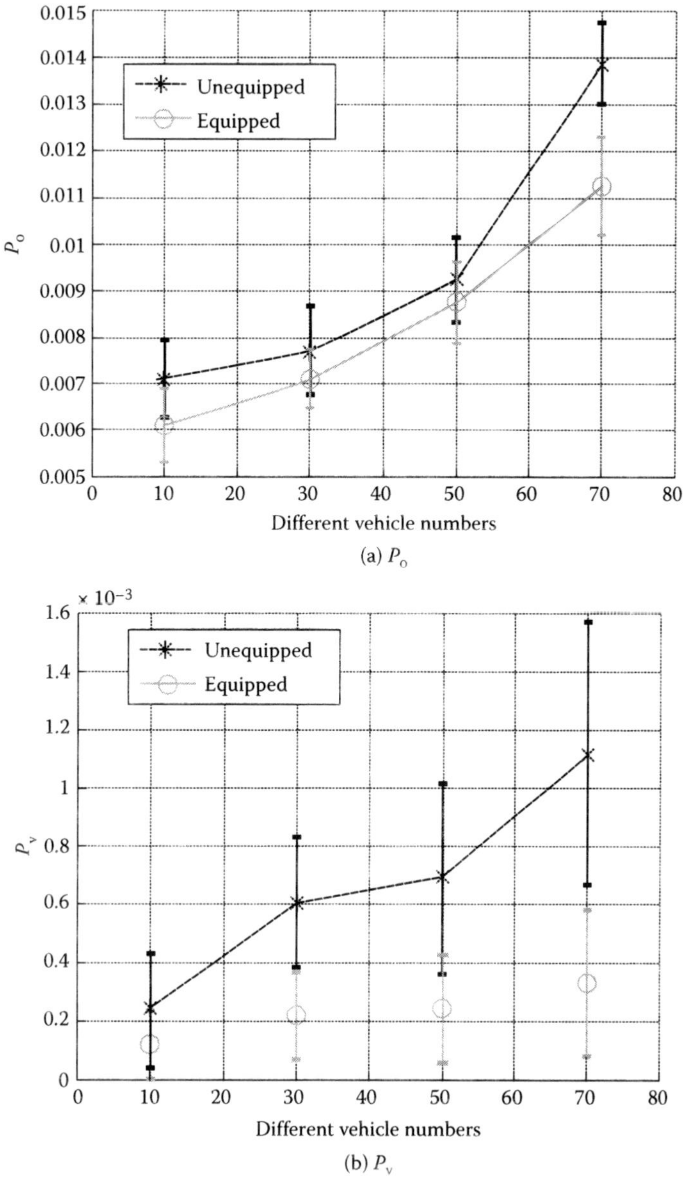

Figure 12.10 The results of P_o and P_v under different vehicle numbers.

set as $\sigma = 1.5$ m/s. The intervehicular communication range varies with $d_i = 200, 250, 300, 400$, and 500 m, and the simulation repeats 30 times at each communication range. Additionally, the traffic density rate is set to be 0.001 veh/m^2. From Figure 12.11a and b, it can be seen that the greater the communication range becomes, the lower the collision probability tends to be. Consequently, the probability of intervehicle collision can be kept at a relatively lower level.

The normal distribution for initializing the vehicle velocity and the traffic density rate are set to be $N(10.73, 2)$ and $\rho = 0.0022$ veh/m^2, respectively, to simulate the peak-hour flow scenario. Similar conclusions can as well be drawn from Figures 12.11c and d that larger communications range can improve vehicle safety. On the other hand, by comparing the results in Figure 12.11a and b with those in Figure 12.11c and d, it can be seen that a smaller collision probability on average is achieved under the dense traffic flow scenario.

12.3.3.2 Analysis of flow rate

To provide a deep insight into the impacts of wireless communications on road traffic efficiency, we would like to conduct other simulations.

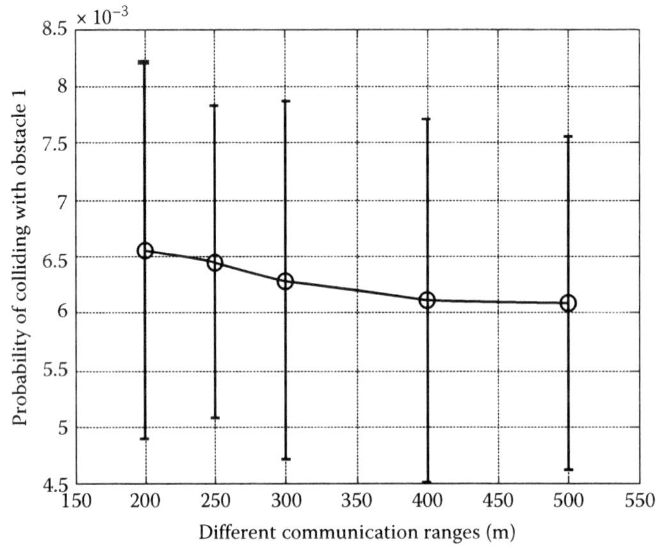

(a) Probability of colliding with obstacle 1, free flow scenario

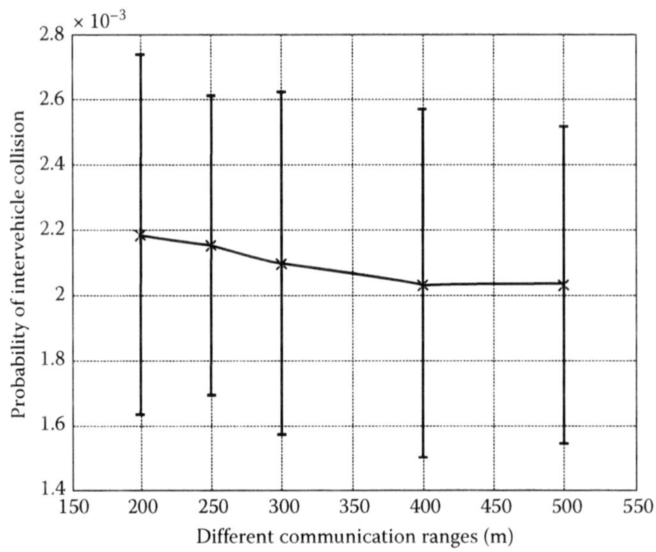

(b) Probability of intervehicle collision, free flow scenario

Figure 12.11 The collision probability in different traffic scenarios. (*Continued*)

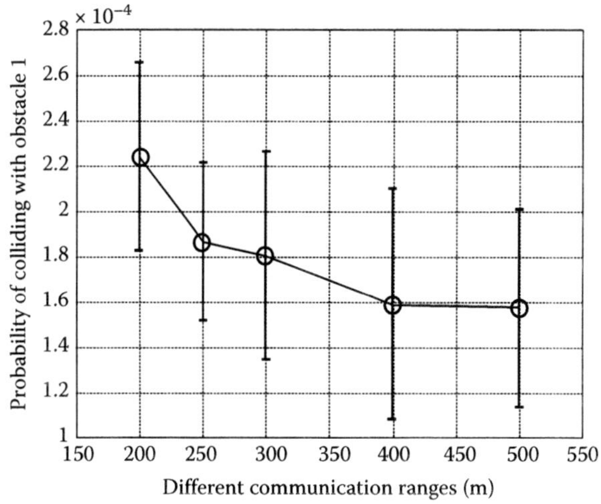

(c) Probability of colliding with obstacle 1, peak-hour flow scenario

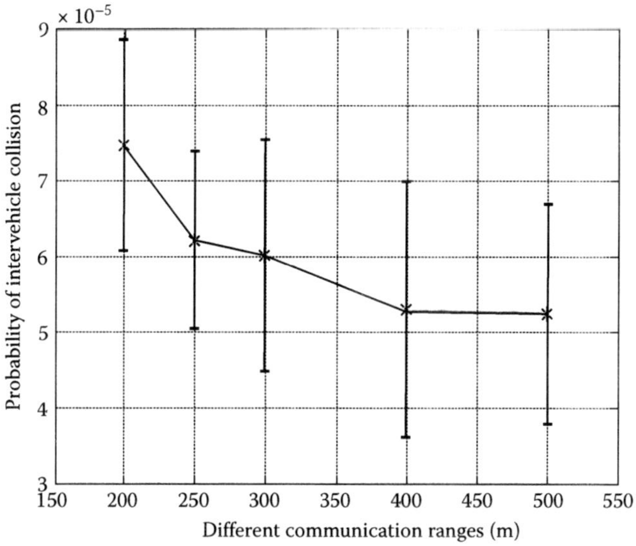

(d) Probability of intervehicle collision, peakhour flow scenario

Figure 12.11 (CONTINUED) The collision probability in different traffic scenarios.

1. Influence of different initial vehicular velocity on flow rate

   First, we fix the total vehicle number as $N = 30$ and then vary the initial vehicular velocity from 10 m/s to 30 m/s. At each velocity point, we repeat the simulation 30 times and calculate the flow rate on average. The flow rate is proposed here, which is measured at the section of the obstacle. Its calculation is defined as follows: in one simulation we can record the time instant t_1 when the first vehicle of the vehicle group passes the section located at the obstacle, and t_2 to indicate the moment when the last vehicle passes this section. The simulation results obtained in the Equipped and Unequipped cases are given in Figure 12.12a. From Figure 12.12a, it can be seen that the flow rate increases

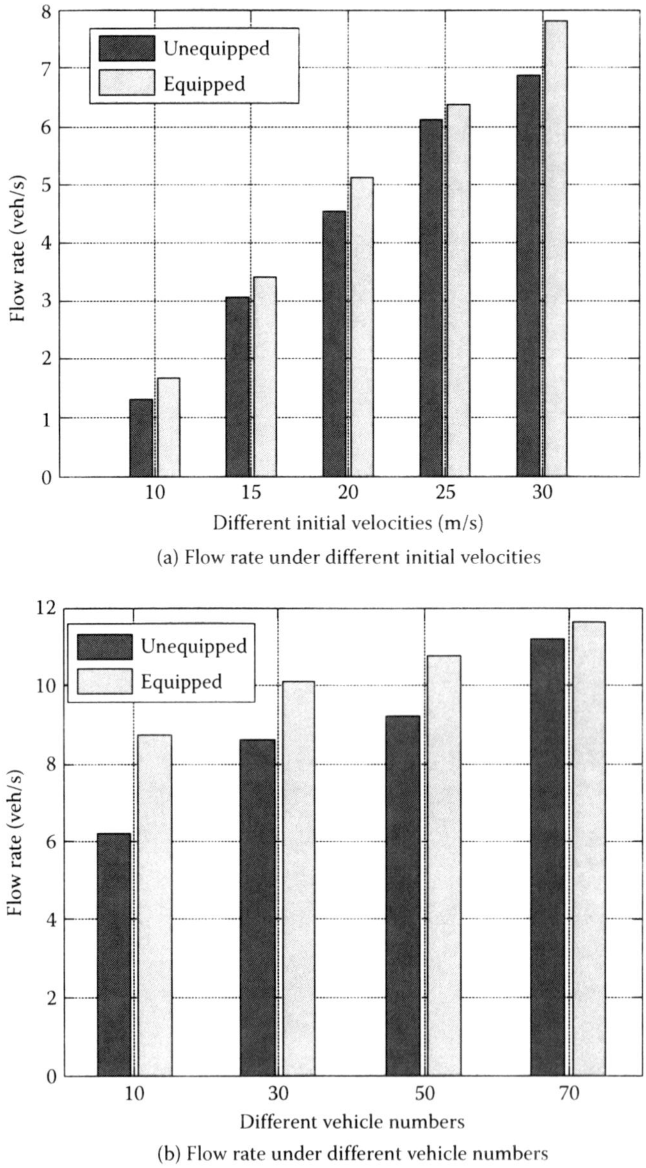

Figure 12.12 The results of flow rate.

along with an increase in the initial vehicular velocity in both compared cases. Logically, the increase in the velocity of each vehicle will lead to an increase in the mean speed of the overall traffic flow. Consequently, whether vehicles are equipped with wireless communications or not, a high vehicular velocity will promote road traffic efficiency. Nevertheless, when comparing the results of the Equipped and the Unequipped cases in Figure 12.12a, the flow rate at each initial velocity setting obtained in the Equipped case is higher than that in the Unequipped case. The reason is that with the assistance of wireless communication, the equipped vehicles are able to sense traffic condition far away from their own positions so that they can smoothly react to potential collisions ahead of time and efficiently avoid collisions in a cooperative manner. Then, the overall mobility of the equipped vehicles can be better maintained than that of unequipped vehicles during the time that vehicles are avoiding a collision.

2. Influence of different vehicle number on flow rates
   In the next experiment, the initial velocity of all the vehicles is fixed at 20 m/s, and the total vehicle number is set to range from 10 to 70. Similarly, the simulations in this experiment have been performed with 30 replications per vehicle number point. The results are shown in Figure 12.13b. The flow rate obtained in the Equipped case at each vehicle number point is obviously larger than that of the Unequipped case. This indicates that a vehicle group of small or large size can benefit more from wireless communication, since those equipped vehicles can coordinate their velocity more smoothly according to the overall movement of the group via wireless communication systems. At this point, this result confirms that the wireless communications promise to improve the overall efficiency of road traffic.

3. Influence of different cross sections of obstacles on flow rate
   Furthermore, the comparative flow rates at different sections are shown in Figure 12.13. Here, section 1, section 2, and section 3 stand for the location of the center of each obstacle on the road separately. According to the concept of the flow rate mentioned in foregoing paragraphs, it is obvious that the total number of vehicles N_i is the same in all these simulations; hence, the results of the flow rate in these three sections are the same approximately. However, there is still a little difference between them. Besides, the focus of these numerical experiments is on comparing the flow rate between the vehicles with V2V communication devices and vehicles without them. When compared to those results obtained under the Unequipped condition, the flow rates obtained under the Equipped condition are relatively larger. For example, in Figure 12.13a, the flow rate is about 26.85% improved by using vehicular communication under the point of 25 m/s. These comparative results demonstrate the effect of vehicular communication in the improvement of traffic efficiency.

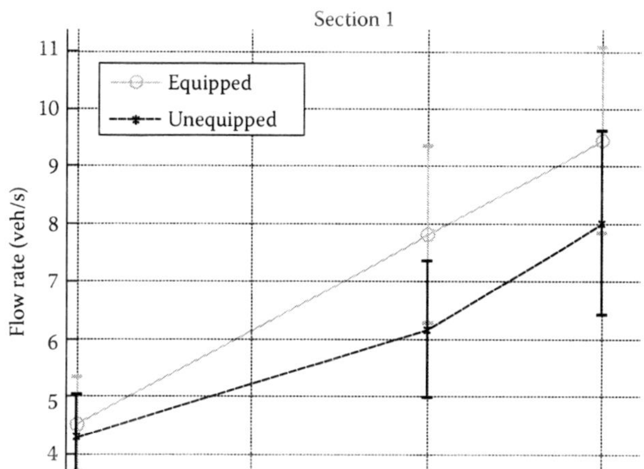

(a) Flow rate of section 1

Figure 12.13 Flow rate. (*Continued*)

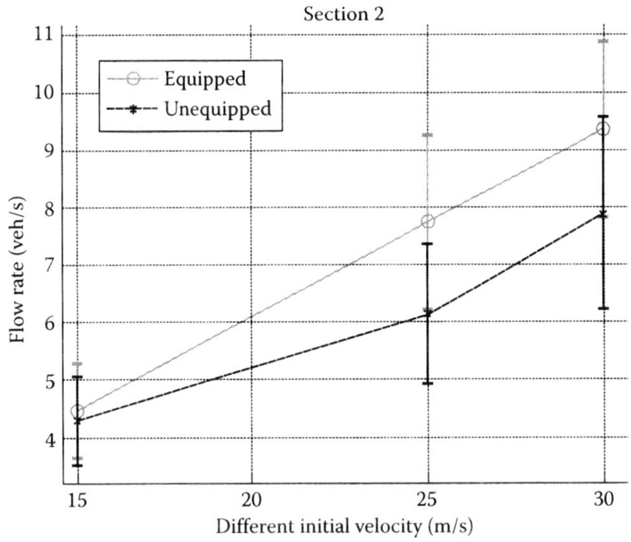

(b) Flow rate of section 2

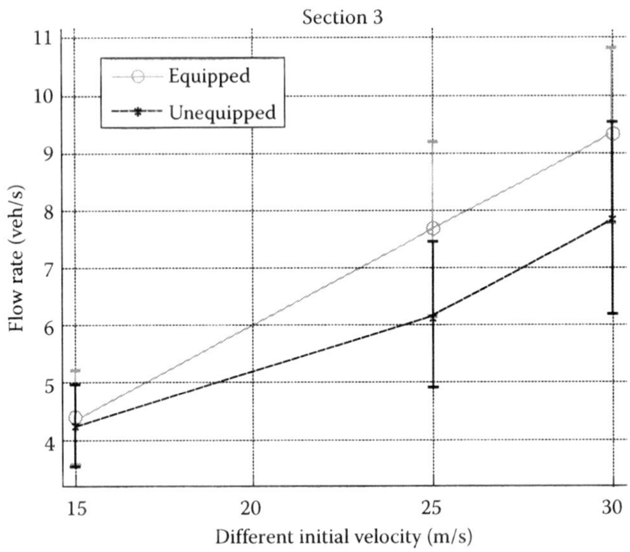

(c) Flow rate of section 3

Figure 12.13 (CONTINUED) Flow rate.

12.4 Conclusion

In this chapter, the connected vehicles are analogized to the fish schools so as to develop a more realistic model to describe the mobility of connected vehicles with wireless communications. The proposed models consist of the attraction of the goal, the repulsion of the obstacles, the constraint of the road, and the interplays of vehicles in the group, including both attractive and repulsive affection. The results prove that the models can be used to analyze the self-organization of multivehicles with V2V communications and the behavior of obstacle avoidance.

References

[1] *R. Horn and C. R. Johnson. Matrix analysis.* Cambridge University Press, New York, 1986.

[2] A. B. Gershman, M. Rübsamen, and M. Pesavento. One-and two-dimensional direction-of-arrival estimation: An overview of search-free techniques. *Signal Processing*, 90:1338–1349, 2010.

[3] A. Mallat, J. Louveaux, and L. Vandendorpe. UWB based positioning: Cramer Rao bound for angle of arrival and comparison with time of arrival. *2006 Symposium on Communications and Vehicular Technology*, IEEE, Liege, Belgium, 65–68, 2006.

[4] M. M. Adankon and M. Cheriet. *Support Vector Machine*. Chemistry World Scientific Pub Co Inc., 1(4):1–28, 2004.

[5] G. Agamennoni, J. I. Nieto, and E. M. Nebot. Robust inference of principal road paths for intelligent transportation systems. *IEEE Transactions on Intelligent Transportation Systems*, 12(1):298–308, 2011.

[6] N. Alam and A. G. Dempster. Cooperative positioning for vehicular networks: Facts and future. *IEEE Transactions on Intelligent Transportation Systems*, 14(4):1708–1717, 2013.

[7] N. Alam, A. Kealy, and A. G. Dempster. Cooperative inertial navigation for GNSS-challenged vehicular environments. *IEEE Transactions on Intelligent Transportation Systems*, 14(3):1370–1379, 2013.

[8] N. Alam, A. T. Balaei, and A. G. Dempster. Relative positioning enhancement in vanets: A tight integration approach. *IEEE Transactions on Intelligent Transportation Systems*, 14(1):47–55, 2013.

[9] N. Alam, A. T. Balaei, and A. G. Dempster. A DSRC Doppler-based cooperative positioning enhancement for vehicular networks with GPS availability. *IEEE Transactions on Vehicular Technology*, 60(9):4462–4470, 2011.

[10] N. Alam, A. T. Balaei, and A. G. Dempster. An instantaneous lane-level positioning using dsrc carrier frequency offset. *IEEE Transactions on Intelligent Transportation Systems*, 13(4):1566–1575, 2012.

[11] A. Albert. Comparison of event-triggered and time-triggered concepts with regards to distributed control systems. In *Proceedings of Embedded World Conference*, Hangzhou, China, 2004.

[12] H. Aly, A. Basalamah, and M. Youssef. Lanequest: An accurate and energy-efficient lane detection system. In *2015 IEEE International Conference on Pervasive Computing and Communications (PerCom)*, 163–171, 2015.

[13] O. Amrani and A. Rubin. Contention detection and resolution for multiple-access power-line communications. *IEEE Transactions on Vehicular Technology*, 56(6):3879–3887, 2007.

[14] O. Arpacioglu and Z. J. Haas. On the scalability and capacity of wireless networks with omnidirectional antennas. In *Proceedings of the Third International Symposium on Information Processing in Sensor Networks*, 2004, Berkeley, CA, USA, April 26–27, ACM, New York, NY, 169–177, 2004.

[15] B. Aygün, M. Boban, and A. M. Wyglinski. ECPR: Environment- and context-aware combined power and rate distributed congestion control for vehicular communications. Computer communications 96: 3–16, 1 November 2016. doi:10.1016/j.comcom.2016.05.015 *CoRR*, abs/1502.00054, 2015.

[16] N. Bahrani and V. Gaudet. Measurements and channel characterization for in-vehicle power line communications. In *Proceedings of 18th IEEE International Symposium on Power Line Communications and its Applications (ISPLC)*, IEEE, Glasgow, UK, 64–69, March 2014.

[17] Z. Bai and G. H. Golub. Bounds for the trace of the inverse and the determinant of symmetric positive definite matrices. *Annals of Numerical Mathematics*, 4:29–38.

[18] S. Barmada, M. Raugi, M. Tucci, and T. Zheng. Power line communication in a full electric vehicle: Measurements, modelling and analysis. In *Proceedings of IEEE International Symposium on Power Line Communications and Its Applications (ISPLC)*, IEEE, Rio Othon Palace, Rio de Janeiro City, 331–336, March 2010.

[19] E. Bassi, F. Benzi, L. Almeida, and T. Nolte. Powerline communication in electric vehicles. In *Proceedings of IEEE International Electric Machines and Drives Conference (IEMDC)*, Prentice-Hall, Englewood Cliffs, NJ, 1749–1753, May 2009.

[20] D. Bertsekas and R. Gallager. *Data Networks*. Hilton Hotel Miami, Miami, FL, Prentice-Hall, 1987.

[21] A. Biryukov. Known plaintext attack. In *Henk C. A. van Tilborg and Sushil Jajodia Encyclopedia of Cryptography and Security*, 704–705. Springer, 2011.

[22] S. Biswas, R. Tatchikou, and F. Dion. Vehicle-to-vehicle wireless communication protocols for enhancing highway traffic safety. *IEEE Communications Magazine*, 44(1):74–82, 2006.

[23] D. M. Blough, G. Resta, and P. Santi. Approximation algorithms for wireless link scheduling with sinr-based interference. *IEEE/ACM Transactions on Networking*, 18(6):1701–1712, 2010.

[24] G. E. P. Box and D. R. Cox. An analysis of transformations. *Journal of the Royal Statistical Society. Series B (Methodological)*, 211–252, 1964.

[25] G. Brar, D. M. Blough, and P. Santi. Computationally efficient scheduling with the physical interference model for throughput improvement in wireless mesh networks. In *Proceedings of the 12th Annual International Conference on Mobile Computing and Networking*, ACM, Los Angeles, CA, 2–13, 2006.

[26] C. M. Breder. Equations descriptive of fish schools and other animal aggregations. *Ecology*, 35(3):361–370, 1954.

[27] L. Cao and J. Krumm. From GPS traces to a routable road map. In *Workshop on Advances in Geographic Information Systems*, ACM, Seattle, WA, 3–12, 2009.

[28] C. Y. Yang, B. S. Chen, and F. K. Liao. Mobile location estimation using fuzzy-based IMM and data fusion. *IEEE Transactions on Mobile Computing*, 9(10):1424–1436, 2010.

[29] P. Chatzimisios, A. C. Boucouvalas, and V. Vitsas. Performance analysis of IEEE 802.11 DCF in presence of transmission errors. In *2004 IEEE International Conference on Communications*, IEEE, Paris, Vol. 7, 3854–3858, June 2004.

[30] B. S. Chen, C. Y. Yang, F. K. Liao, and J. F. Liao. Mobile location estimator in a rough wireless environment using extended Kalman-based IMM and data fusion. *IEEE Transactions on Vehicular Technology*, 58(3):1157–1169, 2009.

[31] C. Chen and Y. Cheng. Roads digital map generation with multi-track GPS data. In *Proceedings of Workshops on Education Technology and Training, and on Geoscience and Remote Sensing*, IEEE, Shanghai, China, 508–511, 2008.

[32] H. Chen and J. Tian. Research on the controller area network. In *Proceedings of International Conference on Networking and Digital Society (ICNDS)*, IEEE, Guiyang, Guizhou, China, Vol. 2, 251–254, May 2009.

[33] J. Chen, H. Zeng, C. Hu, and Z. Ji. Optimization between security and delay of quality-of-service. *Journal of Network and Computer Applications*, 34(2):603–608, 2011.

[34] J. Chen, M. K. Leung, and Y. Gao. Noisy logo recognition using line segment Hausdorff distance. *Pattern Recognition*, 36(4):943–955, 2003.

[35] M. Chen, Y. Ma, J. Song, C. F. Lai, and B. Hu. Smart clothing: Connecting human with clouds and big data for sustainable health monitoring. *Mobile Networks and Applications*, 21(5):825–845, 2016.

[36] M. Chen, Y. Zhang, L. Hu, T. Taleb, and Z. Sheng. Cloud-based wireless network: Virtualized, reconfigurable, smart wireless network to enable 5G technologies. *Mobile Networks and Applications*, 20(6):704–712, 2015.

[37] R. Chen, Z. Sheng, Z. Zhong, M. Ni, V. C. M. Leung, D. G. Michelson, and M. Hu. Connectivity analysis for cooperative vehicular ad hoc networks under nakagami fading channel. *Institute of Electrical and Electronics Engineers*, 18(10):1787–1790, 2014.

[38] S. Chen, A. M. Wyglinski, S. Pagadarai, R. Vuyyuru, and O. Altintas. Feasibility analysis of vehicular dynamic spectrum access via queueing theory model. *IEEE Communications Magazine*, 49(11):156–163, 2011.

[39] N. Cheng, N. Zhang, N. Lu, X. Shen, J. W. Mark, and F. Liu. Opportunistic spectrum access for cr-vanets: a game-theoretic approach. *IEEE Transactions on Vehicular Technology*, 63(1):237–251, 2014.

[40] Y.-C. Cheng, Y. Chawathe, A. LaMarca, and J. Krumm. Accuracy characterization for metropolitan-scale wi-fi localization, 2005.

[41] S. Coleri Ergen, H. S. Tetikol, M. Kontik, R. Sevlian, R. Rajagopal, and P. Varaiya. RSSI-fingerprinting-based mobile phone localization with route constraints. *IEEE Transactions on Vehicular Technology*, 63(1):423–428, 2014.

[42] European Commission. Standardisation mandate addressed to CEN, CENLEC and ETSI in the field of information and communication technologies to support the interoperability of co-operative systems for intelligent transport in the european community. http://www.etsi.org/m453, M453, 2009.

[43] L. Cong and W. Zhuang. Hybrid TDOA/AOA mobile user location for wideband CDMA cellular systems. *IEEE Transactions on Wireless Communications*, 1:439–447, 2002.

[44] MOST Cooperation. MOST Specification Rev. 3.0 E2. http://www.mostcooperation.com/publications/specifications-organizational-procedures/, 2010.

[45] S. Corrigan. Introduction to the controller area network (CAN). s.l. : Texas Instruments, http://rpi.edu/dept/ecse/mps/sloa101.pdf, 2008.

[46] D. Helbing. A mathematical model for the behavior of individuals in a social field. *Journal of Mathematical Sociology*, 19(3):189–219, 1994.

[47] S. Sarvotham D. Baron and R. G. Baraniuk. Bayesian compressive sensing via belief propagation. *IEEE Transactions on Signal Processing*, 58:269–280, 2010.

[48] C. M. Clark, D. Thanh-Son, K. Y. K. Leung and J. P. Huissoon. Markov-based lane positioning using intervehicle communication. *IEEE Transactions on Intelligent Transportation Systems*, 8:641–650, 2007.

[49] J. Daemen and V. Rijmen. The wide trail design strategy. In Springer (ed), *Cryptography and Coding*, 222–238. Berlin, Heidelberg: Springer, 2001.

[50] J. J. Davies, A. R. Beresford, and A. Hopper. Scalable, distributed, real-time map generation. *IEEE Pervasive Computing*, 5(4):47–54, 2006.

[51] V. Degardin, P. Laly, M. Lienard, and P. Degauque. Impulsive noise on in-vehicle power lines: Characterization and impact on communication performance. In *Proceedings of IEEE International Symposium on Power Line Communications and Its Applications (ISPLC)*, IEEE, Orlando, FL, 222–226, 2006.

[52] V. Degardin, M. Lienard, P. Degauque, E. Simon, and P. Laly. Impulsive noise characterization of in-vehicle power line. *IEEE Transactions on Electromagnetic Compatibility*, 50(4):861–868, 2008.

[53] F. F. Digham, M.-S. Alouini, and M. K. Simon. On the energy detection of unknown signals over fading channels. In *Proceedings of IEEE International Conference on Communications (ICC)*, IEEE, Anchorage, AK, Vol. 5, 3575–3579, May 2003.

[54] F. Dressler and O. B. Akan. A survey on bio-inspired networking. *Computer Networks*, 54(6):881–900, 2010.

[55] S. Edelkamp and S. Schrdl. Route planning and map inference with global positioning traces. *Computer Science in Perspective*, 2598(1):128–151, 2003.

[56] A. J. Efron and H. Jeen. Detection in impulsive noise based on robust whitening. *IEEE Transactions on Signal Processing*, 42(6):1572–1576, 1994.

[57] T. ElBatt, C. Saraydar, M. Ames, and T. Talty. Potential for intra-vehicle wireless automotive sensor networks. In *Proceedings of IEEE Sarnoff Symposium*, IEEE, Princeton, NJ, 1–4, March 2006.

[58] M. B. Elowitz, A. J. Levine, E. D. Siggia, and P. S. Swain. Stochastic gene expression in a single cell. *Science*, 297(5584):1183–1186, 2002.

[59] M. Erritali, O. M. Reda, and B. El Ouahidi. A beaconing approach with key exchange in vehicular ad hoc networks. *International Joural of Distributed and Parallel Systems*, 3(6):9–13, 2012.

[60] J. Maslouh, A. Errami, and M. Khaldoun. Resolving the access conflict for shared Ethernet communication channel. *IEEE International Conference on Next Generation Networks and Services*, University of Oxford, 80–87, 2014.

[61] N. Navet, Y. Song, F. Simonot-Lion, and C. Wilwert. Trends in automotive communication systems. *Proceedings of the IEEE*, 93(6):1204–1223, 2005.

[62] T. Nolte, H. Hansson, and L. L. Bello. Automotive communications-past, current and future. *IEEE Conference on Emerging Technologies and Factory Automation*, Catania, Italy, 985–992, September 2005.

[63] T. Nolte et al. Wireless automotive communications. In *Euromicro Conference on Real-Time Systems (ECRTS 05)*, 2005.

[64] W. Fei-Yue. Parallel control and management for intelligent transportation systems: Concepts, architectures, and applications. *IEEE Transactions on Intelligent Transportation Systems*, 11:630–638, 2010.

[65] Y. Fu and Z. Tian. Cramercrao bounds for hybrid TOA/DOA-based location estimation in sensor networks. *IEEE Signal Processing Letters*, 16:655–658, 2009.

[66] I. Fukuyori, Y. Nakamura, Y. Matsumoto, and H. Ishiguro. Control method for a robot based on the adaptive attractor selection model. In *International Conference on Autonomous Robots and Agents*, IEEE, Wellington, New Zealand, 618–623, 2009.

[67] G. Johansson. and K. Rumar. Drivers brake reaction times. *Human Factors: The Journal of the Human Factors and Ergonomics Society*, 13(1):23–27, 1971.

[68] M. Carlin, G. Oliveri, and A. Massa. Complex-weight sparse linear array synthesis by Bayesian compressive sampling. *IEEE Transactions on Antennas and Propagation*, 60:2309–2326, 2012.

[69] S. Galli and O. Logvinov. Recent developments in the standardization of power line communications within the IEEE. *IEEE Wireless Communications Magazine*, 46(7):64–71, 2008.

[70] J. Galtier. Tournament methods for WLAN: Analysis and efficiency. *Graphs and Algorithms in Communication Networks*. Springer, Cham Switzerland, 2010.

[71] Y. Gao and M. K. H. Leung. Line segment Hausdorff distance on face matching. *Pattern Recognition*, 35(2):361–371, 2002.

[72] Z. Gengzhong, L. Sanyang, and Q. Xiaogang. A power control algorithm based on non-cooperative game for wireless CDMA sensor networks. *International Journal of Digital Content Technology and its Applications*, 4(3):137–145, 2010.

[73] Robert Bosch GmbH. CAN Specification. In *BOSCH*, http://docplayer.net/72152-Bosch-can-specification-version-2-0-1991-robert-bosch-gmbh-postfach-30-02-40-d-70442-stuttgart.html, 1991.

[74] A. Goundan, E. Coe, and C. Raghavendra. Efficient broadcasting in delay tolerant networks. In *2008 IEEE Global Telecommunications Conference. IEEE GLOBECOM 2008*, IEEE, New Orleans, LA, 1–5, November 2008.

[75] W. Gouret, F. Nouvel, and G. El-Zein. Powerline communication on automotive network. In *Proceedings IEEE Vehicular Technology Conference (VTC)*, 2545–2549, http://xueshu.baidu.com/s?wd=paperuri%3A%282076e2a3928724f530c973afb4bdc33a%29&filter=sc_long_sign&tn=SE_xueshusource_2kduw22v&sc_vurl=http%3A%2F%2Fieeexplore.ieee.org%2Fxpls%2Ficp.jsp%3Farnumber%3D4212952&ie=utf-8&sc_us=6031007383106161442, April 2007.

[76] O. Goussevskaia, Y. A. Oswald, and R. Wattenhofer. Complexity in geometric SINR. In *Proceedings of the 8th ACM International Symposium on Mobile Ad Hoc Networking and Computing*, ACM, Montreal, Quebec, Canada, 100–109, 2007.

[77] P. D. Grünwald, I. J. Myung, and M. A. Pitt. *Advances in Minimum Description Length: Theory and Applications*. MIT Press, Cambridge, MA, 2005.

[78] B. S. Gukhool and S. Cherkaoui. IEEE 802.11p modeling in NS-2. In *IEEE Conference on Local Computer Networks, 2008. LCN 2008*, Montreal, Quebec, Canada, 622–626, 2008.

[79] E. Gustafsson and A. Jonsson. Always best connected. *IEEE Wireless Communications*, 10(1):49–55, 2003.

[80] R. Tafazolli, R. Kernchen, H. Chong, M. Dianati, and S. Xuemin. Analytical study of the IEEE 802.11p MAC sublayer in vehicular networks. *IEEE Transactions on Intelligent Transportation Systems*, 13:873–886, 2012.

[81] J. Lien, H. Wymeersch, and M. Z. Win. Cooperative localization in wireless networks. *Proceedings of the IEEE*, 97:427–450, 2009.

[82] A. Hagiescu, U. D. Bordoloi, S. Chakraborty, P. Sampath, P. V. V. Ganesan, and S. Ramesh. Performance analysis of FlexRay-based ECU networks. In *Proceedings of ACM/IEEE Design Automation Conference*, ACM/IEEE, San Diego, CA, 284–289, 2007.

[83] M. Haklay and P. Weber. Openstreetmap: User-generated street maps. *IEEE Pervasive Computing*, 7(4):12–18, 2008.

[84] R. Hasseea. A GIS based accident system for reducing road accidents. In *Proceedings of European Transport Conference*, Strasbourg, France, October, 2003.

[85] J. Hasty and J. J. Collins. Noise-based switches and amplifiers for gene expression. *Proceedings of the National Academy of Sciences of the United States of America*, 97(5):2075–80, 2000.

[86] Y. He, M. Chowdhury, P. Pisu, and X. Kang. Vehicle-infrastructure integration-enabled plug-in hybrid electric vehicles for optimizing energy consumption. In *Meeting of the Transportation Research Board*, Washington, DC, 2011.

[87] HomePlug Powerline Alliance. Homeplug Green PHY Specification 1.1.1. 2013. http://groups.homeplug.org/tech/homeplug_gp/.

[88] E. Hossain, G. Chow, V. C. M. Leung, R. D. McLeod, J. Misic, and O. Yang. Vehicular telematics over heterogeneous wireless networks: A survey. *Computer Communicaitons*, 33(7):775–793, 2010.

[89] D. Hristu-Varsakelis and W. Levine. *Handbook of Networked and Embedded Control Systems*. Birkhuser, Basel, Switzerland, 2005.

[90] S.-C. Huang. D2PS: Direction and distance positioning system in wireless networks. *The Computer Journal*, 57(6): 939–951, 2013.

[91] J. M. Huerta, J. Vidal, A. Giremus, and J. Y. Tourneret. Joint particle filter and ukf position tracking in severe non-line-of-sight situations. *IEEE Journal of Selected Topics in Signal Processing*, 3(5):874–888, 2009.

[92] IEEE. 1609.3-2007—IEEE trial-use standard for wireless access in vehicular environments (WAVE)—Networking services. 1–144, http://xueshu.baidu.com/s?wd=paperuri%3A%284742890cf87115184f07f18afef682a8%29&filter=sc_long_sign&tn=SE_xueshusource_2kduw22v&sc_vurl=http%3A%2F%2Fieeexplore.ieee.org%2Fxpls%2Fabs_all.jsp%3Farnumber%3D4167674&ie=utf-8&sc_us=10536129880847936929, 2010.

[93] IEEE. IEEE standard for information technology telecommunications and information exchange between systems local and metropolitan area networks-specific requirements. Part 11: Wireless LAN medium access control layer, 2007.

[94] T. Iwai, N. Wakamiya, and M. Murata. Error-tolerant coverage control based on bio-inspired attractor selection model for wireless sensor networks. In *IEEE International Conference on Computer and Information Technology, Cit 2010*, Bradford, West Yorkshire, UK, June 29–July, 723–729, 2010.

[95] J. Carbaugh, D. N. Godbole, and R. Sengupta. Safety and capacity analysis of automated and manual highway systems. *Transportation Research Part C: Emerging Technologies*, 6(1–2):69–99, 1998.

[96] S. Biswas, J. A. Misener, and G. Larson. Development of V-to-X systems in North America: The promise, the pitfalls and the prognosis. *Computer Networks*, 55:3120–3133, 2011.

[97] G. Rhodes, J. D. Hill, and S. Vollar. *Car Park Designers' Handbook. Drainage*, http://xueshu.baidu.com/s?wd=Car+Park+Designers'+Handbook&rsv_bp=0&tn=SE_ baiduxueshu_c1gjeupa&rsv_spt=3&ie=utf-8&f=8&rsv_sug2=1&sc_f_para=sc_tasktype%3D%7BfirstSimpleSearch%7D&rsv_n=2, 2005.

[98] R. Jain, D. Chiu, and W. Hawe. A quantitative measure of fairness and discrimination for resource allocation in shared computer systems. *Computer Science*, cs.ni/9809099, 1998.

[99] G. Jakimoski and L. Kocarev. Analysis of some recently proposed chaos-based encryption algorithms. *Physics Letters A*, 291(6):381–384, 2001.

[100] K. Jamieson, H. Balakrishnan, and Y. C. Tay. Sift: A MAC protocol for event-driven wireless sensor networks. In *Proceedings of Third European Workshop on Wireless Sensor Networks (EWSN)*, Zurich, Switzerland, February 2006.

[101] S. Jang, T. Kim, and S. Lee. Map generation system with lightweight GPS trace data. In *Proceedings of the 12th International Conference on Advanced Communication Technology*, IEEE, Gangwon-Do, Korea, 1489–1493, 2010.

[102] X. Jiang, K.-K. Wong, Y. Zhang, and D. J. Edwards. Cross-layer design of partial spectrum sharing for two licensed networks using cognitive radios. *Wireless Communications and Mobile Computing*, 15(2):340–353, 2015.

[103] P. Jindal and B. Singh. Performance evaluation of security-throughput tradeoff with channel adaptive encryption. *International Journal of Computer Network and Information Security (IJCNIS)*, 5(1):49, 2013.

[104] J. Jose, C. Li, X. Wu, L. Ying, and K. Zhu. Distributed rate and power control in vehicular networks. *CoRR*, abs/1511.01535, 2015.

[105] F. J. Massey Jr. The Kolmogorov-Smirnov test for goodness of fit. *Journal of the American Statistical Association*, 46(253):68–78, 1951.

[106] A. Kesting and M. Treiber. Calibrating car-following models by using trajectory data: Methodological study. *Transportation Research Record: Journal of the Transportation Research Board*, 148–156, 2008.

[107] K. Li, J. Bigham, E. L. Bodanese and L. Tokarchuk. Outdoor location estimation in changeable environments. *IEEE Communications Letters*, 17:2072–2075, 2013.

[108] S. Kajioka, N. Wakamiya, and M. Murata. Autonomous and adaptive resource allocation among multiple nodes and multiple applications in heterogeneous wireless networks. *Journal of Computer & System Sciences*, 78(6):1673–1685, 2012.

[109] J. M. Kang, J. Strassner, S. S. Seo, and J. W. K. Hong. Autonomic personalized handover decisions for mobile services in heterogeneous wireless networks. *Computer Networks*, 55(7):1520–1532, 2011.

[110] T. H. Kang, C. P. Hong, Y. S. Kim, and S. D. Kim. A context-aware handoff management for seamless connectivity in ubiquitous computing environment. In *International Conference on Pervasive Systems Computing*, PSC 2006, Las Vegas, NV, 128–134, June, 2006.

[111] E. D. Kaplan and C. J. Hegarty. *Understanding GPS: Principles and Applications*. Artech House, Norwood, MA. 2005.

[112] G. Karlsson, V. Lenders, and M. May. Delay-tolerant broadcasting. In *Proceedings of the 2006 SIGCOMM Workshop on Challenged Networks*, CHANTS '06. ACM, New York, 197–204, 2006.

[113] A. Kashiwagi, I. Urabe, K. Kaneko, and T. Yomo. Adaptive response of a gene network to environmental changes by fitness-induced attractor selection. *PLoS One*, 1(1):e49, 2006.

[114] S. A. Kassam and H. V. Poor. Robust techniques for signal processing: A survey. *Proceedings of the IEEE*, 73(3):433–481, 1985.

[115] A. Kenarsari-Anhari and V. C. M. Leung. Multi-carrier medium access control for in-vehicle power line communication with imperfect sensing. In *Proceedings of IEEE Vehicular Technology Conference (VTC Spring)*, 1–5, June 2013.

[116] A. Kenarsari-Anhari, V. C. M. Leung, and L. Lampe. A distributed MAC protocol for in-vehicle power line communication under imperfect carrier sensing. In *Proc. IEEE International Symposium on Personal Indoor and Mobile Radio Communications (PIMRC)*, Vehicular Technology Conference, Dresden, Germany, 1720–1725, September 2013.

[117] M. G. Khoshkholgh, K. Navaie, and H. Yanikomeroglu. On the impact of the primary network activity on the achievable capacity of spectrum sharing over fading channels. *IEEE Transactions on Wireless Communications*, 8(4):2100–2111, 2009.

[118] S. Kim. Adaptive online power control scheme based on the evolutionary game theory. *IET Communications*, 5(18):2648–2655, December 2011.

[119] S. Kirkpatrick, C. D. Gelatt, M. P. Vecchi. Optimization by simulated annealing. *Science*, 220(4598):671–680, 1983.

[120] H. Kitano. Biological robustness. *Nature Reviews Genetics*, 5(11):826–837, 2004.

[121] H. Kitano. Towards a theory of biological robustness. *Molecular Systems Biology*, 3(1):137, 2007.

[122] F. H. Knight. The economics of welfare. *Quarterly Journal of Economics*, 38:582–606, 1924.

[123] Y. Koizumi, T. Miyamura, S. I. Arakawa, E. Oki, K. Shiomoto, and M. Murata. Adaptive virtual network topology control based on attractor selection. *Journal of Lightwave Technology*, 28(11):1720–1731, 2010.

[124] M. Koutny, P. Mlynek, and P. Mrakava. Homeplug simulation model for analysis of the rounding functions. In *Proceedings of 34th International Conference on Telecommunications and Signal Processing (TSP)*, Budapest, Hungary, May 2011.

[125] H. Kremo and O. Altintas. On detecting spectrum opportunities for cognitive vehicular networks in the TV white space. *Journal of Signal Processing Systems*, 73(3):243–254, 2013.

[126] L. Lai and S. Qu. Research on application of fish swarm behaviour in intelligent transportation system. *Advanced Forum on Transportation of China*, 195–200, 2011.

[127] T. Larsson and M. Patriksson. An augmented Lagrangean dual algorithm for link capacity constrained traffic assignment problems. *Transportation Research Part B: Methodological*, 29:433–455, 1995.

[128] Y. Lee and K. S. Park. Meeting the real-time constraints with standard Ethernet in an in-vehicle network. In *Proceedings of IEEE Intelligent Vehicles Symposium (IV)*, Brisbane, Australia, 1313–1318, June 2013.

[129] K. Leibnitz, N. Wakamiya, and M. Murata. A bio-inspired robust routing protocol for mobile ad hoc networks. In *International Conference on Computer Communications and Networks*, Vancouver, Canada, 321–326, 2007.

[130] A. R. Lelewski, J. A. Berenis, and G. M. Pressimone. Analyzing express toll plaza operations using modern simulation models. In *Institute of Transportation Engineers 2003 Annual Meeting*, Seattle, WA, 2003.

[131] M. Z. F. Li. The role of speed flow relationship in congestion pricing implementation with an application to Singapore. *Transportation Research Part B: Methodological*, 36:731–754, 2002.

[132] X. R. Li and V. P. Jilkov. Survey of maneuvering target tracking. part I. Dynamic models. *IEEE Transactions on Aerospace and Electronic Systems*, 39(4):1333–1364, 2003.

[133] Y. B. Li, R. Yang, Y. Lin, and F. Ye. The spectrum sharing in cognitive radio networks based on competitive price game. *Radioengineering*, 225–226(3):632–636, 2012.

[134] Y. Li, H. Long, M. Peng, and W. Wang. Spectrum sharing with analog network coding. *IEEE Transactions on Vehicular Technology*, 63(4):1703–1716, 2014.

[135] M. Lienard, M. O. Carrion, V. Degardin, and P. Degauque. Modeling and analysis of in-vehicle power line communication channels. *IEEE Transactions on Vehicular Technology*, 57(2):670–679, 2008.

[136] LIN and Consortium. LIN Specification v2.2A. http://www.lin-subbus.org/, 2010.

[137] M. Lindhe, P. Ogren, and K. H. Johansson. Flocking with obstacle avoidance: A new distributed coordination algorithm based on voronoi partitions. In *Proceedings of 2005 IEEE International Conference on Robotics and Automation*, IEEE, Barcelona, Spain, 1797–1782, 2005.

[138] R. Lindsey. Do economists reach a conclusion on road pricing? The intellectual history of an idea. *Journal Watch*, 3:292–379, 2006.

[139] X. Liu, J. Biagioni, J. Eriksson, Y. Wang, G. Forman, and Y. Zhu. Mining large-scale, sparse GPS traces for map inference: Comparison of approaches. In *Proceedings of the 18th ACM SIGKDD International Conference on Knowledge Discovery and Data Mining*, Beijing, China, 669–677, 2012.

[140] X. Liu, Y. Zhu, Y. Wang, G. Forman, L. M. Ni, Y. Fang, and M. Li. Road recognition using coarse-grained vehicular traces. *Hp Labs*, IEEE Communications Letters, 2012.

[141] K. Li, P. Jiang, E. L. Bodanese, and J. Bigham. Outdoor location estimation using received signal strength feedback. A survey on position-based routing in mobile ad hoc networks. *IEEE Communications Letters*, 16:978–981, 2012.

[142] V. J. Lumelsky. On fast computation of distance between line segments. *Information Processing Letters*, 21(2):55–61, 1985.

[143] C. A. Lupini. In-vehicle networking technology for 2010 and beyond. s.l. : SAE International, 2010.

[144] H. Wymeersch, M. A. Caceres, F. Penna, and R. Garello. Hybrid cooperative positioning based on distributed belief propagation. *IEEE Journal on Selected Areas in Communications*, 29:1948–1958, 2011.

[145] G. Oliveri, F. Viani, M. Carlin, P. Rocca, and A. Massa. Directions-of-arrival estimation through Bayesian compressive sensing strategies. *IEEE Transactions on Antennas and Propagation*, 61:3828–3838, 2013.

[146] J. Widmer, M. Mauve, and H. Hartenstein. A survey on position-based routing in mobile ad hoc networks. *IEEE Network*, 15:30–39, 2001.

[147] M. Porretta, P. Nepa, G. Manara, and F. Giannetti. Location, location, location. *IEEE Vehicular Technology Magazine*, 3:20–29, 2008.

[148] C. Ma. Techniques to improve ground-based wireless location performance using a cellular telephone network. Geomatics Engineering University of Calgary, 2003.

[149] X. Ma, X. Chen, and H. H. Refai. Unsaturated performance of IEEE 802.11 broadcast service in vehicle-to-vehicle networks. In *Vehicular Technology Conference, 2007. VTC-2007 Fall. 2007 IEEE 66th*, IEEE, Baltimore, MD, 1957–1961, September 2007.

[150] D. B. Maciua. Brake dynamics effect on IVHSLane capacity. https://www.researchgate.net/publication/46439352_Brake_Dynamics_Effect_On_IVHSLane_Capacity, 1995.

[151] B. B. Madan, K. Gogeva-Popstojanova, K. Vaidyanathan, and K. S. Trivedi. Modeling and quantification of security attributes of software systems. In *Proceedings of International Conference on Dependable Systems and Networks*, 2002, IEEE, Washington, DC, 505–514, 2002.

[152] M. Maskery, V. Krishnamurthy, and Q. Zhao. Decentralized dynamic spectrum access for cognitive radios: Cooperative design of a non-cooperative game. *IEEE Transactions on Communications*, 57(2):459–469, 2009.

[153] J. Matsumura, Y. Matsubara, H. Takada, M. Oi, M. Toyoshima, and A. Iwai. A simulation environment based on OMNeT++ for automotive CANEthernet networks. In *Proceedings of International Workshop on Analysis Tools and Methodologies for Embedded and Real-Time Systems*, Paris, France, 2013.

[154] S. Mazuelas, F. A. Lago, J. Blas, A. Bahillo, P. Fernandez, R. M. Lorenzo, and E. J. Abril. Prior NLOS measurement correction for positioning in cellular wireless networks. *IEEE Transactions on Vehicular Technology*, 58(5):2585–2591, 2009.

[155] M. Meisel, V. Pappas, and L. Zhang. A taxonomy of biologically inspired research in computer networking. *Computer Networks*, 54(6):901–916, 2010.

[156] Q. Meng, W. Xu, and H. Yang. A trial-and-error procedure for implementing a road-pricing scheme. *Transportation Research Record*, 1923:103–109, 2005.

[157] J. Mitola. Cognitive radio for flexible mobile multimedia communications. In *IEEE International Workshop on Mobile Multimedia Communications, 1999. (MoMuC'99)*, IEEE, San Diego, CA, 3–10, 1999.

[158] U. Mohammad. Performance analysis of IEEE 802.15.4 for intra-vehicle wireless networks. In *Proceedings of WorldComp 2014*, Las Vegas, NV, USA, 2014.

[159] Y. L. Morgan. Notes on DSRC and WAVE standards suite: Its architecture, design, and characteristics. *IEEE Communications Surveys & Tutorials*, 12(4):504–518, 2010.

[160] Y. L. Morgan. Notes on DSRC and WAVE standards suite: Its architecture, design, and characteristics. *IEEE Communications Surveys & Tutorials*, 12:504–518, 2010.

[161] P. B. Moyle and J. J. Cech. Fishes: An introduction to ichthyology. *International Review of Hydrobiology*, 92(1):98, 2007.

[162] A. T. Balaei, N. Alam, and A. G. Dempster. An instantaneous lane-level positioning using DSRC carrier frequency offset. *IEEE Transactions on Intelligent Transportation Systems*, 13:1566–1575, 2012.

[163] N. Alam, A. T. Balaei, and A. G. Dempster. A DSCRC doppler-based cooperative positioning enhancement for vehicular networks with GPS availability. *IEEE Transactions on Vehicular Technology*, 60:4462–4470, 2011.

[164] N. Patwari, J. N. Ash, S. Kyperountas, A. O. Hero, R. L. Moses, and N. S. Correal. Locating the nodes: Cooperative localization in wireless sensor networks. *IEEE Signal Processing Magazine*, 22:54–69, 2005.

[165] S. S. Nair, S. Schellenberg, J. Seitz, and M. Chatterjee. Hybrid spectrum sharing in dynamic spectrum access networks. In *The International Conference on Information Networking 2013 (ICOIN)*, 324–329. IEEE, 2013.

[166] M. Naor and M. Yung. Public-key cryptosystems provably secure against chosen ciphertext attacks. In *Proceedings of the Twenty-Second Annual ACM Symposium on Theory of Computing*, ACM, Baltimore, MD, 427–437, 1990.

[167] H. P. Ngallemo, W. Ajib, and H. Elbiaze. Dynamic spectrum access analysis in a multi-user cognitive radio network using Markov chains. In *2012 International Conference on Computing, Networking and Communications (ICNC)*, IEEE, Maui, HI, 1113–1117, 2012.

[168] B. Niehoefer, R. Burda, C. Wietfeld, and F. Bauer. GPS community map generation for enhanced routing methods based on trace-collection by mobile phones. In *First International Conference on Advances in Satellite and Space Communications, 2009. SPACOMM 2009*, IEEE, Colmar, France, 156–161, 2009.

[169] D. Niyato, X. Lu, and P. Wang. Adaptive power management for wireless base stations in a smart grid environment. *IEEE Wireless Communications*, 19(6):44–51, 2012.

[170] C.-H. Oh. Location estimation using space-time signal processing in RFID wireless sensor networks. *International Journal of Distributed Sensor Networks*, 2013:8, 2013.

[171] R. Oliveira, L. Bernardo, and P. Pinto. Performance analysis of the IEEE 802.11 distributed coordination function with unicast and broadcast traffic. In *2006 IEEE 17th International Symposium on Personal, Indoor and Mobile Radio Communications*, IEEE, Helsinki, Finland, 1–5, September 2006.

[172] P. Olszewski and L. Xie. Modeling the effects of road pricing on traffic in Singapore. *Transportation Research Part A: Policy and Practice*, 39:755–772, 2005.

[173] R. Ortalo, Y. Deswarte, and M. Kaâniche. Experimenting with quantitative evaluation tools for monitoring operational security. *IEEE Transactions on Software Engineering*, 25(5):633–650, 1999.

[174] D. Haley, P. Alexander, and A. Grant. Cooperative intelligent transport systems: 5.9-GHz field trials. *Proceedings of the IEEE*, 99:1213–1235, 2011.

[175] Y. C. Ho, P. Tientrakool, and N. F. Maxemchuk. Highway capacity benefits from using vehicle-to-vehicle communication and sensors for collision avoidance. *2011 IEEE Vehicular Technology Conference (VTC Fall)*, IEEE, San Francisco, CA, 1–5, 2011.

[176] N. Pavlidou, A. J. H. Vinck, J. Yazdani, and B. Honary. Power line communications: State of the art and future trends. *IEEE Wireless Communications Magazine*, 41(4):34–40, 2003.

[177] A. C. Pigou. *Some Fallacies in the Interpretation of Social Cost.* MacMillan, London, 1920.

[178] O. Pink and C. Stiller. Automated map generation from aerial images for precise vehicle localization. In *2010 13th International IEEE Conference on Intelligent Transportation Systems (ITSC)*, 1517–1522, 2010.

[179] H. Poor. *An Introduction to Signal Detection and Estimation, 2nd Ed.* Springer, New York, 1994.

[180] J. Prieto, S. Mazuelas, A. Bahillo, P. Fernandez, R. M. Lorenzo, and E. J. Abril. Adaptive data fusion for wireless localization in harsh environments. *IEEE Transactions on Signal Processing*, 60(4):1585–1596, 2012.

[181] SPITS project communication stack. https://www.spits-project.com, 2009, 2011.

[182] F. Pukelsheim. The three sigma rule. *The American Statistician*, 48(48):88–91, 1994.

[183] T. Tamaki, T. Uta, N. Matsuzawa, R. Yamasaki, A. Ogino, and T. Kato. TDOA location system for IEEE 802.11 b WLAN. *Wireless Communications and Networking Conference, 2005 IEEE*, IEEE, New Orleans, LA, 2338–2343, 2005.

[184] B. D. Rao and K. Hari. Performance analysis of root-music. *IEEE Transactions on Acoustics, Speech and Signal Processing*, 37:1939–1949, 1989.

[185] T. S. Rappaport. *Wireless Communications: Principles and Practice*, Vol. 2. Prentice Hall, Upper Saddle River, NJ, 1996.

[186] T. S. Rappaport, J. H. Reed, and B. D. Woerner. Position location using wireless communications on highways of the future. *IEEE Communications Magazine*, 34(10):33–41, 1996.

[187] P. D. Robinson and A. J. Wathen. Variational bounds on the entries of the inverse of a matrix. *IMA Journal of Numerical Analysis*, 12(4):463–486, 1992.

[188] R. Roy and T. Kailath. ESPRIT-estimation of signal parameters via rotational invariance techniques. *IEEE Transactions on Acoustics, Speech and Signal Processing*, 37:984–995, 1989.

[189] M. Ruff. Evolution of local interconnect network (LIN) solutions. In *Proceedings of IEEE Vehicular Technology Conference (VTC)*, IEEE, Orlando, FL, Vol. 5, 3382–3389, 2003.

[190] S. Yin, Z. Li, Y. Zhang, D. Yao, Y. Su, and L. Li. Headway distribution modeling with regard to traffic status. In *Intelligent Vehicles Symposium*, Xian, 1057–1062, 2009.

[191] S. C. Ergen, H. S. Tetikol, M. Kontik, R. Sevlian, R. Rajagopal, and P. Varaiya. RSSI-fingerprinting-based mobile phone localization with route constraints. *IEEE Transactions on Vehicular Technology*, 63:423–428, 2014.

[192] S. Ji, D. Dunson, and L. Carin. Multitask compressive sensing. *IEEE Transactions on Signal Processing*, 57:92–106, 2009.

[193] Y. Xue, S. Ji, and L. Carin. Bayesian compressive sensing. *IEEE Transactions on Signal Processing*, 56:2346–2356, 2008.

[194] S. Mazuelas, F. A. Lago, J. Blas, A. Bahillo, P. Fernandez, R. M. Lorenzo, and E. J. Abri. Prior NLOS measurement correction for positioning in cellular wireless networks. *IEEE Transactions on Vehicular Technology*, 58:2585–2591, 2009.

[195] S. Ullah, W. Alsalih, A. Alsehaim, and N. Alsadhan. A review of tags anti-collision and localization protocols in RFID networks. *Journal of Medical Systems*, 36:4037–4050, 2012.

[196] F. Sabahi. The security of vehicular ad hoc networks. In *2011 Third International Conference on Computational Intelligence, Communication Systems and Networks (CICSyN)*, IEEE, Bali, Indonesia, 338–342, 2011.

[197] R. O. Saber. Flocking for multi-agent dynamic systems: Algorithms and theory. *IEEE Transactions on Automatic Control*, 51(3):401–402, 2006.

[198] R. O. Saber and R. M. Murray. Flocking with obstacle avoidance: Cooperation with limited communicaiton in mobile networks. *Proceedings of 42nd IEEE Conference on Decision and Control*, IEEE, Maui, HI, 2:2022C–2028, 2003.

[199] SAE. Class A Application Definition. J2507-1, 1997. http://standards.sae.org/j2057/1_200609/.

[200] A. H. Sayed, A. Tarighat, and N. Khajehnouri. Network-based wireless location: Challenges faced in developing techniques for accurate wireless location information. *IEEE Signal Processing Magazine*, 22(4):24–40, 2005.

[201] A. Schiffer. Statistical channel and noise modeling of vehicular DC-lines for data communication. In *Proceedings of IEEE Vehicular Technology Conference (VTC)*, IEEE, Boston, MA, Vol. 1, 158–162, 2000.

[202] T. M. Schmidl and D. C. Cox. Robust frequency and timing synchronization for OFDM. *IEEE Transactions on Communications*, 45(12):1613–1621, 1998.

[203] R. Schmidt. Multiple emitter location and signal parameter estimation. *IEEE Transactions on Antennas and Propagation*, 34:276–280, 1986.

[204] R. Sengupta, S. Rezaei, S. E. Shladover, D. Cody, S. Dickey, and H. Krishnan. Cooperative collision warning systems: Concept definition and experimental implementation. *Journal of Intelligent Transportation Systems*, 11(3):143–155, 2007.

[205] S. Sengupta, M. Chatterjee, and K. Kwiat. A game theoretic framework for power control in wireless sensor networks. *IEEE Transactions on Computers*, 59(2):231–242, 2010.

[206] C. E. Shannon. Communication in the presence of noise. *Proceedings of the IRE*, 37(1):10–21, 1949.

[207] C. E. Shannon. A mathematical theory of communication. *ACM SIGMOBILE Mobile Computing and Communications Review*, 5(1):3–55, 2001.

[208] S. A. Sharna and M. M. Murshed. Performance analysis of vertical handoff algorithms with QoS parameter differentiation. In *IEEE International Conference on High Performance Computing and Communications, HPCC 2010, 1–3 September 2010*, 623–628, Melbourne, Australia, 1–3 September, 2010.

[209] W. Shi, S. Shen, and Y. Liu. Automatic generation of road network map from massive GPS, vehicle trajectories. In *12th International IEEE Conference on Intelligent Transportation Systems, 2009. ITSC '09*. 1–6, 2009.

[210] K. P. Shih, Y. D. Chen, and C. C. Chang. A physical/virtual carrier-sense-based power control MAC protocol for collision avoidance in wireless ad hoc networks. *IEEE Transactions on Parallel and Distributed Systems*, 22(2):193–207, February 2011.

[211] F. Shih-Hau. Cross-provider cooperation for improved network-based localization. *IEEE Transactions on Vehicular Technology*, 62:297–305, 2013.

[212] S. E. Shladover and S.-K. Tan. Analysis of vehicle positioning accuracy requirements for communication-based cooperative collision warning. *Journal of Intelligent Transportation Systems*, 10(3):131–140, 2006.

[213] F. Simonot-Lion and Y. Q. Song. Design and validation process of in-vehicle embedded electronic systems. s.l. : Online material, https://www.researchgate.net/publication/228622171_Design_and_validation_process_of_in-vehicle_embedded_electronic_systems, 2005.

[214] B. Sklar. Rayleigh fading channels in mobile digital communication systems. I. characterization. *IEEE Communications Magazine*, 35(7):90–100, 1997.

[215] GCDC. Grand Cooperative driving challenge communication stack. https://www.gcdc.net, 2011.

[216] G. Staple and K. Werbach. The end of spectrum scarcity [spectrum allocation and utilization]. *IEEE Spectrum*, 41(3):48–52, 2004.

[217] P. Stoica and R. L. Moses. *Spectral Analysis of Signals*. Pearson/Prentice Hall, Upper Saddle River, NJ. 2005.

[218] P. Stoica and K. Sharman. Maximum likelihood methods for direction-of-arrival estimation. *IEEE Transactions on Acoustics, Speech and Signal Processing*, 38:1132–1143, 1990.

[219] G. Sun, J. Chen, W. Guo, and K. J. R. Liu. Signal processing techniques in network-aided positioning: A survey of state-of-the-art positioning designs. *IEEE Signal Processing Magazine*, 22(4):12–23, 2005.

[220] J. Sun, C. Zhang, Y. Zhang, and Y. Fang. An identity-based security system for user privacy in vehicular ad hoc networks. *IEEE Transactions on Parallel and Distributed Systems*, 21(9):1227–1239, 2010.

[221] J. A. K. Suykens, J. Vandewalle, and B. De Moor. Optimal control by least squares support vector machines. *Neural Networks*, 14(1):23–35, 2001.

[222] N. Taherinejad, R. Rosales, L. Lampe, and S. Mirabbasi. Channel characterization for power line communication in a hybrid electric vehicle. In *Proceedings of IEEE International Symposium on Power Line Communications and Its Applications (ISPLC)*, 328–333, IEEE, Beijing, China, March 2012.

[223] N. Taherinejad, R. Rosales, S. Mirabbasi, and L. Lampe. A study on access impedance for vehicular power line communications. In *Proceedings of IEEE International Symposium on Power Line Communications and Its Applications (ISPLC)*, 440–445, Udine, Italy, April 2011.

[224] A. Tang and A. Yip. Collision avoidance timing analysis of DSRC-based vehicles. *Accident Analysis and Prevention*, 42(1):182–195, 2010.

[225] H. G. Tanner, A. Jadbabaie, and G. J. Pappas. Flocking in fixed and switching networks. *IEEE Transactions on Automatic Control*, 52(5):863–868, 2007.

[226] G. T. Taoka. Break reaction times of unalerted drivers. *ITE Journal*, 59(3):19–21, 1989.

[227] T. S. Dao, K. Y. K. Leung, C. M. Clark, and J. P. Huissoon. Markov-based lane positioning using intervehicle communication. *IEEE Transactions on Intelligent Transportation Systems*, 8(4):641–650, 2007.

[228] P. Tientrakool, Y. C. Ho, and N. F. Maxemchuk. Highway capacity benefits from using vehicle-to-vehicle communicaiton and sensors for collision avoidance. In *Vehicular Technology Conference (VTC Fall)*, IEEE, San Francisco, CA, 1–5, 2011.

[229] National Transportation Statistics. 2012. Bureau of transportation statistics. U.S. Department of Transportation. Washington, DC, 103–109, 2012.

[230] Y.-C. Tseng, S.-Y. Ni, Y.-S. Chen, and J.-P. Sheu. The broadcast storm problem in a mobile ad hoc network. *Wireless Networks*, 8(2–3):153–167, 2002.

[231] S. Tuohy, M. Glavin, C. Hughes, E. Jones, M. Trivedi, and L. Kilmartin. Intra-vehicle networks: A review. *IEEE Transactions on Intelligent Transportation Systems*, 99:1–12, 2014.

[232] H. Urkowitz. Energy detection of unknown deterministic signals. *Proceedings of the IEEE*, 55(4):523–531, 1967.

[233] A. B. Vallejo-Mora, J. J. Sánchez-Martínez, F. J. Cañete, J. A. Cortés, and L. Díez. Characterization and evaluation of in-vehicle power line channels. In *Proceedings of IEEE Global Telecommunications Conference (GLOBECOM)*, 1–5, December 2010.

[234] H. L. Van Trees. *Detection, Estimation, and Modulation Theory*. John Wiley & Sons, New York, 2004.

[235] T. Vanderbilt. Heading for the cloud. *ITS Magazine*, 2011(3):10–11, 2011.

[236] W. S. Vickrey. Point of view: Principles and applications of congestion pricing. *TR News*, 167:4–5, 1993.

[237] W. Viriyasitavat, M. Boban, H. M. Tsai, and A. Vasilakos. Vehicular communications: Survey and challenges of channel and propagation models. *IEEE Vehicular Technology Magazine*, 10(2):55–66, 2015.

[238] I. Moerman, W. Vandenberghe, and P. Demeester. Approximation of the IEEE 802.11p standard using commercial off-the-shelf IEEE 802.11a hardware. *2011 11th International Conference on ITS Telecommunications*, IEEE, Graz, Austria, 21–26, 2011.

[239] A. A. Walters. The theory and measurement of private and social cost of highway congestion. *Econometrica*, 29:676–699, 1961.

[240] B. Wang, Z. Ji, and K. J. R. Liu. Primary-prioritized Markov approach for dynamic spectrum access. In *IEEE Symposium on New Frontiers in Dynamic Spectrum Access Networks (DySPAN07)*, IEEE, Dublin, Ireland, 507–515, 2007.

[241] D. C. Wang and J. P. Thompson. Apparatus and method for motion detection and tracking of objects in a region for collision avoidance utilizing a real-time adaptive probabilistic neural network, http://xueshu.baidu.com/s?wd=paperuri%3A%2850eafe9949d733cf729e68d96f643acd%29&filter=sc_long_sign&tn=SE_xueshusource_2kduw22v&sc_vurl=http%3A%2F%2Fwww.freepatentsonline.com%2F5613039.html&ie=utf-8&sc_us=11714964232604303703, 1997.

[242] J. Wang and D. Katabi. Dude, where's my card?: RFID positioning that works with multipath and non-line of sight. *Proceedings of the ACM SIGCOMM 2013 Conference on SIGCOMM*, Hong Kong, China, 2013.

[243] J. Wang, Y. Liu, and K. Deng. Modelling and simulating worm propagation in static and dynamic traffic. *Intelligent Transport Systems*, 8(2):155–163, 2014.

[244] J. Wang, L. Zhang, D. Zhang, and K. Li. An adaptive longitudinal driving assistance system based on driver characteristics. *IEEE Transactions on Intelligent Transportation Systems*, 14(1), 2013.

[245] J. Wang, F. Adib, R. Knepper, D. Katabi, and D. Rus. RF-compass: Robot object manipulation using RFIDs, 2013.

[246] J. B. Wang, M. Chen, X. Wan, and C. Wei. Ant-colony-optimization-based scheduling algorithm for uplink CDMA nonreal-time data. *IEEE Transactions on Vehicular Technology*, 58(1):231–241, 2009.

[247] S. Wang, C. Fan, C.-H. Hsu, Q. Sun, and F. Yang. A vertical handoff method via self-selection decision tree for internet of vehicles. *IEEE System Journal*, 10(3):1183–1192, 2016.

[248] Y. Wang, Y. Zhu, Z. He, Y. Yue, and Q. Li. Challenges and opportunities in exploiting large-scale GPS probe data. *Hp Labs*, http://www.hpl.hp.com/techreports/2011/HPL-2011-109.pdf, 2011.

[249] Y. Wang, X. Duan, D. Tian, G. Lu, and H. Yu. Throughput and delay limits of 802.11 and its influence on highway capacity. *Procedia—Social and Behavioral Sciences*, 96:2096–2104, 2013.

[250] J. G. Wardrop. Some theoretical aspects of road traffic research. In *Proceedings of Institution of Civil Engineers Part II*, Vol. 1, 325–378, 1952.

[251] A. Wasef, R. Lu, X. Lin, and X. Shen. Complementing public key infrastructure to secure vehicular ad hoc networks [security and privacy in emerging wireless networks]. *IEEE Wireless Communications*, 17(5):22–28, 2010.

[252] D. Washington. Highway capacity manual. Special Report, 1:5–7, 1985.

[253] W. Li and Y. Jia. Consensus-based distributed multiple model UKF for jump Markov nonlinear systems. *IEEE Transactions on Automatic Control*, 57(1):227–233, 2012.

[254] W. Li and Y. Jia. Location of mobile station with maneuvers using an IMM-based cubature Kalman filter. *IEEE Transactions on Industrial Electronics*, 59(11):4338–4348, 2012.

[255] W. Li, Y. Jia, J. Du, and J. Zhang. Distributed multiple-model estimation for simultaneous localization and tracking with NLOS mitigation. *IEEE Transactions on Vehicular Technology*, 62(6):2824–2830, 2013.

[256] T. L. Willke, P. Tientrakool, and N. F. Maxemchuk. A survey of inter-vehicle communication protocols and their applications. *IEEE Communications Surveys & Tutorials*, 11(2):3–20, 2009.

[257] M. Z. Win, A. Conti, S. Mazuelas, Y. Shen, W. M. Gifford, D. Dardari, and M. Chiani. Network localization and navigation via cooperation. *IEEE Communications Magazine*, 49(5):56–62, 2011.

[258] Ath5k wireless driver. https://linuxwireless.org/en/users/Drivers/ath5k, 2010.

[259] J. Wu and F. Dai. Broadcasting in ad hoc networks based on self-pruning. In *INFOCOM 2003, Twenty-Second Annual Joint Conference of the IEEE Computer and Communications. IEEE Societies*, IEEE, San Franciso, CA, Vol. 3, 2240–2250, March 2003.

[260] M. Y. Wu, S. J. Ma, and W. Shu. Scheduled video delivery a scalable on-demand video delivery paradigm. *IEEE Transactions on Multimedia*, 8(1):179–187, 2007.

[261] T. Wu, J. Wang, Y. Liu, W. Deng, and J. Deng. Image-based modeling and simulating physical channel for vehicle-to-vehicle communicaitons. *Ad Hoc Networks*, 19:75–91, 2014.

[262] D. Zhao, X. Huang, and H. Peng. Empirical study of DSRC performance based on safety pilot model deployment data. *Transactions on Intelligent Transportation Systems*, 99:1–10, 2016.

[263] Z. Wang, X. Wang, and B. O'Dea. A TOA-based location algorithm reducing the errors due to non-line-of-sight (NLOS) propagation. *IEEE Transactions on Vehicular Technology*, 52:112–116, 2003.

[264] Y. Xing, R. Chandramouli, and S. Mangold. Dynamic spectrum access in open spectrum wireless networks. *IEEE Journal on Selected Areas in Communications*, 24(3):626–637, 2006.

[265] J. Xiong and K. Jamieson. Arraytrack: A fine-grained indoor location system. *ACM HotMobile*, http://xueshu.baidu.com/s?wd=paperuri%3A%280ff9cbd225b1689045300871cf325195%29&filter=sc_long_sign&tn=SE_xueshusource_2kduw22v&sc_vurl=http%3A%2F%2Fdl.acm.org%2Fcitation.cfm%3Fid%3D2482626.2482635&ie=utf-8&sc_us=10168236717634003154, 2012.

[266] J. Xu, Y. Liu, J. Wang, W. Deng, and T. Ernst. Vike: Vehicular ike for context-awareness. *Wireless Networks*, 21(4):1343–1362, 2015.

[267] A. LaMarca, Y.-C. Cheng, Y. Chawathe, and J. Krumm. Accuracy characterization for metropolitan-scale Wi-Fi localization. *Proceedings of the 3rd International Conference on Mobile Systems, Applications, and Services*, Seattle, Washington, DC, 2005.

[268] Y. Gu, A. Lo, and I. Niemegeers. A survey of indoor positioning systems for wireless personal networks. *IEEE Communications Surveys & Tutorials*, 11:13–32, 2009.

[269] Y. Ma, M. Chowdhury, A. Sadek, and M. Jeihani. Real-time highway traffic condition assessment framework using vehicle-infrastructure integration (VII) with artificial intelligence (AI). *IEEE Transactions on Intelligent Transportation Systems*, 10:615–627, 2009.

[270] Y. Xie, Y. Wang, B. Wu, X. Yang, P. Zhu, and X. You Localization by hybrid TOA, AOA and DSF estimation in nlos environments. In *Vehicular Technology Conference Fall (VTC 2010-Fall), 2010 IEEE 72nd*, 1–5, 2010.

[271] Y. Yao, L. Rao, and X. Liu. Performance and reliability analysis of IEEE 802.11p safety communication in a highway environment. *IEEE Transactions on Vehicular Technology*, 62:4198–4212, 2013.

[272] Y. Yabuuchi, D. Umehara, M. Morikura, T. Hisada, S. Ishiko, and S. Horihata. Measurement and analysis of impulsive noise on in-vehicle power lines. In *Proceedings of IEEE International Symposium on Power Line Communications and Its Applications (ISPLC)*, IEEE, Rio de Janeiro, 325–330, March 2010.

[273] X. Yan, Y. A. Şkercilu, and S. Narayanan. A survey of vertical handover decision algorithms in fourth generation heterogeneous wireless networks. *Computer Networks*, 54(11):1848–1863, 2010.

[274] H. Yang and H. J. Huang. Principle of marginal-cost pricing: How does it work in a general network? *Transportation Research Part A: Policy and Practice*, 32:45–54, 1998.

[275] H. Yang and H. J. Huang. *Mathematical and Economic Theory of Road Pricing*. Elsevier, Oxford, UK, 2005.

[276] R. Yang, G. W. Ng, and Y. Bar-Shalom. Tracking/fusion and deghosting with Doppler frequency from two passive acoustic sensors. In *2013 16th International Conference on Information Fusion (FUSION)*, IEEE, Istanbul, Turkey, 1784–1790.

[277] X. Yang and J. Rosdahl. Throughput and delay limits of IEEE 802.11. *IEEE Communications Letters*, 6:355–357, 2002.

[278] X. Yang, J. Liu, N. F. Vaidya, and F. Zhao. A vehicle-to-vehicle communication protocol for cooperative collision warning. In *International Conference on Mobile and Ubiquitous Systems: NETWORKING and Services, 2004. MOBIQUITOUS*, IEEE, Istanbul, Turkey, 114–123, 2004.

[279] Z. Yang, Z. Zhou, and Y. Liu. From RSSI to CSI: Indoor localization via channel response. *ACM Computing Surveys (CSUR)*, 46(2):25, 2013.

[280] L.-Y. Yeh, Y.-C. Chen, and J.-L. Huang. PAACP: A portable privacy-preserving authentication and access control protocol in vehicular ad hoc networks. *Computer Communications*, 34(3):447–456, 2011.

[281] S. Yuan, S. Mazuelas, and M. Z. Win. Network navigation: Theory and interpretation. *IEEE Journal on Selected Areas in Communications*, 30(9):1823–1834, 2012.

[282] S. Yuan and M. Z. Win. Fundamental limits of wideband localization—Part I: A general framework. *IEEE Transactions on Information Theory*, 56(10):4956–4980, 2010.

[283] S. Zeadally, R. Hunt, Y.-S. Chen, A. Irwin, and A. Hassan. Vehicular ad hoc networks (VANETs): status, results, and challenges. *Telecommunication Systems*, 50(4):217–241, 2012.

[284] R. Zekavat and R. M. Buehrer. *Handbook of Position Location: Theory, Practice and Advances*, Vol. 27. John Wiley & Sons, New York, 2011.

[285] M. Zekri, B. Jouaber, and D. Zeghlache. A review on mobility management and vertical handover solutions over heterogeneous wireless networks. *Computer Communications*, 35(17):2055–2068, 2012.

[286] G. Zhang, S. Yan, and Y. Wang. Simulation-based investigation on high occupancy toll lane operations for Washington State Route 167. *ASCE Journal of Transportation Engineering*, 135(10):677–686, 2009.

[287] Y. Zhang. GroRec: A group-centric intelligent recommender system integrating social, mobile and big data technologies. *IEEE Transactions on Services Computing*, 9(5):786–795, 2016.

[288] Y. Zhang, M. Chen, S. Mao, L. Hu, and V. C. M. Leung. CAP: Community activity prediction based on big data analysis. *IEEE Network*, 28(4):52–57, 2014.

[289] Y. Zhang, Y. K. Ji, W. Shu, and M.Y. Wu. Capacity-aware mechanisms for service overlay design. *IEEE Globecom*, 135(10), 2008.

[290] Z. Zhang, D. Yang, T. Zhang, Q. He, and X. Lian. A study on the method for cleaning and repairing the probe vehicle data. *IEEE Transactions on Intelligent Transportation Systems*, 14(1):419–427, 2013.

[291] Q. Zhao and B. M. Sadler. A survey of dynamic spectrum access: Signal processing and networking perspectives. *IEEE Signal Processing Magazine*, 24(3):79–89, 2007.

[292] C. Zheng and D. C. Sicker. A survey on biologically inspired algorithms for computer networking. *IEEE Communications Surveys & Tutorials*, 15(3):1160–1191, 2013.

[293] K. Zheng, H. Meng, P. Chatzimisios, L. Lei, and X. Shen. An SMDP-based resource allocation in vehicular cloud computing systems. *IEEE Transactions on Industrial Electronics*, 62(12):7920–7928, 2015.

[294] K. Zheng, Q. Zheng, P. Chatzimisios, W. Xiang, and Y. Zhou. Heterogeneous vehicular networking: A survey on architecture, challenges, and solutions. *IEEE Communications Surveys & Tutorials*, 17(4):2377–2396, 2015.

[295] W. Zhonghai and S. A. Zekavat. A novel semidistributed localization via multinode TOA-DOA fusion. *IEEE Transactions on Vehicular Technology*, 58:3426–3435, 2009.

[296] X. Zhou, R. K. Ganti, J. G. Andrews, and A. Hjorungnes. On the throughput cost of physical layer security in decentralized wireless networks. *IEEE Transactions on Wireless Communications*, 10(8):2764–2775, 2011.

[297] H. Zinner, O. Kleineberg, C. Boiger, and A. Grzemba. Automotive requirements and definitions. s.l. : Online presentation slides, http://www.ieee802.org/1/files/public/docs2012/new-avb-zinner-boiger-kleineberg-automotive-requirements-0112-v01.pdf, 2012.

Index

Page numbers followed by *f* indicate figures; those followed by *t* indicate tables.

A

ABC concept, see Always best connected concept
ABS, see Antilock braking systems
ADC, see Analog-to-Digital Converter
Additive white Gaussian noise (AWGN), 32
Advanced Encryption Standard (AES) encryption, 189, 194
AES encryption, see Advanced Encryption Standard encryption
Always best connected (ABC) concept, 72
Analog-to-Digital Converter (ADC), 32
Angular scope discretization, 149*f*
Antilock braking systems (ABS), 23
Arduino, 13, 14*f*
Automotive networking applications, 4–7
  Class A, 4, 5*f*
  Class B, 4–5, 5*f*
  Class C, 5–6, 6*f*
  functional classes, comparisons of, 6–7, 6*t*
Automotive networking protocols, 7–10
  automotive communication development roadmap, 12*f*
  automotive network architecture, 10*f*
  controller area network (CAN), 7–8
  Ethernet, 9–10
  FlexRay, 8–9
  local interconnect network (LIN), 7
  media-oriented system transport (MOST), 9
  technical comparisons, 10–12, 11*t*
Automotive network test bed implementation, 12–15, 13*f*
  aim of, 12
  head/dashboard unit, 14–15, 14*f*
  key components of, 13
  onboard/V2X unit, 14, 14*f*
  performance outcomes from, 15–19
    Ethernet link test results, 16–17, 16–17*f*
    further work areas for test bed, 19
    Wi-Fi link test results, 17–19, 18–19*f*
  roadside unit, 15
  ZigBee gateway, 14*f*, 15
  ZigBee units, 14*f*, 15
AWGN, see Additive white Gaussian noise

B

Backbone network, 12–13
Backoff period, 10
Band-pass filter (BPF), 32
Bayesian compressive sensing (BCS) technique, 135
BCS technique, see Bayesian compressive sensing technique
Beacon to Beacon InterFrame Space (B2BIFS), 50
BEB algorithm, see Binary Exponential Backoff algorithm
BER, see Bit error rate
Binary Exponential Backoff (BEB) algorithm, 10
Binary phase shift keying (BPSK) modulation, 98

Bio-inspired QoS-oriented handover model
  experiment and analysis
    applications settings, 88*t*
    model settings, 89*t*
    network conditions settings, 88*t*
    numerical evaluation, 89–93, 90–92*f*
    settings on weights, 89*t*
    simulation scenario, 87, 87*f*
    simulation settings, 87–89
  in heterogeneous wireless networks, 76–84
  QoS-oriented handover decision, 81–84
    extended attractor selection model, 82, 82*f*
    quantification of terminal's QoS satisfaction, 83–84, 84*f*
  system model, 76–81
    basic attractor selection model, 76–77
    extended attractor selection model, 77–78
    impact of random noise variability, 81*f*
    model validation and discussion, 78–81
    problem formulation, 76
    varying activity in simulations, 80*f*
Bit error rate (BER), 50
BPF, see Band-pass filter
BPSK, see Binary phase shift keying modulation
Broadcom's BroadR-Reach 10/100 PHY Ethernet module, 9
B-spline fitting, 116, 127

C

CAM packet, see also Cooperative awareness message packet
  accuracy of, 101, 102, 107
  bit error rate of, 98
  data in, 101–102
  data transmission method in, 97–98
  packet header, 98, 99
  packet transmission delay (T_{delay}), 100*f*, 101, 103
  price of, 96
  purpose of, 101
CAN-based media access control protocol, 21–45
  impulse noise filtering, 31–33
  introduction to, 21–23
  MAC layer, 25–26
  multichannel contention resolution method, 26–27
  numerical results, 36–45
  performance analysis
    channel selection probabilities, 30*f*
    parameters used for, 28*t*
    under perfect sensing, 27–31
    under presence of sensing errors, 31–36, 34*f*
    success probability, maximizing, 30*t*
  physical layer, 23–25, 24*f*
  sensing errors, impact of, 33–36
  system assumption, 26–27
  throughput evaluation, 31
CAN, see Controller area network
Carrier frequency offset (CFO), 133
Carrier sense multiple access (CSMA) approach, 48, 61
Carrier sense multiple access/collision detection (CSMA/CD), 8, 9, 26
Carrier sense multiple access with collision avoidance (CSMA/CA), 22, 97, 132
Carrier sensing, 9
CCH, see Control channel interval
CCo, see Central coordinator
CDFs, see Cumulative distribution functions
CDMA, see Code division multiple access
CDR, see Contention detection and resolution
Centerline fitting, 116, 126–130
  B-spline, fitting result of, 127*f*
  final fitting centerlines of test area, 130*f*
  fitting algorithm, evaluating, 126
  leftmost marked part, 128*f*
  lines fitted by two methods, deviation of, 129*f*
  LS-SVM, fitting result of, 127*f*
  middle marked part, 128*f*
  nearest point on road for each node, 126*f*

original lines, deviation of, 129*f*
rightmost marked part, 129*f*
Central coordinator (CCo), 49
CFO, see Carrier frequency offset
Channel capacity model, 191–193
Charging station (CCo), 50
CIFS, see Contention InterFrame Space
Class A automotive application, 4, 5*f*
Class B automotive application, 4–5, 5*f*
Class C automotive application, 5–6, 6*f*
Code division multiple access (CDMA), 190
Collision probability, analysis of, 239–242
communication ranges on, influence of, 240–242
in different traffic scenarios, 242–243*f*
initial vehicle velocity on, influence of, 239, 240*f*
vehicle numbers on, influence of, 239–240, 241*f*
Connected vehicles (CVs), 132, 210
Connected vehicles-enhanced DIMM-KF (CV-DIMM-KF), for mobile vehicle localization, 172–174
algorithm for vehicular positioning, 175
combination, 174
interaction, 173
mixing probabilities calculation, 172–173
mode update and prediction steps, 173–174
Connected vehicles-enhanced IMM-KF (CV-IMM-KF), for vehicular positioning, 163–165, 164*f*
algorithm for vehicular positioning, 165
estimates combination, 165
for estimating the TVs state, 164–165
and GPS error comparison, 167*t*
and GPS performance in positioning error, 168*f*
mixing probabilities, 164
mode update and prediction steps, 164–165
Contention detection and resolution (CDR), 22
Contention InterFrame Space (CIFS), 50
Contention window (CW) size, reducing, 49*t*
Continuous-time Markov chains (CTMC), 60, 61
Control channel (CCH) interval, 72, 72*f*
Controller area network (CAN), 7–8, 14, 25, 48, 49
Cooperative awareness message (CAM) packet, 96, 102
accuracy of, 101, 102, 107
bit error rate of, 98
data in, 101–102
data transmission method in, 97–98
packet header, 98, 99
packet transmission delay (T_{delay}), 100*f*, 101, 103
price of, 96
purpose of, 101
Cooperative multivehicle mobility model, 215–222
intervehicular communication
on mobility of multiple vehicles, 235–238, 236–237*f*
on vehicle safety and traffic efficiency, 238–246
simulation and verification, 231–235, 233*f*, 234*f*, 235*t*
Cooperative positioning (CP) method, 160
Cooperative vehicle infrastructure systems (CVIS), 132, 160
Cooperative vehicles, model of, 215–231
bump function with different parameters, 218*f*
cooperative multivehicle mobility model, 215–222
impact of, 214
improved cooperative multivehicle mobility model, 226–231
interaction range, 228*f*
potential functions, variation of, 220*f*
sigmoid function with different parameters, 229*f*
stability analysis of, 222–226
Cramér-Rao lower bound (CRLB), 175
CRC, see Cyclic redundancy check
CRLB, see Cramér-Rao lower bound

Cross-layer multihop broadcasting, 95–111
IEEE 802.11 broadcast service, analytic model for, 96–97
introduction to, 95–96
meanings and values of parameters, 104*t*
numerical results, 103–111
system model, 97–103
analytic model, 97–100, 100*f*
decision-making process, 100–103, 102*f*
VANET, multicast methods for, 97
CSI, see Wireless channel state information
CSMA, see Carrier sense multiple access approach
CTMC, see Continuous-time Markov chains
Cumulative distribution functions (CDFs), 156, 156*f*
CVIS, see Cooperative vehicle infrastructure systems
CVs, see Connected vehicles
Cyber space
CPS architecture, 206–207, 206*f*
social-network-enabled architecture, 207–209
activity manager, 208
cyber space and physical space, interaction between, 208–209
edges, mapping of, 208
m-nodes, mapping of, 207–208
s-node, mapping of, 208
transportation manager, 208–209
Cyclic redundancy check (CRC), 8, 48

D

Data cleaning, 117
DCC, see Decentralized Congestion Control
DCF interframe space (DIFS), 98
DCF mode, see Distributed coordination function mode
Decentralized Congestion Control (DCC), 188
Decision-making multicast, 100–103, 102*f*
under different multicast ranges (r), 106–107*f*
under different packet lengths (L_P), 108*f*
under different transmission data rates (R_d), 109*f*
under different vehicle densities, 105*f*
under different vehicle speeds (v), 110*f*
flow diagram of, 104*f*
Dedicated short-range communication (DSRC), 71–72, 131–158, 160, see also DSRC-based vehicular positioning enhancement model
Bayesian compressive sensing (BCS), 135
embedded self-adaptive configuration scheme, 75–76
energy threshold analysis, 149–151
highway capacity limits of condition using V2V communication, 138–143, 140–142*f*
hybrid TOA/DOA
Cramér-Rao bounds for, 155–156
data model, 146–147
estimation, 134–135, 135*f*
location procedure, 148–149
MT-BCS–based location method, 151–155
implementation and field testing, 143–146
introduction to, 132–134
multitask BCS, based DOA estimation, 147–148
network performance of, 136–143
performance evaluation test bed, 144–146
self-adaptive V2V communication system with, 73–76, 74*f*
simulation for real-time vehicular positioning, 156–158, 157*f*
throughput and delay limits of 802.11p, 136–138
traffic capacity, effect on, 136–143
V2X communication system, 143
vehicular positioning, 134
Delay lower limit (DLL), 132
Delay tolerant networks (DTNs), 97
DFS, see Doppler frequency shift
Dial-a-Ride system, 205
Differential cryptanalysis, 194
DIFS, see DCF interframe space
Direction of arrival (DOA), 133, 160
Discrete-time Markov chain, 98

Distributed coordination function (DCF) mode, 97, 136
Distributed interactive multiple-model-based Kalman filter (DIMM-KF), 161
Distributed multiple-model Kalman filter (KF), 168–185
  CV-enhanced DIMM-KF for mobile vehicle localization, 172–174, 172*f*
  general performance analysis, 174–180
  Monte Carlo results, 181–185
  simulation setup, 180–181
  single-trial analysis, 181
  system model and localization enhancement, 168–174
DLL, see Delay lower limit
DOA, see Direction of arrival
Doppler effect, 162
Doppler frequency shift (DFS), 133, 160
DSAN, see Dynamic Spectrum Access Networks
DSRC-based vehicular positioning enhancement model
  using distributed multiple-model KF, 168–185
    general performance analysis, 174–180
    Monte Carlo results, 181–185
    simulation setup, 180–181
    single-trial analysis, 181
    system model and localization enhancement, 168–174
  using multiple-model KF, 161–168
    connected vehicles–enhanced IMM-KF (CV-IMM-KF), 163–165, 164*f*
    simulation results, 167–168
    simulation scenario, 166–167, 166*f*
    vehicular positioning enhancement model, 161–163
DTNs, see Delay tolerant networks
Dynamic spectrum, 59–61, 60*f*
Dynamic Spectrum Access Networks (DSAN), 60

E

ECDF, see Empirical cumulative distribution function
ECUs, see Electronic control units
Electric vehicles (EVs), 49
Electronic control units (ECUs), 23
  multiplexing, 7
Empirical cumulative distribution function (ECDF), 32, 33*f*
Energy threshold analysis, 149–151, 149*t*
ESPRIT, see Estimation of signal parameters via rotational invariance techniques
Estimation of signal parameters via rotational invariance techniques (ESPRIT), 133
Ethernet, 9–10
  10BASE-T Ethernet, 9
  100BASE-T Ethernet, 9
  1000BASE-T Ethernet, 9
  transmission medium for, 9–10
Euclidean distance, 119, 126
European Telecommunications Standards Institute/International Organization for Standardization (ETSI/ISO), 187–188
EVs, see Electric vehicles
Experimental data, 125–126

F

FCC, see Federal Communications Commission
FEC, see Forward error correction
Federal Communication Commission (FCC), 59, 132
FIM, see Fisher information matrix
Fingerprinting technique, 160
First-order Markov chain, 166
Fisher information matrix (FIM), 161, 176
Flexible time division multiple access (FTDMA), 8
FlexRay, 7, 8–9, 26, 49
Forward error correction (FEC), 132
FTDMA, see Flexible time division multiple access

G

Gale–Shapley algorithm, 210
Gaussian distribution, 127
Gigabit Ethernet, 9
Global navigation satellite system (GNSS), 132, 160
Global positioning system (GPS), 132, 160
  devices, 116, 125–126

Global System for Mobile communications (GSM), 133
GNSS, see Global navigation satellite system
GPS, see Global positioning system
GPS-TOA positioning technology, 134
Green transportation system framework
congestion pricing strategies, 204
cyber space social-network-enabled architecture, 207–209
Dial-a-Ride system, 205
global optimal pricing strategies, 204
incentives and pricing formulation, 209
introduction to, 203–204
microscopic traffic simulation, 205–206
problem definition and related work, 204–206
public vehicle (PV) assignment and hyper simulator, 210–211
scenario-driven traffic operations, 204
social-network-enabled, 203–211
social-network-enabled transportation CPS architecture, 206–207, 206*f*
traffic congestion, reducing, 205
GSM, see Global System for Mobile communications

H

HEV, see Hybrid electric vehicle
High-Speed ROBO (HS-ROBO) modulations, 48, 49
HomePlug Green PHY protocol
background of, 48
compatibility with, 54
extended, topology of, 51*f*
improvements on delay performance, 54–55
introduction to, 47–48
proposed real-time MAC protocol, 48–50
simulation results, 50–54
100BASE-T Ethernet, 9
Hybrid electric vehicle (HEV), 24
Hybrid time of arrival (TOA)/direction of arrival (DOA)
actual and estimated DOAs, 153*f*, 154*f*
Cramér-Rao bounds for, 155–156
cumulative distribution functions (CDFs), 156, 156*f*
data model, 146–147, 147*f*
DOA estimation based on MT-BCS and MSE values, 151*t*, 153*t*, 154*f*
estimation, 134–135, 135*f*
location procedure, 148–149, 149*t*
MT-BCS–based location method, 151–155
root mean square error (RMSE) of, 156*f*
Hyper simulator, 210–211, 211*f*

I

IEEE 1609.3 standard, 72, 74
IEEE 1609.4 standard, 72, 72*f*, 74
IEEE 1609 standard, 70, 71
IEEE 802.11a standard, 132
IEEE 802.11 broadcast service
analytic model for, 96–97
performance evaluation of, 96
performance indices for, 97
IEEE 802.11p standard, 70, 71, 73–74, 132
delay lower limit (DLL) of, 138*f*
parameters of, 136*t*
throughput upper limit (TUL), 137*f*
IEEE 802.3 standard, 9, 25
Impulse noise filtering, 31–33
Inertial navigation system (INS), 160
INS, see Inertial navigation system
Intelligent transport systems (ITS), 96, 132, 188
Interaction range, 228*f*
Interactive multiple-model-based Kalman filter (IMM-KF), 161
Internal combustion engine (ICE) vehicle, 24
International Telecommunication Union Telecommunication Standardization Sector (ITU-T), 96
Internet-of-Vehicle (IoV), 54
Internet protocol version 6 (IPV6) layer, 71*f*, 72
Intervehicle communication
cross-layer multihop broadcasting, 95–111
on mobility of multiple vehicles, influence of, 235–238, 236–237*f*

physical channel modeling and sharing, 59–68
short-range communication (SRC) technology, 70–93
on vehicle safety and traffic efficiency, influence of, 238–246
Intra-vehicle communications
automotive networking applications, 4–7
automotive networking protocols, 7–10
automotive network test bed implementation, 12–15
buses using HomePlug GP, worst delay analysis of, 47–55
HomePlug Green PHY protocol, 47–55
real-time media access control (MAC) protocol, 48–50
IoV, see Internet-of-Vehicle
ITS, see Intelligent transport system

K

Kalman filter (KF), 161
KDE, see Kernel density estimation
Kernel density estimation (KDE), 116
KF, see Kalman filter

L

LANs, see Local area networks
LBS, see Location-based services
Least-squares support vector machine (LS-SVM) fitting
borders and centerline of the roads, 125*f*
centerline fitting, 126–130
data cleaning, 117
experimental data, 125–126
introduction to, 116
maps, generating, 121–125
mechanism of, 124
nearest point on road for each node, 126*f*
road recognition based on, 115–130
roads centerline generation by, 123
subtrajectory partitioning, 117–121
Lightweight and adaptive security mechanism, 187–201
channel capacity model, 191–193
introduction to, 187–189
meanings and values of parameters, 197*t*
numerical results, 197–201
optimal transmit power, 195–196
problem statement, 189–190
related work, 190–191
security cost model, 193–195
utility model, 195
Limit Theorem, 137
Linear cryptanalysis, 194
Line-of-sight (LOS), 135, 160
LIN, see Local interconnect network
Local area networks (LANs), 9
Local Interconnect Network (LIN), 7, 25
"Local warning scope" concept, 215, 216*f*
Location-based services (LBS), 160
Logic link control (LLC) layer, 71, 71*f*

M

MANETs, see Mobile ad hoc networks
Maps, generating, 121–125
algorithm of, 121
borders and centerline of the roads, 125*f*
by data, 122*f*
key idea of, 121
merged map, 123*f*
Open Street Map (OSM), 122*f*
raw road lines, merging, 122
roads centerline generation by LS-SVM, 123
Markov chain model, 61
Markov localization algorithm, 133
MATLAB®, 66, 103, 231
Maximum-likelihood estimation (MLE), 133
Maximum throughput (MT), 136, 137*f*
Mean relative error (MRE), 150
Measurement of distance, defined, 119, 119*f*
Media access control (MAC), 97
layer, 25–26
protocol, 8
Media-oriented system transport (MOST), 9
Merged map, 123*f*
Microscopic traffic simulation, 205–206
Minimum delay (MD), 136, 138*f*

Mini-ROBO modulation, 48
Mini-slots, 8
Mobile ad hoc networks (MANETs), 97
Mobile terminal (MT), 133
Mobility model, for connected vehicles, see Cooperative vehicles, model of
MOST, see Media-oriented system transport
MRE, see Mean relative error
mSPEB, 161, 176–178, 182*f*
Mulitple-model Kalman filter (KF), 161–168
  connected vehicles-enhanced IMM-KF (CV-IMM-KF), 163–165, 164*f*
  simulation results, 167–168
  simulation scenario, 166–167, 166*f*
  vehicular positioning enhancement model, 161–163
Multichannel contention resolution method, 26–27
Multiple signal classification (MUSIC), 133
Multitask BCS–based DOA estimation, 147–148
Multivariate Taylor Expansion theorem, 171
MUSIC, see Multiple signal classification

N

Negotiated deceleration rates, 139*f*
Network Interface Card (NIC), 10
NIC, see Network Interface Card
Non-line-of-sight (NLOS), 133, 160
Non-return-to-zero (NRZ)-bit-encoding scheme, 8

O

OBUs, see Onboard units
OFDM, see Orthogonal frequency-division multiplexing
OMNeT++, 50–52
Onboard units (OBUs), 143, 143*f*, 146*f*, 162
Open Street Map (OSM), 121, 122*f*
Optimal scheduling, 210
Optimal transmit power, 195–196
  attacker's capability on node utility, effect of, 201*f*
  information security strength on node utility, effect of, 197*f*
  key length on node utility, effect of, 200*f*
  relative density of eavesdropper nodes on node utility, effect of, 198*f*
  total node density on node utility, effect of, 198*f*
  transmit power of other nodes on node utility, effect of, 199*f*
  wireless channel bandwidth on node utility, effect of, 200*f*
Orthogonal frequency-division multiplexing (OFDM), 132
OSM, see Open Street Map
Overlay system model
  capacity of secondary user, 63–64, 66
  of n>3 secondary users, 64–66
  secondary system throughput
    comparison between two and three secondary users, 67*f*
  of three secondary users, 62–64, 63*f*
  throughput comparison between two and three secondary users, 68*f*

P

PAACP, see Portable privacy-preserving authentication and access control protocol
Packet delivery ratio (PDR), 103, 106
PDF, see Probability Density Function
PDR, see Packet delivery ratio
Performance evaluation test bed, DSRC, 144–146, 144*f*
PGF, see Probability generating function
Physical block (PB), 48, 49
Physical channel modeling and sharing, 59–68
  dynamic spectrum access taxonomy, 60*f*
  introduction to, 59–60
  meanings and values of parameters, 66*t*
  numerical results, 66–68
  overlay system model of n>3 secondary users, 64–66
  overlay system model of three secondary users, 62–64, 63*f*
  related work, 60–61
  system models, 61–66

Physical space, CPS architecture, 206–207, 206*f*
Plastic optical fiber (POB), 9
PLC, see Power line communication
POB, see Plastic optical fiber
Point of Interest (PoI), 206
POINT problem, 190
PoI, see Point of Interest
Portable privacy-preserving authentication and access control protocol (PAACP), 190
Power line communication (PLC), 21
Priority resolution slot (PRS), 48
Probability density function (PDF), 167, 189
Probability generating function (PGF), 97, 99
Probability mass function, 27, 29
 of number of transmission cycles, 39–40*f*
PRS, see Priority resolution slot
Public vehicle (PV) assignment, 205, 210–211

Q

QoS, see Quality-of-service
Quality-of-service (QoS), 25, 48–49, 187, 188–190, 195

R

Radio frequency (RF) spectrum, 59
Random arbitration register (RAR), 22
RAR, see Random arbitration register
Raspberry Pi, 13, 14*f*, 15
Real-time media access control (MAC) protocol, 48–50
 contention window (CW) size, reducing, 49*t*
 frame structure of, 50*f*
 hypothesis of, 48–49
 length of data message, reducing, 49
 message length information, altering, 49
 slot time, reducing, 49
 summary of differences, 50*t*
Received signal strength (RSS), 133, 160
Receiver operating characteristic (ROC) curves, 33
Response InterFrame Space (RIFS), 50–51
RF, see Radio frequency spectrum
RIFS, see Response InterFrame Space
Rijndael algorithm, 194
Riverbed's Packet Analyzer, 15
RMSE, see Root mean square error
Road recognition based on least-squares support vector machine (LS-SVM) fitting
 centerline fitting, 126–130
 data cleaning, 117
 experimental data, 125–126
 introduction to, 116
 maps, generating, 121–125
 subtrajectory partitioning, 117–121
Roadside units (RSUs), 15, 133, 143, 143*f*, 146*f*
Robust sensing module, 32
ROC curves, see Receiver operating characteristic curves
Root mean square error (RMSE), 153–155, 156*f*, 167, 182
Round-trip time (RTT), 16–17, 17*f*, 18, 19*f*
RSS, see Received signal strength
RSUs, see Roadside units
RTT, see Round-trip time

S

SAE, see Society of Automotive Engineers
SAFESPOT, 132
SA, see Simulated annealing
Security cost model, 193–195
Self-adaptive V2V communication system
 with dedicated short-range communication (DSRC), 73–76
 experiment and analysis, 85–86
 host, 74–75
 packet loss rate *vs.* relative distance, 86*f*
 real testing scenario, 85*f*
 router, 75
 system architecture, 74*f*
 system model, 75*f*
 transmission delay *vs.* relative distance, 86*f*
Sensing errors
 false alarms, 33–35
 impact of, 33–36, 34*f*
 miss-detection, 33–35
 receiver operating characteristic (ROC) curves, 33
Service channel (SCH) interval, 72, 72*f*

Short-range communication (SRC) technology, in Internet of Vehicles
bio-inspired model, for making handover decision, 72–73
bio-inspired QoS-oriented handover model, in heterogeneous wireless networks, 76–84
dedicated short-range communication (DSRC), 71–72
embedded self-adaptive configuration scheme, 75–76
experiment and analysis, 85–93
introduction to, 70
Signal-to-noise ratio (SNR), 98, 101, 133, 191
Simulated annealing (SA), 120–121
Simulation and verification, of cooperative multivehicle mobility model, 231–235
of cooperative multivehicle mobility model, 231–232
improved cooperative multivehicle mobility model, 232–235
road situation for simulation, 234*f*
traveling trajectory of vehicles, 233*f*
values of parameters, 235*t*
SNR, see Signal-to-noise ratio
Society of Automotive Engineers (SAE), 4
Spectrum sharing, 59–61
SPITS, see Strategic Platform for Intelligent Traffic Systems
Square position error bound (SPEB), 175, 182*f*
Standard deviation (STD), 133, 161
Standard ROBO (STD-ROBO) modulations, 48, 49
State probabilities
of channel occupancy in overlay, 67*f*
of primary and secondary users, 62
STD, see Standard deviation
Steady-state probability, of primary and secondary users, 62–64
Strategic Platform for Intelligent Traffic Systems (SPITS), 132
Subtrajectory partitioning, 117–121
algorithm of, 121
definition of, 117
ideal partitioning, 118*f*
key procedure of, 118
MDL principle, 118
measurement of distance, 119, 119*f*
potential problem, 120*f*
purpose of, 118
simulated annealing (SA), 120–121
Success probability, 29
CDR *vs.* number of contending nodes, 41*f*
maximizing, 30
vs. number of channels, 37*f*
vs. number of contending nodes, 36*f*, 43–44*f*
of the proposed protocol, 41
vs. threshold of energy detector, 42*f*
vs. total number of nodes, 38*f*

T

Target vehicle (TV), 162
TCP, see Transmission Control Protocol
TDMA, see Time division multiple access
TDOA, see Time difference of arrival
10BASE-T Ethernet, 9
1000BASE-T Ethernet, 9
3 principle, 125
Throughput evaluation, 31
CDR *vs.* number of contending nodes, 42*f*
of proposed protocol, 42*f*
system throughput *vs.* number of channels, 37*f*
Throughput upper limit (TUL), 132, 137*f*
Time difference of arrival (TDOA), 133, 160
Time division multiple access (TDMA), 8
functionality, 54–55
Time of arrival (TOA), 133, 160
TOA, see Time of arrival
Traffic capacity, effect on, 136–143
Traffic congestion, reducing, 205
Transmission control protocol (TCP), 14, 15, 19, 71*f*, 72
Transmission control protocol/Internet protocol (TCP/IP), 74–75, 93
TUL, see Throughput upper limit

U

UDP, see User datagram protocol
Universal asynchronous receiver/transmitter and SCI (UART/SCI) interface hardware, 7

Unlocalized mobile nodes, 134
User datagram protocol (UDP), 71*f*, 72
Utility model, 195

V

VANETs, see Vehicular ad hoc networks
Vehicle Infrastructure Integration (VII), 132
Vehicle Miles Traveled (VMT), 204
Vehicle-to-Everything (V2X) communication, 14, 15, 143, 143*f*
Vehicle-to-infrastructure (V2I) communication, 132, 160, 210, 211*f*
Vehicle-to-roadside (V2R) communications, 70, 95, 214
Vehicle-to-vehicle (V2V) communications, 70, 95, 132, 160, 189, 189*f*, 210, 211*f*, 214
Vehicular ad hoc networks (VANETs), 95–96, 132, 160, 187, 214
  broadcast method for, 97
  CAM application in, 101
  multicast CAM messages in, 97
  multicast methods for, 97
Vehicular positioning enhancement, 159–185
  connected vehicles-enhanced DIMM-KF (CV-DIMM-KF), 172–174
  connected vehicles-enhanced IMM-KF (CV-IMM-KF) for, 163–165, 164*f*
  by DFS and RSSI measurements, 168*f*, 182–185*f*
  DSRC-based model, see DSRC-based vehicular positioning enhancement model
  introduction to, 160–161
Vehicular positioning, for traffic systems, 134
Vehicular power line communication (VPLC) channel
  CDR protocol for, 22
  challenges, addressing, 24
  collision resolution algorithm, 27*f*
  MAC protocol for, 22
  multiple frequency channels, use of, 22
  single transmission cycle, 26*f*
Verkehr In Stdten-SIMulationsmodell (VISSIM), 205–206, 210
VISSIM COM, 210
VISSIM, see Verkehr In Stdten-SIMulationsmodell
VMT, see Vehicle Miles Traveled
VPLC, see Vehicular power line communication channel

W

WAVE MAC layer (WAVE MAC), 71, 71*f*
WAVE physical layer (WAVE PHY), 71, 71*f*
WAVE, see Wireless access in vehicular environments
WAVE short message protocol (WSMP) layer, 71, 71*f*
WAVE stack, 71, 71*f*
Wide trail strategy, 194
Wi-Fi-based positioning system, 133
WiMAX, 70
Wireless access in vehicular environments (WAVEs), 70, 160
Wireless channel state information (CSI), 160
Wireless LANs, 70
Wireless sensor networks (WSNs), 190
Wireshark, 15
World Telecommunication Standardizing Assembly (WTSA), 96
WSMP, see WAVE short message protocol layer
WSNs, see Wireless sensor networks
WTSA, see World Telecommunication Standardizing Assembly

X

XBEE Arduino shield, 15
XBEE ZigBee modules, 15
X-by-wire systems, 48

Y

Y-splitting method, 121

Z

ZigBee gateway, 14*f*, 15
ZigBee units, 14*f*, 15
ZigBee wireless technology, 12